互聯互通的
金融大時代

小加隨筆

The Great Financial Era
of Connectivity

Blogs by Charles Li

李小加 著
Charles Li Xiaojia

商務印書館

互聯互通的金融大時代 —— 小加隨筆

作　　者：李小加

責任編輯：張宇程

封面設計：趙穎珊

出　　版：商務印書館（香港）有限公司

　　　　　香港筲箕灣耀興道 3 號東滙廣場 8 樓

　　　　　http://www.commercialpress.com.hk

發　　行：香港聯合書刊物流有限公司

　　　　　香港新界大埔汀麗路 36 號中華商務印刷大廈 3 字樓

印　　刷：美雅印刷製本有限公司

　　　　　九龍觀塘榮業街 6 號海濱工業大廈 4 樓 A 室

版　　次：2019 年 8 月第 1 版第 2 次印刷

The Great Financial Era of Connectivity —— Blogs by Charles Li

Author: Charles LI Xiaojia
Executive editor: Chris CHEUNG
Cover design: Cathy CHIU
Publisher: The Commercial Press (H.K) Ltd.,
8/F, Eastern Central Plaza, 3 Yiu Hing Road, Shau Kei Wan, Hong Kong
Distributor: The SUP Publishing Logistics (H.K.) Ltd.,
3/F, C & C Building, 36 Ting Lai Road, Tai Po, New Territories, Hong Kong
Printer: Elegance Printing & Book Binding Co., Ltd.
Block A, 4th Floor, Hoi Bun Building
6 Wing Yip Street, Kwun Tong, Kowloon, Hong Kong

© 2018 Hong Kong Exchanges and Clearing Limited
First edition, Second printing, August 2019
ISBN: 978 962 07 6606 0

Printed in Hong Kong

目錄

CONTENTS

第三章　互聯互通與共同市場

第四章　定息及貨幣產品市場與人民幣國際化

Chapter 3   Connectivity and Mutual Market

Chapter 4   Fixed Income & Currency Market and RMB Internationalisation

第五章　大宗商品市場

第六章　其他

Chapter 5 Commodities Market

Chapter 6 Others

自 序

不是我不明白，這世界變得太快。

變化，大概是這個世界唯一不變的趨勢。與時俱進，我們就能處於不敗之地；稍有疏忽，我們就可能被這個時代邊緣化。對於香港金融市場，尤其如此。

縱觀香港經濟和金融市場的歷史，可以説是一部隨時代變化不斷反覆運算與升級的歷史。尤其是過去幾十年，憑藉大膽實幹、靈活應變的精神，香港成功把握中國改革開放的機遇迅速轉型，從一個區域性證券市場，成長為舉世矚目的國際金融中心。

過去幾十年，香港為中國的改革開放主要做了三件大事：第一，轉口貿易；第二，直接投資（FDI）；第三，資本市場的大發展。轉口貿易給內地帶來了第一桶金，FDI 直接投資把中國變成了世界的工廠，而香港資本市場的大發展則為中國內地源源不斷地輸送了發展經濟的寶貴資本，從 1993 年 H 股誕生開始，一家又一家內地公司在香港上市募集來自全球的資金，發展成了今天世界上最大的電訊公司、能源公司、銀行和保險公司。

這三件大事都有一個核心主題，那就是為中國內地輸入資本，因為那個時候內地很缺錢。隨着經濟的發展和國家的富裕，現在內地已經很不缺錢，甚至正在逐漸走向一個資本輸出國。未來幾十年，香港應當如何找到自己新的定位？

在我看來，香港未來也有三件大事要做：

第一件事是幫助內地國民財富實現全球配置。從幾年前開始，中國內地的國民財富逐漸開始從房地產和銀行儲蓄走向股市和債市、從單一的國內資產配

置走向全球分散配置的歷史性「大搬家」。一些「勇敢的人們」已經開始將資產直接投資海外，但絕大多數中國投資者還暫時沒有能力直接「闖世界」。以滬港通和深港通為代表的互聯互通，正是為了把世界帶到中國門口的香港，讓中國內地居民可以「坐在家裏投世界」。

第二件事是幫助中外投資者在離岸管理在岸金融風險。人民幣已經成功加入國際貨幣基金特別提款權（SDR）的貨幣籃子，未來人民幣國際化程度還會加快。今天的內地市場，利率與匯率尚未完全開放，內地的債券與貨幣衍生品市場也還沒有建立成熟的風險管理機制。內地衍生品市場的缺失使得大量有意持有中國資產（包括股票、債券與貨幣）的國際投資者在國門外望而卻步。香港擁有國內外投資者都認可和熟悉的法治和語言環境，完全有條件開發符合雙方投資者需求的產品，發展成為亞洲時區內最主要的國際風險管理中心。

第三件事是幫助中國實現商品與貨幣的國際定價，為內地的資金定價海外資產提供舞台。未來幾十年，中國的投資者將學着走向世界、用中國的錢來「定價」海外的公司股權和大宗商品；同時，也要用中國的購買力讓越來越多的國際權益與商品以人民幣定價，這樣，中國就能在全球範圍內逐步掌握人民幣匯率與利率的定價權。

在這個互聯互通的金融大時代，在不斷變化的全球格局中，香港應該如何做好這三件大事？這是香港交易所近幾年需要思考的一個重要問題。

作為香港證券與衍生品市場的營運者和監管者，香港交易所集團最重要的使命就是緊跟時代步伐、引領和推動市場的發展。在此過程中，我們必須向市場耐心解釋我們的初衷，我們必須仔細聆聽市場各方的聲音，我們既要有擔當，勇於決策，更要尊重程序正義、凝聚市場共識，充分兼顧各方的利益。而在凝聚市場共識的過程中，大家總會有很多思想火花的碰撞和激烈的討論。

為了讓更多人能夠了解我們的所思所想，也為了讓更多的人能夠參與這些激烈的討論，我嘗試着寫一些大白話的網誌來分享我對金融市場的一些淺見，以拋磚引玉，集思廣益。幾年下來，竟然也拋了幾十塊磚。今天，我把這些「磚頭」收集起來，整理成一本書冊，希望能對大家了解香港金融市場有所幫助。

　　在撰寫網誌的過程中，我得到了香港交易所許多部門和同事的幫助，因此，這些網誌是我們大家集體智慧的結晶，我衷心感謝所有為網誌提出寶貴建議的同事。陳涓涓女士率領的企業傳訊部為網誌的創作和發佈傾盡全力，尤其是錢傑女士和 Cam MacMurchy 先生對於網誌的潤色和修改付出了大量心血，在此，我向她們表示衷心的感謝。

　　由於本人學識有限，網誌中難免會出現一些錯誤，説得不對的地方，還請讀者指正和原諒。

李小加

香港交易及結算所有限公司　集團行政總裁

2018 年 5 月 31 日於香港

Preface

The world is changing fast.

"Change" is perhaps the only thing in the world that is eternal. We know that we must continue to evolve — if we are complacent, we risk being marginalised. This is especially true of the financial market of Hong Kong.

The Hong Kong economy and the Hong Kong financial market have continued to grow and change with each generation over time, which has been the key to our success. Hong Kong has successfully captured opportunities offered by China's reform and opening up in the last few decades by being bold, pragmatic and versatile, transforming itself from a regional securities market into a prominent international financial centre.

Hong Kong has traditionally served Mainland China in its reform and opening up in three main ways: first, entrepot trade; second, foreign direct investment (FDI); and third, big development of its capital market. While entrepot trade gave the Mainland its first pot of gold and FDI turned China into the world's factory, the big development in the Hong Kong capital market channelled much-needed capital to China's developing economy. Since the birth of H shares in 1993, Mainland companies have come to list in the city one after another, raising funds from across the globe that have nurtured these companies to become among the world's largest in telecommunications, energy, banking and insurance.

These three big deeds share the same core theme — directing capital into the Mainland in the days when Mainland China was short of money. Now, the Mainland has plenty of money after years of economic growth and prosperity and is becoming an exporter of capital. So how should Hong Kong reposition itself in the decades to come?

In my opinion, there are three things Hong Kong must do in the days ahead:

The first thing is to facilitate the global allocation of Mainland assets. Mainland Chinese began shifting their wealth from properties and bank savings to stocks and bonds a few years ago. Historically, onshore asset allocation has increasingly given way to diversified global asset allocation. Some Mainland investors are even bold enough to invest directly overseas, though the majority do not yet have the ability to do so. Stock Connect, which links the Shanghai and Shenzhen exchanges with the Stock Exchange of Hong Kong, is exactly the kind of programme designed to bring the world to China's doorstep and give Mainland investors safe and convenient access to international investments, all through Hong Kong.

The second thing is to enable both Mainland and overseas investors to manage their Mainland financial risks in an offshore environment. The pace of the Renminbi's (RMB) internationalisation is set to accelerate now that the currency has been admitted into the International Monetary Fund's Special Drawing Rights (SDR) currency basket. However, Mainland interest rates and exchange rates are not yet fully liberalised, and a mature risk management system is not yet well established in the onshore bond and currency derivatives markets. Such inadequacies in the Mainland derivatives market deter many international investors from holding Mainland assets (including stocks, bonds and the currency). Hong Kong, with a legal, regulatory and language environment respected by both Mainland and overseas investors, is in an excellent position to develop products that will meet the needs of both types of investors. This could lead to Hong Kong's transformation into a leading global risk management centre in the Asian time zone.

The third thing is to help China exercise international pricing power in commodities and currencies, and to provide a platform for Mainland investors to price overseas assets. In the decades ahead, Mainland investors will learn to step into the world, using their own capital to price overseas equities and commodities and using their purchasing power to price international equity rights and commodities in RMB. This will give China global influence over the RMB's exchange and interest rates.

As we enter an era of rapid change and enhanced financial connectivity, what should Hong Kong do to achieve these three missions? This is a critical question HKEX has been asking itself over the past few years.

HKEX, as the operator and regulator of the Hong Kong securities and derivatives markets, has a key role to play in leading and promoting market development so we stay ahead of the curve. Along the way, we must explain our vision clearly and carefully and listen to voices from all sectors. Due process, market consensus and the interests of all stakeholders are of paramount importance, in addition to a strong sense of commitment and bold decisions. But in the process of reaching consensus, we cannot avoid diverse opinions and debates.

To let a wider spectrum of the market understand our thinking and to involve more people in these discussions, I periodically share my views on topics of concern in Charles Li Direct, my blog. There are now dozens of posts since the first one was published some years ago. I hope this compilation will be of assistance to understanding our vision of the local financial market.

While I am the one who penned the articles, many departments and colleagues at HKEX have offered their insights. The articles are, I must say, the collective wisdom of HKEX staff. I am grateful to each and every one of them who has assisted me in composing the articles, not least Corporate Communications led by Lorraine Chan. Lorraine's team has spared no effort in sparking my creativity and getting it released. Qian Jie, in particular, has devoted wholeheartedly in polishing the Chinese articles and Cam MacMurchy for the English version. I fully appreciate their dedication and professionalism.

Bound by my own knowledge, the blog posts are anything but perfect. If there is any inadequacy, do not hesitate to let us know. As always, your suggestions and valuable opinions are most welcome.

Li Xiaojia, Charles
Chief Executive
Hong Kong Exchanges and Clearing Limited
31 May 2018, Hong Kong

第一章

香港交易所戰略發展

香港新年開市有派利是（紅包）的習俗。

01 | 香港交易所新年展望： 我們是否過於專注中國？

恭喜發財！希望各位新春與家人共聚天倫，同享佳節氣氛！

一年之計在於春。香港交易所正為蛇年積極籌劃，磨拳擦掌，迎接這一個滿載機遇和希望的新一年！今後，我希望能通過網誌這個平台來跟大家交流，講述我們為市場做的事，詳解我們的某些觀點、想法以及方針的理據。

目前，香港正處於向國際金融中心發展的重要轉折期。過去 20 年，中國迅速崛起並成為了世界第二大經濟體。通過抓住中國在發展過程中的商機並服務於中國經濟整體需求，香港亦取得了驕人的增長。自上世紀九十年代起，香港一直是中國企業走向國際的首要融資平台，幫助當時的中國邁出了改革開放的重要一步。

不過，中國發展的需求不斷演變，我們亦必須相應作出調整。香港交易所 2013-2015 年戰略規劃闡述了我們對未來幾年中國發展的方向的判斷以及在其影響下香港發展的可能路徑。在這關鍵時刻，市場上有些聲音質疑我們似乎過於專注中國。有人曾問我：「我們為何如此重視中國？我們是否把太多的雞蛋放在了同一個籃子裏？我們為何不先走向國際化，然後再回頭關注中國？」對於類似的問題與疑惑，我想談談個人的一些淺見，以希望得到更多市場人士的理解，並以此拋磚引玉，聽取大家的意見反饋。

我們是否過於專注中國？

香港是中國的一部份，但同時也是全世界最國際化的大都市之一；香港實

施「一國兩制」。香港在過去把這個獨一無二的「雙重特性」發揮得淋漓盡致，取得了輝煌成績；如果將來也能恰當運用，香港就能延續輝煌。「中國」雖然只是這個「雙重特性」的其中一面，卻是我們整體計劃的關鍵一環。請允許我在下面打一比方說明。

設想你在籌辦一個派對，而且你想把它辦成一個全城矚目的盛事，關鍵問題是怎樣才能邀請到城中最重要、最多的客人。我們都知道，人們參加派對，通常是為了遇見那些在其他場合中不輕易碰到的人。作為派對的主人，你需要仔細考量邀請的優先次序，分辨出那些對其他客人極具吸引力的「核心」人物，首先邀請並落實他們的參與。

在香港，我們所做的就是籌辦一個資本市場的「派對」，讓無論是來自中國還是海外的投資者能在這裏相遇、相識，讓這個「派對」成為全球必去的盛會之一；而中國便是這個「派對」的「核心」嘉賓。對於中國這位嘉賓而言，香港的「派對」提供了其試驗金融與資本市場開放的平台，讓它能夠「摸着香港這塊熟悉的石頭」邁出踏出國門的第一步；對於我們這個「籌辦方」而言，中國的到來是對其他海外客人的莫大吸引力。正所謂「物以類聚，人以羣分」，在我們的市場，獨特的投資產品與機會吸引最多的投資者，而投資者又吸引更多的投資者，流動性吸引更大的流動性。

香港交易所過去 20 年的發展歷程驗證了上述「派對」模式的成功性。20 年前，香港交易所從一個本土交易所起步，看準了中國企業來港上市的浪潮，不遺餘力地吸引中國發行人來港進行首次公開發行與上市；與此同時，被這些中國企業與香港這個國際平台所吸引，一批又一批的國際投資者慕名而來，其數量前所未見。最終，香港由一個本土交易所成功轉型為在股票及其衍生產品方面領先的國際交易所。對我們的市場參與者和股東來說，香港交易所過去 20 年的成功就是最好的「派對」。更重要的是，香港已經成為籌辦這類中外結合「派對」的最有經驗的能手；而香港過去 20 年中累積的成功經驗，將令我們在未來籌辦更多、更大型中外結合「派對」時更有優勢。我們若不能充分發揮利用這項已經建立起來的獨特優勢，就等於將過去 20 年的心血白白浪費掉。

為什麼我們開始潛心發展定息產品、貨幣及商品市場，
而不只專注於開發股票證券這個傳統核心業務？

基於上述原因，我們在首個「派對」取得空前成功之後，責無旁貸地應該着手籌辦下一個更大的盛會，邀請更多的、不同的嘉賓。香港今天面對的最大機遇是中國資本、金融及商品市場開放的提速。隨着中國開始由資本進口國轉型為資本出口國，香港需要從中國離岸集資中心這個傳統角色，加速轉型成為一個能夠提供全方位股本、定息、貨幣及商品產品的綜合性國際金融中心。在這個千載難逢的良機面前，我們不僅有責任而且有能力確保香港成為中國走向世界的第一站。

在股票證券方面，除了進一步發展我們的 IPO 及二級市場增發業務以外，我們下一個「派對」的焦點將轉為邀請中國本土投資者前赴香港，並以此吸引更多國際企業來港上市融資，而這恰恰是香港過去 20 年股票市場故事的另一面。假以時日，內地投資者在香港市場中的結構性增長將會吸引更多的國際投資者。而對於那些已在香港的本地與國際投資者，我們會致力於降低他們參與內地資本市場的門檻。若我們能成功促成香港、內地、國際市場的互聯互通，香港將無愧於其東西匯合的樞紐的稱號。所以説，我們不僅沒有忽略我們傳統的股票核心業務，而是在更積極地強化核心業務，同時推動市場結構轉型，為日後資金跨境雙向流動的大勢作好準備。

在商品產品方面，港交所是一個新手，我們充分意識到由零開始籌辦一個「派對」將何其艱巨。香港並沒有很多籌辦這類「派對」的經驗，但歷史的巨輪給我們呈現了這個稍縱即逝的機會。我們知道，作為世界許多大宗商品最大的生產國和消費國，中國需要加速開放內地商品市場、促進內地與國際商品市場的互通，以獲取與其經濟實力相匹配的國際影響力。同時，國際商品市場的參與者也能夠通過香港為中介，基於國際最佳慣例和標準，以更公平、公開、可持續的方式來參與內地在岸商品市場的發展。

收購倫敦金屬交易所（LME）是我們構建商品平台的重要一環。雖然它不

能解決我們在發展商品交易與結算平台中的所有挑戰，但此舉是我們的第一槍。它告訴世界，香港能夠成為世界級的商品中心，能夠為中國商品界賓客提供大膽邁向國際的第一站。

在定息與貨幣產品方面，我們知道，中國熱切地希望並大力推進人民幣國際化。內地與國際通過香港實現股票市場與商品市場上的互聯互通，一定會以人民幣產品為主旋律。這必將帶來人民幣大規模的增量出境，並以投資貨幣、交易貨幣等形式更長久地沉澱於離岸市場。這將為離岸人民幣定息與貨幣產品的交易與結算帶來巨大的發展空間與機遇。

香港當前所面臨的巨大挑戰和機遇，就是克服自己信心與經驗的不足，抓住歷史機遇，勇敢承辦這個「奧運會」級的歷史性「派對」，既能切合中國和國際投資者各自的戰略目標和需求，又能使雙方賓至如歸、共同創造價值。借鑒歷史，我們需要再次把目光聚焦中國，務求邀請中國成為我們的「核心」嘉賓。

那麼回到面前的問題：為何要專注中國而不是國際？我們的目標是兩者兼得，因為兩者是環環相扣、互為因果。我們的目標並不是成為純粹的中國交易所、區域交易所或者只關注國際市場的國際交易所；如我們在三年戰略規劃中所說的，我們的願景是要成為中國客戶走向世界以及國際客戶進入中國的首選全球交易所。

在推行各項規劃之際，我熱切期盼借用這個網誌跟大家討論分享更多話題，包括 LME 的整合、經紀業界的發展、市場互聯互通，以及建設香港成為頂尖國際金融中心的發展路徑。我期待聆聽您的聲音，歡迎您通過電郵方式（ceo@hkex.com.hk）分享您的寶貴意見。

謹此祝願大家蛇年進步、龍馬精神、事事亨通！

2013 年 2 月 14 日

02 | 對香港交易所 近期舉措的回顧

　　香港交易所在過去幾年投資、收購、改革與求變，動作頻頻，引起不少關注。經常有朋友問我，為什麼要做這麼多？為什麼要現在做？它們之間有什麼內在的聯繫？不做是不是不行，做了是不是一定有成效？成效何時來？我想借這篇博客「溫故而知新」。

　　在我看來，我們過去幾年的努力就像是翻新鞏固一棟「大房子」。我們的第一件要務便是加固房子的根基。為此，我們投資於核心基礎設施，建成了將軍澳新數據中心（其中包括設備託管服務）、提升了交易平台（AMS／3.8），並且還將繼續投資於領航星系統工程等。這些措施可能投入大，回報較慢，但香港市場在巨大發展機會面前，必須加固與更新其業務發展的核心基礎設施。在此，香港交易所並無其他選擇。這些措施將壘實香港市場的根基、支撐香港市場的長遠發展。

　　我們做的第二件事便是搭建新平台，這相當於為房子搭建新的樓層，不僅是適應今天的需求，更加是為馬上到來的明天做準備。我們的措施遠的如金磚五國交易所聯盟、近的如中華交易服務公司（與內地兩大交易所的合資公司）、自建的場外結算所以及收購而來的倫敦金屬交易所（LME），這些新的業務、地區與資產發展平台能使香港交易所跨越傳統股票業務，讓它在中國資本市場逐步開放的大潮中獲得更加廣闊和可持續發展的平台。

　　一所牢固的「房子」不僅要有堅實的根基、廣闊與多層次的平台，而且要有一道安全的防火牆。我們在前兩年花了大力氣改革香港交易所的清算風險管理體系，大大提高了香港金融市場的穩健水平和抗壓能力，相當於為香港交易

所加固了防火牆。雖然這並不是讓人激動的增長點，卻使市場的長遠發展少了後顧之憂。

有了前面的基礎、平台、防火牆，香港交易所這棟「大房子」就能基本屹立不倒，但我們不能停步於此。所以在今年我們還開始仔細審視「房子」的微結構，並作必要的「裝修」與改造，使其運營得更加合理和有效。例如剛剛推出的股票期權市場改革，從交易費、莊家服務、合約設計等幾個方面，審視一些採用多年但未必再適應市場發展的微結構。這些微結構如果改好了，就能與世界上通用的「房子」標準更一致，釋放出來的能量可以更大。當然，還有不少市場微結構的改革在不同市場參與者之間存在着較大的爭議與質疑，譬如收市競價、隱名交易等，我們會多傾聽市場的意見與聲音再考慮改革措施的方式和最佳時機。

光把自己房子修整得煥然一新還是不夠，我們的最終目標是讓各方賓客齊集香港，客似雲來，所以我們要搭建與內地和國際市場的聯通。這方面，我們從人民幣功能與交易時間入手。一方面，我們推出了人證港幣交易通、人民幣港幣雙櫃檯交易、人民幣期貨等，為的就是迎接不斷加速的人民幣國際化的進程，為日後內地資金來港而做準備；另一方面，我們修改交易時間、加開期貨夜市，為的就是與內地、國際市場的交易時間接軌，為互聯互通的加快實現提供條件。

這些措施對我們財務又有什麼影響呢？這些舉措有着不同的特徵。數據中心投入大，屬成本中心；建成後，折舊成本將會在收入增長來臨之前對利潤率產生壓力。相反，金磚聯盟和滬港深合資公司的投資很小，但品牌影響和戰略意義重大。對於 LME，它是香港交易所目前最大的一筆戰略投資。中長期來看，它會是香港交易所未來發展最重要的發動機之一。但由於 LME 收費還未充分商業化，加上 LME 正經歷最重要的建設投資階段，LME 對香港交易所財務貢獻還很有限。它對香港交易所的最有形貢獻，應會體現在促進香港交易所與內地市場大幅提前實現互聯互通。

以上措施，如果每個單獨來看，可能不全是驚天動地的創舉，甚至可能給

人繁多、雜亂的感覺。但事實上，許多的舉措是一環扣一環的，放在一起、特別是在市場機會來臨時便能釋放出巨大能量。例如，當初修改交易時間、讓香港開市時間與內地同步，因為交易量並未同步增加，市場反響很大。現在回過頭看，近兩年與 A 股相連產品、特別如人民幣 A 股 ETF 的湧現，相當大程度歸功於開市時間的同步。若未來 A 股期權期貨產品相繼推出，互聯互通逐漸形成，兩地市場若仍要承受開市時間錯位的風險，情況實在難以想像。

　　無疑，在如此短的時間裏集中推出如此多的措施，也對市場特別是我們的中小經紀帶來相當的影響與困惑。對於業界的難處，我們深深理解，我們也在決策、執行與實際選擇上盡量減少可能的負面影響。我深信，有一些成果可能比大家想像中來得快、來得多。

　　過去幾年裏，大家給了香港交易所許多耐心和莫大幫助，我深表謝意。我的目標是，把香港交易所這個「房子」建得更好、更大、更穩固、實用率更高、吸引四方賓客，最終給我們的市場和投資者創造巨大的商機。

2013 年 5 月 8 日

03 自貿區來了，香港怎麼辦？

　　備受關注的上海自貿區終於揭開了神秘的面紗。在內地，上海自貿區引發了人們對於進一步改革開放的無限遐想與憧憬。在香港，大家對上海自貿區的關注絲毫不遜於內地，各種討論早已沸騰。

　　最常聽到的議論是，「上海自貿區來了，香港的領先地位能否保持」，「自貿區將幫助上海成為國際金融中心，直接威脅香港金融中心的地位」……無論怎麼說，這樣的議論總給我一種「既生瑜何生亮」的酸楚感覺。在我看來，這樣的討論根本問錯了問題，因為它們幾乎都是在問「上海自貿區成立會令上海還是香港最受益？」這一點其實無需討論，上海自貿區給上海帶來的好處當然大過給香港帶來的好處。但是，作為香港人，其實我們更應該關心的問題是，有了上海自貿區的中國和沒有上海自貿區的中國，哪個對香港更有利？

　　我想肯定是前者對香港更有利。因為，有了上海自貿區的中國一定是更加開放、更加國際化的中國，而在所有國際金融中心中，香港最有條件將中國的開放轉化為自身的機遇，中國經濟和金融開放的節奏越快，香港的機遇也越大。

　　過去 30 年，香港是中國改革開放最大的受益者之一。在中國經濟和金融仍處封閉狀態的時候，彈丸之地的香港憑藉「一國兩制」創造的優勢成為中國內地與世界經濟一體化的最主要橋樑。在金融領域，香港幫助內地實現了資本項下開放的第一步，即中國企業的股權境外上市，香港成為中國最主要的境外資本聚集中心，也成為海外投資者參與中國投資最重要的市場與基地。

　　作為國家級的自由貿易試驗區，上海自貿區的主要任務是制度創新，是為中國經濟的進一步開放探路，在自貿區試驗成功的改革未來將被複製到全國。

換言之，上海自貿區將成為中國進一步改革開放的催化劑和加速器，它的成功將意味着中國經濟的開放力度加大、速度加快，相當於內地和海外之間貿易與資金的「直航」限制有可能以自貿區為突破而逐漸放開。

這一發展對香港之影響主要體現在兩個方面。首先，香港資本市場前 20 年的發展，已使全球資金、機構與人才大量沉積香港，香港已成為全球投資中國的最主要前沿陣地。未來上海自貿區的發展會從香港吸引越來越多的人、財、物北上，但一個更加開放的中國又會使香港成為更有吸引力的「中轉站」。從這個意義上說，上海自貿區對香港一定是一個淨利好，但還不是質的提升。

上海自貿區迅速發展對香港具有質變的利好體現在中國開放本地資本市場的速度與幅度。香港很快會從「單向」變成「雙向」的中轉站和終點站。一方面，海外的資金會繼續通過香港這個設施完備又方便的中轉站進入中國內地。另一方面，國內資本市場對外開放，允許國內投資者進軍國際市場的步伐加快，香港市場憑藉其法制、信用、國際化和開放的營商環境等優勢，必將成為中國資本國際化進程中初期的主要終點站和長期的中轉站。

我相信，內地經濟改革開放的規模越大、速度越快，給香港這個「中轉站」和「終點站」的機會也就越大。以金融業為例，中國目前的金融資產總值已經超過 100 萬億人民幣。在過去 20 年間，H 股公司從香港市場募集的資金僅 1.5 萬億港元，還不到中國金融總資產的 2%，已經為香港資本市場帶來了有目共睹的巨大機遇。可以預見，中國金融業的進一步開放將加快資本的流動，為香港金融業帶來更大的蛋糕。

至於上海自貿區會否加劇香港與上海之間的競爭，我想說的是，其實香港自開埠以來，無時無刻不在競爭當中。香港所面臨的競爭，不僅是與上海或內地其他城市的競爭，更是在增量中的競爭。增量越大，增速越快，香港的機會越大。中央發展上海自貿區的力度越大，意味着開放速度會越快；上海自貿區的成功來得越快，意味着內地市場的開放來得越快，香港的成功也來得越快。從這個角度看，香港應是為上海自貿區發展喝彩最響亮的拉拉隊，也是最好的發展夥伴。

因此，作為一個在香港工作多年的金融從業者，我由衷地為上海自貿區成立的消息感到高興，並祝福它能夠取得成功，為香港帶來更多機遇！

2013 年 10 月 14 日

04 ┃ 十五歲再出發

今天是香港交易所集團上市的 15 歲生日，感謝各位好友前來參加我們的生日派對，和我們一起慶祝生日。15 歲是人生最美好的青春，也是最多夢的季節。此時此刻，我們的心中充滿了喜悅和感激，當然，更多的還是對明天的憧憬和嚮往，因為我們懷揣着一個新的遠大夢想，即將踏上新的征程。

這個夢想就是**連接世界與中國、重塑全球市場格局**。具體而言，就是我們將致力於在香港建設一個巨大的**境外人民幣生態圈**，努力使香港發展成**中國的離岸財富管理中心**，把世界帶到中國門前，讓中國人通過香港投資世界，讓世界可以通過香港投資中國。

眾所周知，二十幾年前，香港與內地開創性地聯手推出 H 股機制，為內地的改革開放提供了寶貴的資金來源，也為香港金融業迎來了歷史性的發展機遇。如今，香港已經成為中國企業最主要的境外融資中心，香港交易所也成長為全球市值最高的上市交易所集團。

不過，我們深知，成績只屬於過去，奮鬥成就未來，尤其是在這個日新月異的時代。世界在變，市場在變，我們也必須及時調整自己的定位和步伐，與時俱進。放眼當下和未來，對於香港交易所而言，這個時代最大的改變是什麼？是中國已經由一個「渴求資本」的國家變為「資本充裕」的國家，中國金融體系的主要目標，也已經從單純強調大規模動員儲蓄、吸引外資，轉變為高效率、多元化、國際化地配置金融資源，來促進和支持中國經濟的進一步升級與發展。

在此背景下，中國內地的國民財富也開始了歷史性的「大搬家」，以前大

家的財富主要是趴在銀行賬戶上吃利息，主要追求保本；或者是大量集中在房地產等實物資產投資上，房地產投資的市值比居民銀行儲蓄還要大；而今天，中國的國民財富必須開始追求保值與增值。大家的財富紛紛從房地產投資市場、從銀行儲蓄開始往股票、債券等各種金融資產轉移和配置，也開始迅速走向海外配置分散投資風險，從全球範圍內來把握投資機會——套用一句現在的流行語，世界那麼大，誰不想去看看呢？這一財富「大搬家」的趨勢才剛剛開始，隨着中國經濟總量的不斷增長，未來十到二十年這一趨勢還將加速。

與此同時，隨着中國資本市場的對外開放和人民幣國際化的步伐加快，海外投資者對於投資中國本土市場的興趣也日益濃厚，現在國際貨幣基金組織（IMF）正在考慮將人民幣加入 SDR（特別提款權），MSCI 也正在考慮將 A 股市場納入其全球指數系列，相信這些發展都已不是加入與否的問題，而是多快、多大的問題。目前海外投資者對中國資產的配置在其投資組合中佔比仍然較低，未來增加配置也是大勢所趨。

面對這樣的歷史性大機遇，我們的角色也需要轉變，我們不能再僅僅局限於為內地發行人提供國際資本，還應該為內地投資者提供投資國際金融產品的機會，同時也為國際投資者提供渠道，直接投資內地資本市場。擁有「一國兩制」優勢的香港完全有能力「華麗轉身」，把連接世界與中國的角色再次升級。打個簡單的比喻，過去香港只需要服務兩位客戶，一位是需要融資發展企業的中國發行人，另一位是需要尋找財富增值機會的國際投資人；未來的香港需要服務好四位客戶：中國發行人、國際投資人、中國投資人、國際發行人。而發行人與投資人的定義也從狹義的上市公司與股民延伸到各類金融產品的發行者、使用者和投資者。

這種歷史性的角色轉變也是我們萌生建設境外人民幣生態圈想法的初衷，我們希望這個境外人民幣生態圈能夠充分利用香港作為「一國兩制」的國際金融中心的優勢，發揚光大滬港通開啟的交易模式，盡量採用統一的價格發現、依靠本地清算結算、以本地交易習慣為主，交易對方市場產品，倡導兩地監管機構之間共同監管、執法合作，為海外的人民幣提供豐富多樣的投資產品，建

立離岸和在岸人民幣對接的市場機制，以期用最小的制度成本，獲取最大的市場開放效果。只要這個生態圈裏的產品足夠豐富、交易足夠方便，願意持有人民幣的海外投資者就會越來越多，人民幣國際化也就能走得越來越遠，反過來又會提高香港這個境外人民幣生態圈的人氣，從而形成良性循環。而一個健康發展的強大的境外人民幣生態圈也為中國國民財富多元化、全球化的投資配置提供最安全可靠和可持續的國際財富管理中心，並與內地資本市場的其他對外開放措施（譬如自貿區發展）相輔相成、相得益彰。

滬港通朝這個方向邁出了第一步，我們也很快會邁出深港通這第二步，但是還遠遠不夠，未來我們還要與監管機構、市場參與者及內地合作夥伴攜手不斷拓展這個生態圈的內涵和規模，爭取早日把香港建設成為中國的離岸財富管理中心，衍生產品、大宗商品、固定收益類產品等等都是我們未來需要努力的方向。

身逢這樣的大時代，是香港交易所的幸運，未來的日子裏，我們會把成績歸零，向着新的夢想奮力前行，絕不辜負這個時代。希望在下一個 15 年回首往日的時候，我們可以昂起頭驕傲地說，我們真的改變了世界，香港的繁榮有我們的一份貢獻！

2015 年 6 月 22 日

05 淺談中國金融市場的開放路徑

隨着滬港通的順利推出，中國資本市場已經邁入了雙向開放的新紀元，開放不僅為兩地投資者都帶來了全新的機遇，推動了人民幣國際化的進程，其本身也已經成為人民幣國際化的重要組成部份之一。在剛剛落幕的陸家嘴金融論壇上，大家都在熱議如何穩妥擴大中國金融市場的開放和加快人民幣國際化進程，我們在準備滬港通和深港通的過程中也有一些體會和淺見，在此拋磚引玉，希望能與朋友們一起探討。

為什麼雙向開放已是必然趨勢？

縱觀人類經濟發展史，任何一個大國的崛起，必然離不開一個發達的金融市場，而金融市場的發展壯大又離不開「開放」二字。只有一個開放的金融市場才會有真正的廣度和深度，形成的價格信號才真正有代表性和影響力，才能夠吸引更多資金和投資者，支持實體經濟的發展。

當前中國經濟已與世界經濟幾乎完全接軌，所處的發展階段必然需要一個穩步開放的金融市場，逐步融入國際金融秩序。中國經濟的增長引擎正在從出口和投資轉向內需和創新，中國已經從資本進口國變為資本輸出國，一個開放和國際化的金融市場體系，有利於推動經濟體系中的創新活動，有利於支持中國企業以更高的效率在全球範圍內收購資產以提升競爭力，有利於居民和機構投資者優化金融資產結構來提高消費能力。

中國的人民幣國際化進程也需要進一步開放金融市場，為海外人民幣提供

豐富的投資機會。人民幣要成為國際儲備貨幣，必須首先成為世界廣泛接受的
貿易貨幣和投資貨幣。如今，隨着人民幣匯率步入均衡區間，人民幣單邊升值
的趨勢已經結束，人民幣國際化的主要驅動力，將越來越依賴於可供海外人民
幣投資的金融產品的豐富程度。

　　此外，金融市場的雙向開放有助於優化國家資產負債表，提高整個國家的
海外淨資產總收益，同時還可以通過採用多樣化的融資渠道和產品來降低一個
國家從國際市場融資的成本，鼓勵創業和創新。由於多年的貿易順差，中國累
積了巨額的外匯儲備，但大部份集中在政府手中，以外匯儲備的形式投資在外
國國債等流動性高但收益率低的資產，儘管中國的外匯儲備經營收益水平在國
際上已經算是比較高的，但是外匯儲備「安全第一」的特性決定了其風險偏好
很低，不可能追求高風險高收益。一般來說，私人部門的投資效率通常高於公
共部分，所以未來中國需要藏匯於民、逐漸拓寬企業和居民海外投資渠道，鼓
勵他們自己走出去投資高收益的資產，同時鼓勵企業充分利用國外金融市場上
的不同金融工具來靈活融資，降低融資的成本。

雙向開放可以有哪些路徑？

　　中國金融市場的開放和國際化無外乎兩種路徑：一是直接請進來，直接走
出去；二是「組團」請進來，「組團」走出去。

（1）直接請進來，直接走出去

　　直接請進來包括把海外流量（投資者）請進來和把海外產品（上市公司或
投資產品）請進來兩種方式，比如 QFII／RQFII 試點和上海自貿區即將推出的
國際原油期貨準備向國際投資者開放，屬於前者；而正在討論中的國際板則屬
於後者。無論哪一種「請進來」，都是將海外的流量和產品完全納入中國的監
管範圍之內，讓他們完全遵守中國的規則，就地國際化。

　　這是最大限度的延伸中國監管機構的監管半徑，也是最大限度的輸出中國標準，自然是最終極最理想的開放路徑，通過這一路徑實現雙向開放的時間與程度取決於內地市場與國際市場的制度和結構完全融合和接軌的速度與程度。鑒於目前內地金融市場與海外市場的制度差異，海外流量和產品要徹底融入內地市場需要進行不小的制度改革，因此，「請進來」的速度相對比較慢，想達到一定規模也需要時間。舉個簡單的例子，如果國內的監管理念是以保護中小投資者為先，就必須在交易結算各環節中預先設定防止「壞人」為非作歹的一系列防範措施（包括各種審批和限制），而很多海外市場監管機構通常是假定大家都是好人，在交易結算環節也是以保障市場自由為先，只有在「壞人」犯事以後才開始狠狠追責。再譬如乘火車，我們國家是要求乘客檢票進站再上車，車上驗票，出站再檢票，這樣自然可以完全杜絕無票乘車；但如果為了方便乘客和提高效率，也可以像國外很多火車站那樣進站和出站都不檢票，在車上檢票，一旦發現逃票嚴懲。兩種監管理念很難說孰優孰劣，但習慣了海外市場的監管理念的機構和發行人要融入內地市場肯定需要付出不小的轉換成本。

　　直接走出去包括允許中國的投資者和中國的產品直接走出國門，比如目前的 QDII 和即將推出的 QDII2 試點制度是讓中國的投資者走出去，而中國公司在海外發行股票和債券屬於中國的產品走出去。這種開放模式下，國內的流量和產品一旦走出去，就完全脫離了內地監管機構的監管，需要徹底接受海外市場的規則。這一開放模式的速度和規模很大程度上取決於人民幣資本項目開放的速度，而且，選擇這一路徑「走出去」的機構和個人需要足夠的知識儲備和風險承受能力，並不適合所有人，比如，絕對不適合風險承受力較弱的普通散戶。

　　直接請進來與直接走出去就好比旅遊市場上的自助遊：那些對中國充滿好奇、又會點中文、喜歡獨立和冒險的外國遊客不喜歡跟團旅遊，他們喜歡獨來獨往、自助來華旅行；同理，越來越多的中國人（尤其是那些會英語、自理能力強的）更願意自由行出去看世界。

（2）組團請進來、組團走出去

相對於以上所説的「自助遊」，還有一種「組團遊」。全球有海量的國際投資者都十分關注中國，在過去 20 年裏，他們都只能投資在海外上市的中國公司（如 H 股等），他們越來越嚮往直接參與中國資本市場投資，可是，由於交易習慣、治理規則、投資文化等多種因素，他們在相當長一段時期內還不能或不願來中國「自助遊」，他們更習慣使用美國旅行社統一安排來華。同理，未來 10 年裏，越來越多中國投資者希望通過投資國際市場來分散風險和多元化配置個人資產，但他們不懂英文，不熟悉海外規則，還不能或不敢直接走出去，就如同很多中國遊客更喜歡參加旅行團出國遊，因為可以聽中文講解、吃中餐、在中國導遊的悉心照顧下走世界。

在今天的中國資本市場上，滬港通就是這種「組團遊」的一個最佳實踐。在滬港通機制下，投資者並沒有扛着錢過境，只有交易訂單通過兩家交易所總量過境撮合，實現最大價格發現；而股票結算與資金交收則通過本地結算所先在本地進行對減，然後再以淨額方式與對方結算所作最終結算，實現最小跨境流動。滬港通採用人民幣境外換匯，實現全程回流，其結算交收全程封閉，實現了風險全面監控。

由於它最大限度地轉化了兩邊市場的制度差異、保障了投資者原有的投資習慣，這一開放路徑將在未來很長的時間裏成為中外投資者互聯互通的重要方式。在監管層面，「組團遊」的方式使境內外的監管當局能夠延伸其監管半徑，對處於中間地帶的「共同市場」進行聯合審批、共同監管。而且，由於它已經部份處於內地監管機構的監管範圍內，風險可控，非常適合短期內還不願或者不敢直接走進來／走出去的投資者和發行人。對於中國的中介機構而言，這一開放路徑不僅為其帶來了增量客戶與業務，也提供了難得的國際化機遇。

滬港通這個「組團遊」試點首先面對兩地二級市場上的股票，下一個「組團遊」的目的地是深港通，將來它們的投資標的也可以延伸到股指期貨、一級市場、商品和固定收益產品等一系列資產類別。就像一般的遊客出國遊可能會

先從周邊國家的短線遊開始，然後再嘗試比較長線的目的地自由探索深度遊。

在我看來，中國資本市場一直以來採取的都是多層次、多維度的開放政策。以上這些不同形式的開放路徑分別針對不同類型的投資者，可協同發展，而且，它們都是中國的經濟轉型所必需的，能夠滿足投資者多樣化的全球資產配置需求。

作為一種漸進式的開放路徑，基於滬港通模式的境外人民幣生態圈與前兩種開放方式不僅沒有衝突，而且能相輔相成、互為補充，共同促進人民幣國際化。我們相信，在充分控制風險的前提下，開放程度越高，市場主體的選擇越多，金融市場的效率就越高，對經濟增長和轉型的支持就更為顯著。

香港可以貢獻怎樣的價值？

經常有朋友問我，在中國金融市場開放的進程中，香港能夠貢獻什麼價值？作為中國的門戶和一個國際金融中心，香港最大的價值在於它可以利用「一國兩制」的獨特優勢，緩衝兩邊市場的差異，連接世界與中國。滬港通就是一個典型的例子，通過滬港通的創新安排，只需最低的制度成本，海外投資者可以以自己的交易習慣在香港投資中國，內地的投資者也可以通過內地券商以近似A股的交易習慣來投資港股。

香港還有一大價值就是充當人民幣匯率和利率改革的「試驗田」。當前，中國的利率市場化改革接近收官階段，匯率市場化也在穩步推進。作為金融市場基礎價格標竿的匯率和利率的改革面臨着「牽一髮而動全身」的複雜性，必須審慎推進；在中國的利率管制基本完全放鬆之後，其實更大的挑戰還在於在金融市場上形成能引導市場預期的利率信號。在此背景下，香港這塊離岸人民幣市場「試驗田」的作用將更加重要。在香港這個離岸市場上形成的利率和匯率信號，是在充分開放的市場條件下的形成的，它既可避免風險向內地市場的直接傳導，又可為繼續穩步推進人民幣匯率和利率改革提供參考。

我們正在香港建設的境外人民幣生態圈正是希望充分利用香港的獨特優

勢，與內地的合作夥伴和監管機構一起開創共贏。香港和國際市場上的人民幣金融資產越豐富，人民幣進行國際化的基礎就越堅實。我們期望，能在不久的將來，把香港建設成為中國的離岸財富管理中心，為中國的金融市場開放和人民幣國際化進程做出應有的貢獻！

2015 年 6 月 28 日

06 | 詳解 《戰略規劃 2016-2018》

不久前，我們公佈了最新的 3 年戰略規劃，勾劃出香港交易所未來的發展藍圖。不少朋友表示很感興趣，想了解整個規劃的來龍去脈，也有朋友表示困惑，香港為什麼要發展貨幣產品和大宗商品呢？當時限於時間，無法一一當面解答，在此，請聽我慢慢道來。

簡單來說，《戰略規劃 2016-2018》的主題就是三個「通」字：「股票通」、「商品通」與「貨幣通」。定價能力是一個金融中心的核心競爭力，具體而言，就是為公司、為商品、為貨幣定價的能力。

之所以強調「通」字，是因為中國雖已成為世界第二大經濟體，但它的金融市場還未與全球完全互聯互通，中國的公司、商品和貨幣還不能取得完全有效的國際定價。因此，我們與我們在內地的合作夥伴今天的歷史使命就是共同攜手建立中國與國際市場的互聯互通機制，連接中國與世界，重塑全球市場格局。

「通」的過程，是滿足市場需求的過程，也是做大整個市場蛋糕的過程，在這個過程中，所有的參與者都會不同角度地受益。中國資本市場的對外開放是多維度的，也是漸進式的，在這個開放的過程中，香港交易所也僅僅是眾多的「修橋者」之一，肯定會有無數的「大橋」落成造福兩邊的投資者，但我們願意成為最早的探索者；我們不怕失敗、尋求合作、創造共贏。

股票通

首先來說說「三通」中最基礎層面的股票通，股票業務是香港交易所的核心傳統業務，我們一直在思考如何與時俱進強化這一傳統優勢。股票通可以分別從滬深港通、「股指期貨通」，以及「新股通」（「三小通」）三個方面實現：

第一，滬港通的順利推出，算是我們邁出了股票通的第一步，首次實現了內地和香港股票二級市場的互聯互通，為兩地投資者都帶來了全新的投資機遇，也引發了全球金融市場對於交易所互聯互通模式的關注。接下來，我們將進一步優化和拓展滬港通，儘早推出深港通。

第二，實現「股指期貨通」的可能路徑有三種：一是複製滬港通模式，與內地期貨交易所合作建立互聯互通；二是兩地交易所產品互掛；三是在香港發展以 A 股指數為標的的衍生產品，滿足國內外投資者管理 A 股波動風險的需求，為大量外資進入 A 股現貨市場提供必要的配套設施。第一種路徑和第二種路徑原本是「股指期貨通」最簡單的實現方式，但由於目前內地市場正處於休養生息時期，短期內這兩條路徑都充滿挑戰，關於第三種路徑，我們也在探討其可行性。

第三，我們也在研究如何將股票通模式從二級市場延伸到一級市場推出「新股通」，讓兩地市場的投資者都可以在未來申購對方市場的新股。香港一直希望吸引更多國際大型企業來港上市，豐富上市公司的來源和類別。但香港市場目前投資者基礎和歐美市場相近，難以吸引海外名企大規模來上市，如果「新股通」可行，中國投資者可以以滬港通模式更大規模地投資國際公司。對於內地市場而言，「新股通」可以幫助其引入更多國際機構投資者，參與 A 股一級市場新股發行或定向增發，改善投資者結構，完善新股定價機制。

有朋友可能會問，這樣的「新股通」與上海或深圳的「國際板」是什麼關係？很明顯，如果內地市場的法律框架能夠在短期內迅速與國際接軌，適應國際公司在 A 股的上市融資，那麼香港再搞「新股通」的意義就只是為國際發行人引入中國投資者提供了多一個選擇。但如果中短期內實現是有困難的，「新

股通」就意義重大。在這裏，大家不應把這個問題簡單狹隘地看成兩地交易所爭奪上市資源、此消彼長的競爭問題，更應把它看成兩地市場共同為中國國民財富尋求更多元、更有效全球資產配置的機會，因為這才是交易所使命的真諦。如果從這個出發點看問題，儘早允許一級市場的互聯互通，為中國投資者提供在中國市場框架下投資海外的機會，可以方便他們在全球市場低迷時利用滬港通安全可控的運營模式，儘早佈置海外資產配置。同時，內地公司若有機會吸引國際投資者成為其股東，也可大大改變內地投資者結構，促進發行機制的改革。而兩地交易所若能攜手聯合發行、協調監管，「新股通」將成為雙方互惠共贏、共同發展的大好契機。

在香港，不時有些朋友問我，滬港通啟動之後去年香港市場似乎比以前波動更大，進一步的互聯互通不會輸入更多內地風險嗎？你們不怕得不償失嗎？我想說的是，全球市場一體化的進程是不可逆的，如今中國經濟感冒了，美國市場都得打噴嚏，更何況是香港市場。無論是否與內地互聯互通，我們作為一個開放型的市場都得承受來自中國經濟波動的風險。換句話說，如果不與內地互聯互通，我們好事遇不上，壞事也躲不開。

商品通

「三通」中的第二通是商品通，主要也包括「三小通」：

一是「倫港通」：把亞洲的流動性尤其是中國的流動性導向倫敦金屬交易所（LME），讓 LME 更加金融化。「倫港通」的核心是將在 LME 交易的合同帶到亞洲時段，通過香港交易所的平台進行交易結算，擴大 LME 的亞洲投資者基礎，特別是在香港的中國投資者的參與。

二是「現貨通」：通過在內地構建大宗商品現貨交易平台，幫助內地大宗商品市場實體化；

三是「內外通」：在「倫港通」與「現貨通」成功運作一段時期後，我們可以通過產品互掛、倉單互換、價格授權、指數開放等方式有機地打通內地現貨

市場和期貨市場，實現內地商品市場與國際商品市場的連通，輸出中國的商品定價權。

不少朋友對我們商品通中的「現貨通」充滿好奇或疑惑。國內外大型商品交易所都是期貨交易所，香港交易所為什麼要去做現貨交易？會不會得不償失？

自從 2012 年收購 LME 之後，我們對國際大宗商品市場有了更深刻的認識。而熟悉股票業務的香港朋友也可以從股票的視角來看大宗商品。

大宗商品市場的交易通常可以分為三個層面：

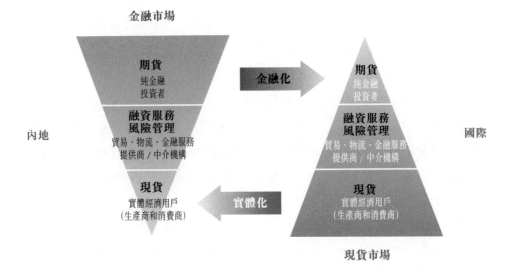

第一層是實體經濟用戶（生產商和消費商）：他們通常是現貨市場的參與者，他們交易出的現貨價格就好像股票市場的 IPO 發行，是商品市場定價啟動的原動力，為其他交易提供價格基準。

第二層是服務實體經濟用戶的貿易、物流、金融服務提供商/中介機構：他們交易的目的是融資服務、對沖，與風險管理，他們參與現貨、遠期以及期貨交易，這些交易就類似股票上市後的二級市場交易。這些交易不斷為第一層實體經濟交易提供價格發現與交易對手方，使一級市場活動持續發展。

　　第三層是純金融投資者，包括投資基金與散戶：他們參與交易的動機主要是投機。他們通常只參與非實物交割的標準期貨產品，但他們為商品交易提供重要的流動性，他們的交易就如同股票市場的期權、期貨等衍生產品交易。總之，這三層的結構有點像金字塔，第一層和第二層交易所形成的價格基準是第三層交易的基礎，就如同股票市場，沒有 IPO，就沒有二級市場交易；沒有二級市場，就不可能有衍生品市場。一個健康的多層次商品市場，應該從滿足實體經濟需求出發，擁有這樣的正金字塔形架構。

　　LME 控制了全球近 80% 的基礎金屬交易，三個層面的用戶均有覆蓋，但主要參與者是第一層與第二層，第三層的純金融投資者參與不如國內市場充分，還有很大的發展空間。它是個典型的正金字塔。我們希望通過倫港通、在香港開發更多類 LME 產品等方式，方便更多的純金融投資者（尤其是亞洲金融投資者）參與 LME 的交易，讓 LME 更加「金融化」。

　　另一方面，內地商品市場的現狀與 LME 有很大不同：儘管中國已是世界上主要大宗商品的消費國，但第一層與第二層用戶（即實體經濟用戶和貿易、物流、金融服務提供商/中介機構）參與的現貨市場還不太發達和規範，市場結構鬆散，價格發現效率不高，倉儲認證增信環節薄弱，流通領域融資困難，整個現貨生態系統效率低下，成本高企，風險叢生。

　　與此同時，第三層的期貨市場監管充分，結構完整，交易異常活躍，且由於期貨市場投資羣體以散戶為主，實物交割有限，實體經濟用戶參與較少，交易量遠遠大於持倉量。正因為此，它與現貨市場的相關度還不夠高，大宗商品市場的金融服務遠遠不能滿足實體經濟需求。我們相信，中國內地需要一個更有效率的大宗商品市場，來真實反映其於全球大宗商品生產及消費方面的影響力。國際市場同樣也需要這樣一個有效率的大宗商品市場，來觀察中國這樣一個具有巨大影響力的市場的變動趨勢。

　　作為香港交易所內地商品戰略的重要一環，現貨通可以將 LME 的成功模式和歷史經驗移植到內地市場，在內地打造一個能夠有效服務實體經濟的商品現貨交易平台，幫助內地商品市場「實體化」。假以時日，我們希望中國的現

貨市場可以產生一系列真正具有全球影響力和代表性的「中國價格」基準，而這些價格基準也將為向第三層發展提供堅實的基礎與持續的發展動力。

也許有朋友會問，香港發展商品業務為什麼一定要去內地建現貨平台而不是直接在香港做期貨？香港一向缺乏商品交易的土壤，即使在收購了 LME 之後，香港交易所在亞洲時段也不具備上述三個層面中的任何一層。要構建一個穩固結實的商品金字塔，香港首先要做的是打地基 —— 建現貨交易平台，這有點類似當年香港先發展 H 股上市現貨平台再開發股指期貨。但香港面積有限，並不具備建設商品現貨交易平台包括倉儲物流系統的條件。與發展股票業務不同的是，大宗商品不會像 H 股上市公司一樣自己可以主動「跑」來香港上市。我們必須去內地開創一個大宗商品的「IPO 平台」。因此，有現貨平台需求的內地成為香港交易所發展第一層與第二層市場的必然選擇。

貨幣通

「三通」中的最後一通是貨幣通，就是大力發展定息及貨幣產品，借助互聯互通把香港打造成利率和匯率衍生產品中心，具體可以分為「匯率通」、「債券通」和「利率通」三步走。

幾年前，我們推出了管理美元兌人民幣匯率風險的人民幣貨幣期貨合約，算是邁出了「匯率通」的一小步。去年以來，隨着人民幣匯率雙向波動風險加大，人民幣貨幣期貨合約的成交量穩步上升。未來，我們還將推出人民幣兌其他貨幣匯率的期貨產品，滿足更多市場需求。

此外，我們還準備從「債券通」入手，豐富香港的貨幣類現貨產品。「債券通」的初步構思為雙向模式：境內和境外。境內方面，內地債券交易大多集中在場外交易（即銀行間債券市場），因此，方便海外投資者參與國內銀行間債券市場是我們的主要着眼點。境外方面，香港本身欠缺大型國債發行人，恐怕吸引力不足。所以，我們希望與國際平台合作，方便內地投資者投資更多不同類型的國際債券，而交易及結算則仍然在香港進行。

　　我們希望借助「債券通」為香港引來足夠的投資者和流動性，為將來的利率通做準備。隨着人民幣國際化進程的深入，香港的人民幣資金池會不斷擴大，管理人民幣利率風險的需求可能會逐漸增加，我們將適時推出相關的利率產品，完善香港的人民幣生態圈。

　　在「債券通」方面，香港交易所過去沒有發展經驗與基礎，也不具備獨特優勢。但從金融中心的長期發展來看，我們必須邁出這一步，希望能找到好的合作模式，以可控的投資、合理的定位，取得事半功倍的效果。

　　以上就是我們未來三年的發展計劃。從今年的開局來看，2016 年注定仍是充滿挑戰的一年，中國資本市場對外開放的步伐也可能時快時慢。因此，我們必須尋求一個多元平衡的戰略發展規劃。我們的某些發展計劃（如股票通），與內地市場的改革開放息息相關，需要與內地的夥伴合作才能創造共贏，其推進程度可能取決於內地市場開放的進程，但也有一些發展計劃（如商品通），它們的成功將更多取決於我們自己的努力。

　　總而言之，未來的 3 年要做的事情很多，要克服的困難的也不少。有的朋友也許會說：你這些目標太宏大了，簡直就是夢想清單。是的，這就是我們未來 3 年的夢想。回想香港金融市場的歷次轉型，其實也大多起源於前輩們在洞察市場需求之後的大膽設想。敢想才能敢拼，沒有夢想，何有未來？套用一句香港電影的經典對白：「做人如果沒有夢想，跟鹹魚又有什麼分別呢？」

　　還有幾天，就是農曆猴年了。我在此恭祝大家身體健康，投資有道，事事亨通！

2016 年 2 月 3 日

07 | 香港金融市場的新征途

　　今年以來，國際市場對於中國經濟和金融穩定的憂慮似乎傳導到了香港。先是國際評級機構調低了香港的評級展望，之後不久，在一家英國調查機構公佈的全球金融中心指數排名中，香港跌出三甲，落後新加坡。一時間有不少悲觀情緒彌漫，有些朋友開始對香港國際金融中心的地位和獨特優勢產生懷疑，我一些在香港的朋友質疑香港的發展為什麼非要依靠內地，我亦有不少內地朋友認為強大的中國今天已經不再需要香港。

　　香港的優勢還在嗎？未來 30 年，香港還能延續東方之珠的璀璨嗎？

過去三件事

　　要回答這些問題，不妨先回望一下香港的過去。過去 30 年是中國改革開放的 30 年，也是內地經濟發展最快的 30 年。在這個 30 年中，香港為中國的改革開放主要做了三件大事：第一，轉口貿易；第二，直接投資（FDI）；第三，資本市場的大發展。轉口貿易給中國帶來了第一桶金，FDI 直接投資把中國變成了世界的工廠，而香港資本市場的大發展則為中國源源不斷地輸送了發展經濟的寶貴資本，從 1993 年 H 股誕生開始，一家又一家中國的公司在香港上市募集來自全球的資金，發展成了今天世界上最大的電訊公司、能源公司、銀行和保險公司。在改革開放這三大潮流中，香港都憑藉自己獨特的優勢為中國做出了巨大貢獻，起到了不可替代的先鋒作用。當然，在此過程中，香港自己也

收獲了繁榮富強，成為全球認可的國際金融中心。

香港過去為改革開放所做的三件大事都有一個核心主題，那就是為中國輸入資本，因為那個時候中國很缺錢，香港一直是中國最可靠的海外資本聚集中心。

細心的朋友也許會問，現在中國已經成為一個資本充裕的國家，國內到處是尋找投資渠道的資金，不差錢的中國還會像過去一樣需要香港嗎？

我的答案是肯定的。香港的開放市場、法治環境、與國際接軌的監管制度與市場體系、專業人才和中英雙語環境在今天和未來仍然是中國非常寶貴的「軟實力」，它們是成就一個國際金融中心必不可少的條件，需要經過幾代人的積累與努力。內地現在也已經越來越開放，不少城市的硬件環境也已經趕超香港，但香港作為國際金融中心的獨特優勢依然十分突出。只不過，時移世易，香港需要與時俱進。未來的 30 年是中國資本雙向流動的時代。除了發揮其傳統的融資功能外，香港還須盡快調整定位和轉型，做好新的三件事。

未來三件事

第一件事是幫助中國國民財富實現全球配置。隨着中國經濟的崛起，中國的國民財富進行全球性分散配置的需求已經出現。從幾年前開始，中國內地的國民財富逐漸開始了從房地產和銀行儲蓄走向股市和債市、從單一的國內資產配置走向全球分散配置的歷史性「大搬家」。國內現在所謂的「資產荒」，其實質就是目前國內的產品遠遠不能滿足國民的資產配置需求。一些「勇敢的人們」已經開始將資產直接投資海外，但絕大多數中國投資者還暫時沒有能力直接「闖世界」。與此同時，中國在相當一段時間裏仍須對資本外流進行管制，我們若能加快發展類似滬港通和深港通這種安全、可控的互聯互通模式，把世界帶到中國門口的香港，讓中國人「坐在家裏投世界」，中國就會再次通過香港找到開放和繁榮的捷徑。

第二件事是幫助中外投資者在離岸管理在岸金融風險。在今天的內地市

場，利率與匯率尚未完全開放，因此內地的債務與貨幣衍生品市場也還沒有建立成熟的風險管理機制。內地的股市雖已經完全市場化，但具有雙刃劍效果的衍生品市場仍會在股市大幅動盪時受到限制，難以完全發揮風險對沖功能。衍生品市場的缺失使得大量有意持有中國資產（包括股票、債券與貨幣）的國際投資者在國門外望而卻步；而國內大量高風險偏好的資金也因找不到合適的高風險高回報投資目標而在內地市場中「東奔西竄」，形成風險隱患。

從這個角度看，香港完全可以幫助內地市場「另闢戰場」，讓中外資金在與內地市場有一定安全隔離的境外市場充分博弈、對沖風險。香港擁有國內外投資者都認可和熟悉的法治和語言環境，在這裏，海內外的投資者最容易各取所需、形成良好的互動。因此，只要充分了解雙方投資者的需求，提供適合的產品，香港完全有條件發展成為亞洲時區內最主要的國際風險管理中心。內地的期貨交易所也可以在內地股市休養生息時與香港市場合作，在境外繼續發揮風險管控的功能。

在匯率方面，人民幣匯率已經進入雙向波動時代，無論是國內的進出口貿易商、QDII 基金還是國外的 QFII 基金和使用滬港通的投資機構，都有管理人民幣匯率風險的需求，我們推出的人民幣匯率期貨，就是為他們度身定做的風險管理產品。事實上，從人民幣匯率期貨今年以來的成交量分析，不難發現一些洞悉先機的投資者已開始利用我們的產品去管理人民幣匯率波動風險。今年，我們計劃推出更多人民幣貨幣期貨，包括歐元兌人民幣、日圓兌人民幣、澳元兌人民幣以及人民幣兌美元期貨等品種，更好地滿足市場需求。

第三件事是幫助中國實現商品與貨幣的國際定價，為中國的資金定價海外資產提供舞台。未來 30 年，是中國資金不斷向外走的 30 年，中國的投資者不能再像中國的外匯儲備一樣，只當「債主」（特別是美國的債主），我們也要學着走向世界、用我們的錢來「定價」我們要買的海外公司股權、我們要買的大宗商品；同時，也要用我們的購買力讓越來越多的國際權益與商品以人民幣定價，這樣，我們就能在全球範圍內逐步掌握人民幣匯率與利率的定價權。

要想成為一個真正的國際定價中心，一個市場必須具備中外各方均能接受

和認可的規則體系與制度安排，也需要高度國際化和專業化的市場服務環境。擁有「一國兩制」優勢的香港既是內地投資者的「主場」，也是外國投資者的「友場」，具備以上所有條件，完全可以成為中國首選的海外定價中心，讓中國的資本通過這個定價中心取得在全球金融體系中應有的話語權和影響力。

如果能夠做好以上這三件大事，香港一定能夠再次為中國的發展創造不可替代的價值，也自然可以鞏固自己國際金融中心的地位。

「一國兩制」是關鍵

香港做好這三件事的關鍵在於「一國兩制」。「一國」使內地可以放心地讓香港去努力創造這三個新的中心；「兩制」可以讓國際社會充分擁抱香港這三個新的功能。「一國兩制」是一個並不輕鬆的政治話題。作為香港交易所的集團行政總裁，我觸及這個話題的唯一原因是它將直接影響上述香港金融市場未來發展的新方向、新功能。換句話說，上述願景的實現取決於內地是否相信「一國」不會受到挑戰；也取決於香港及國際社會是否相信「兩制」會長期持續。

從我在香港二十多年的生活體驗來看，絕大多數的香港人從未對「一國」有過異議。有的內地朋友會說：香港人既然擁護「一國」，那為什麼沒有表現出更多的愛國熱情？為什麼沒有更堅決地反對損害「一國」的雜音？恰恰就是因為「兩制」，香港在處理這些問題的方式上與內地就會有所不同。只要反對「一國」的聲音屬雜音，而不是主旋律，就應該允許香港以自己的方式去處理雜音。在某種意義上，今天小角落裏的雜音沒有被主旋律淹沒，可能也是因為香港的主流開始對「兩制」能否持續產生了一些憂慮。

反過來看，香港人也應有自信，內地的主流從未想要損害「兩制」，因為維護「兩制」符合中國發展的根本利益。中國已很富強，她並不需要再多一個上海、深圳或廣州，她需要一個與眾不同的香港。香港人若沒有這份自信，我們就會與「兩制」漸行漸遠。所以說，香港人保證「兩制」的關鍵是擁抱「一國」。對內地來說，讓香港人擁抱「一國」的關鍵是保障「兩制」。

正確的問題

我們再次回到開篇提及的質疑與負面情緒。今天的香港充滿挑戰，共同走出困境需要大家的集體智慧。我們可以嘗試的第一步是拒絕回答錯誤的問題，努力提出正確的問題：

香港的繁榮能否離得開中國的發展？這是錯誤的問題。正確的問題應是：香港的繁榮為什麼要離開中國的發展？

強大的中國為什麼還需要香港？這也是錯誤的問題。正確的問題應是：有一個繁榮、穩定、自信的香港是不是對中國更好？

如果我們問對了問題，我們就會得到正確的答案：香港的繁榮不應離開中國的發展；中國的發展更應利用香港的獨特優勢。過去如此，今後亦然。

2016 年 4 月 20 日

08 香港可以成為「一帶一路」的新支點

今年是香港回歸祖國懷抱的第 20 個年頭，香港和內地都舉辦了一系列的慶祝活動來回顧香港的昨天，展望香港的明天。慶賀之餘，大家不約而同關心一個問題：中國經濟已經騰飛，未來香港如何延續過去的輝煌？

在我看來，「一國兩制」下的香港擁有獨特的優勢，是連接中國與海外的轉換器。只要與時俱進、找準自己的定位，香港一定能在中國未來的發展中繼續扮演重要的角色，比如，在中國牽頭的「一帶一路」發展戰略中，香港就有機會成為最重要的槓桿支點和權益轉換器。

為什麼這麼說呢？首先讓我們來看看「一帶一路」到底是什麼。不同的人對於「一帶一路」戰略有着不同的理解。在很多香港朋友看來，「一帶一路」是非常遙遠的事情，跟香港好像沒有什麼關係。在我看來，「一帶一路」既不是中國版的馬歇爾計劃，也不是簡單的產能輸出，而是一項以基礎設施互聯互通為突破口、帶動歐亞大陸經貿合作、從而實現互利互惠共同發展的重大國家發展戰略。在當前發達經濟體的逆全球化風潮下，「一帶一路」實際上提出了一個在新的環境下推動全球化的大方向。香港因把握全球化趨勢而興，在當前國際經濟金融格局變幻的大背景下，如果我們能夠充分認識「一帶一路」對全球經濟和政治格局的深遠影響，如果我們能夠有效發揮香港「一國兩制」的制度優勢，我們就會找準香港的定位，就能成就下一個 20 年的輝煌。

這樣的信心來自哪裏呢？請允許我在此分享一些思考。首先，既然不是馬歇爾計劃，「一帶一路」就是遵循市場規律的投資，不是無償援助，因此我們必須關心錢從哪裏來；第二，既然不是簡單的產能輸出，那麼沿線國家就必

須「購買」中國的產能進行基礎設施建設。因此我們也必須關心錢從哪裏回的問題；第三，既然是大規模的基礎設施建設，就必然需要用到大宗商品，如果全球大宗商品價格迎來新一輪牛市、或者出現價格劇烈波動，我們必須關心如何影響全球大宗商品定價、並且管理其波動風險的問題；第四，既然是中國領投「一帶一路」建設，人民幣必將有重要的戰略機會，努力成為「一帶一路」建設中最重要的國際貨幣，因此，我們必須關心人民幣國際化進程中離岸價格的形成和風險管控。在這四個既然與必須中，我們都可以清晰地看到「一國兩制」下的香港擁有的優勢。

一、錢從哪裏來？

目前很多「一帶一路」沿線經濟體的發展程度差異性巨大，不少經濟體尚不發達，無力購買中國產能，雖然有些經濟體有礦產資源或市場潛力，但由於政局不穩、投資環境欠佳，資源變現很困難，投資回報很難保證，可謂花錢容易掙錢難。因此，「一帶一路」建設能否成功推進，一定要解決投資的「錢從哪裏來」與「錢從哪裏回」這兩個問題。

先看看「錢從哪裏來」，由於中國是「一帶一路」的宣導者和推動者，「一帶一路」基礎設施建設項目多以中國主導的亞投行和中國政策性銀行提供貸款的形式啟動先期投資。但是，建設「一帶一路」不可能是政府一家的事，必須引導市場的資金，也不能單靠中國一國之力，必須爭取更多的國家和市場加入「一帶一路」的「朋友圈」。也就是說，「一帶一路」應該爭取以儘量少的政府資金來撬動儘量多的市場資金，以儘量少的中國資金撬動儘量多的國際資金。在這裏，政府是領投人，市場是跟投人；中國是領投人，世界是跟投人。

既然要撬動最大的資源，「一帶一路」投資必須找到一個最有效的槓桿支點，這個支點可以放在國內，用中國政府的錢來撬動中國市場的錢，但這樣一來，「一帶一路」建設的風險就全部鎖定在了國內，擔子完全落在了中國人的肩膀上。「一帶一路」的成功將為世界經濟帶來巨大的增長機遇，全球資金

中不乏願意共擔風險享受高回報的積極參與者，從這個意義上講，以香港為支點，可以撬動更海量的國際資金參與到「一帶一路」的建設之中。

二、錢從哪裏回？

如果「一帶一路」建設的資金是靠槓桿撬動市場而來的，那市場需要的收益又如何回籠呢？答案是國際資本市場。資本市場就是各種金融需求進行權益互換的場所，有人用今天的錢去換明天的收益，也有人願意用明天的收益換取今天的投資；有人願意把錢借給有信用的人，也有人願意把錢借給有抵押品的人；有的抵押品是實物資產，未來的收益可以證券化作為抵押品來換取今日的資金。跨地區、跨時間、跨行業進行資源配置，正是金融中心可以大顯身手的地方。

具體到「一帶一路」，我們應該讓沿線有資源的經濟體將其資源在一個大家公認的市場打包上市，形成有價證券，使其資源迅速形成今天就可以利用的市場價值，以補充其欠缺的償還能力，用抵押或其他形式來保證中國和國際投資者的回報安全。由於抵押權益的市場價值直接影響資源國的償付能力和融資功能，這樣的安排可以將資源國的市值管理與基礎設施投資者的投資安全放在一個利益共同體中。

這樣的利益共同體實際就是一個權益轉換器，它用一個安全有效的機制，把今天的錢換成明天的收益，用今天的股權抵押了未來需要償還的債務，把貨幣投資換成資源的未來收益。這裏的權益人包括中國產能、中國資本、世界資本、資源國，能夠保證這一權益互換的是一個大家都信任的資本市場。

「一國兩制」下的香港就是這樣一個神奇的轉換器，「一國」決定了香港是值得中國信任的地方，「兩制」下與國際接軌的法治和市場環境則為香港贏得了海外投資者的信任，吸引了來自全球的資金。

三、如何管理大宗商品價格波動？

中國目前已是全世界最大的原材料消費國，但是並未取得與其經濟實力相匹配的國際大宗商品定價權。隨着「一帶一路」建設的推進，必然需要消耗更多的大宗商品，如果大宗商品價格迎來新一輪牛市，中國勢必處於比較被動的位置，應該及早打算，提升對於國際大宗商品的定價權。

儘管目前內地期貨交易所的成交量已位居世界前列，但由於尚未對外開放，暫時未能形成有國際影響力的基準價格。「一帶一路」建設橫跨歐亞大陸，必然需要使用有國際影響力的基準價格。在這方面，連接東方與西方的香港可以貢獻獨特的價值。香港近幾年已大步發展大宗商品業務，通過海外併購與在岸建設，香港有望在不久的將來建設大宗商品的深港倫互聯互通。

四、如何讓人民幣國際化與「一帶一路」比翼雙飛？

「一帶一路」建設將會面臨很多風險，包括政治風險、債務風險、匯率風險等等，這些風險中最容易被管控但也最容易被忽視的是匯率風險。如果「一帶一路」建設仍然使用美元或其他外幣結算，匯率風險很難管控。如果「一帶一路」建設儘量使用人民幣作為結算貨幣，相對而言，更容易把風險的管控權掌握在我們自己手中。

目前，香港已經是世界上最大的離岸人民幣中心，不僅擁有大量的離岸人民幣存款，而且開發出了多種多樣的離岸人民幣投資工具和風險管理工具，包括人民幣兌美元期貨、以人民幣計價的金屬期貨、人民幣債券、滬港通、深港通、債券通等等，可以滿足不同的海外機構對於人民幣的資產配置和風險管理需求。香港完全有能力成為「一帶一路」建設中人民幣國際化進程的助推器。

總而言之，無論是在解決錢從哪裏來、錢從哪裏回的問題上，還是在解決大宗商品與人民幣價格風險管控方面，香港都能為「一帶一路」建設做出獨特的貢獻。

對於金融圈的朋友來說，以上道理不難理解。但對於普通市民來說，這些道理並不一定那麼顯而易見，我 80 歲的老母親就特別關心國家大事，不止一次地問我：「習大大要搞一帶一路建設了，你們香港有份兒參與嗎？」我就給她打了個通俗的比喻：地球好比是一個鄉鎮；中國是這鎮上的一個大戶人家，經過幾十年的努力，現在家大業大，有錢有設備，勞力充裕。這個大戶人家的鄰居有住在鎮郊田野裏的農戶（「一帶一路」區域沿線的中西亞國家）和鎮上的其他富裕人家（歐洲經濟圈）。住在鎮郊田野裏的農戶仍很貧窮，家裏未通水、未通電，路也坑坑窪窪，但他們家裏有大棗樹（資源），他們靠賣當年收穫的棗勉強生存。作為大戶人家的中國雖然很希望帶領這些農戶共同致富，但是也不能為這麼大一片田野無償地修路、引水、通電。

在這種情況下，大戶投資人家必須動員全鎮其他富裕人家共同參與投資、共擔風險、共同成功，同時，今天沒錢但有棗樹的農戶可以將棗樹未來的收成拿到鎮上的「錢莊」估值，形成大家都能接受的價格，富戶們都可以認購一部分「棗證」。農戶們不僅可拿到一些錢培育棗樹多產，保證棗樹的價值增長，還可以用自己的「棗證」抵押，換來大戶們出錢出力出設備，讓鄉野之間通路通水通電，為了贖回自己的「棗證」，或者為了融到更多資金，農戶們自然會把棗樹養得越來越大，實現共同發展。

無論如何給農戶家的棗樹定價，都需要一個交易定價中心，好在富人們在鎮上開了不少錢莊（資本市場）。可是，到哪家「錢莊」估值定價好呢？由於大戶人家是領投，自然希望估值在自家錢莊，而其他富戶雖然是跟投，但也想用自家的錢莊估值，而有棗樹的人家則希望找第三方開的錢莊，大家爭執不下。其實，大戶人家的海歸兒子（香港）開的那家錢莊比較容易得到大家認可。由於「一國兩制」，它有可能成為大家都可以接受的「錢莊」。

那到「錢莊」為什麼東西定價呢？首先是拿來做抵押品的棗樹（資源）收益，然後是建設中所需要的材料（大宗商品）以及建設資金（人民幣匯率與利率）。由於參與各方的利益和風險不同，政治、經濟、法律制度也存在差異，

大家都會對於錢莊的選擇十分在意。如何找到一家大家都能接受與認可的錢莊至關重要。

「一國兩制」下的香港就是這樣一個神奇的「錢莊」。「一帶一路」是國家戰略，無論有沒有香港的參與，都會推進。但有了香港，「一帶一路」戰略可以獲得更大的成功，成功的概率也會更高。

眾所周知，在過去 20 年裏，香港一直是連接中國與世界的轉換器，在中國改革開放的大潮中，香港為內地企業籌集來自世界的資金，並成功轉型為一個國際金融中心。在今後的 20 年裏，香港除了繼續發揮融資中心的作用，還可成為中國的全球財富管理中心、領先的離岸風險管理中心和中國的全球資產定價中心。

阿基米德曾經説過，「給我一個支點，我能撬動整個地球。」中國今天宣導的「一帶一路」戰略，即將引領新一輪的全球化進程，香港完全可以成為「一帶一路」的新支點，幫助中國撬動整個世界。

2017 年 6 月 30 日

09 | 新時代的交易所運營

今天，我們身處於一個創新大爆炸的新時代。新科技和新經濟正在以驚人的速度湧現，改變着我們的生活，也顛覆着各個行業。很多人每天都要思索同一個問題，如何創新才能讓我們的公司活得更好、更有競爭力？

也許，在你們看來，交易所似乎跟創新扯不上關係。其實不然，金融全球化的時代早已來臨，全球各大交易所都要憑實力來爭取投資者和發行人資源，競爭激烈，不進則退。尤其是對於我們香港來說，更是如此，香港本地的經濟體量較小，單憑本地的經濟是難以支撐起一個國際交易所和國際金融中心的。因此，我們不得不永遠行走在創新的路上，從二十多年前的 H 股上市，到連接內地與香港資本市場的滬港通、深港通，再到今年推出的債券通，香港交易所一次又一次創新和升級，為香港市場連接中國與世界創造了獨特的價值。

在此，我想跟大家分享一下香港交易所近年來的三大創新探索：

一、通過互聯互通交易機制的創新來改善我們的市場結構

2014 年，我們攜手上海證券交易所推出了滬港通。2016 年，在滬港通成功的基礎上，我們聯合深圳證券交易所推出了深港通。滬港通和深港通採用的互聯互通模式，通過交易總量過境、結算淨量過境的獨特交易機制創新，讓兩地市場可以在充分保留各自市場監管規則、市場結構和交易習慣的前提下實現完全市場化的交易互聯互通，以最小的制度成本為內地資本市場取得了最大效果的開放，也豐富了香港的流動性，大大提升了香港作為國際金融中心的吸引

力。我們欣喜地看到，滬深港通機制自推出以來，一直運作平穩順暢，交易穩步增長，逐漸贏得了兩地市場監管者和投資者的信賴。實踐證明，滬港通開創的互聯互通機制具有高度透明、封閉運行、靈活可控、可複製、可延伸的特性。受此啟發的債券通今年也已經成功推出，我堅信，未來互聯互通機制可以延伸到更多資產類別，例如 ETF、交易所債券和新股。

二、通過上市機制的改革支持創新型公司融資

二十多年前，我們的前輩開創性地推出了 H 股上市機制，為內地的改革開放提供了寶貴的資金來源，也讓香港成長為國際金融中心和全球一大首選上市地。儘管成績斐然，但香港要保持今天的優勢地位並不容易，需要克服不少挑戰，其中一大挑戰就是我們的上市公司中低增長行業佔比過高，高增長行業少，極有可能讓投資者喪失興趣、影響我們市場的活力。因此，香港交易所今年 6 月就完善香港上市機制展開公眾諮詢，希望能夠吸引更多高增長的創新型公司。感謝市場各界的廣泛參與，此次諮詢達成了積極的共識，探明了上市機制改革的大體方向，很快我們將會公佈諮詢結論。下一步，我們將積極推進第二輪諮詢和《上市規則》的細化改革，與市場各方一起找到最有利於香港市場發展、最能保持香港國際競爭力的方案。

三、探索應用金融科技來提升我們的系統和服務

近年來，以大數據、雲計算、人工智能、區塊鏈為代表的現代訊息技術蓬勃發展，廣泛應用到了包括金融業在內的各領域，金融科技正在深刻改變全球金融業的發展格局。面對這些令人震撼的新科技，我們作為金融市場的營運者和監管者，經常思考這樣的問題：我們的系統夠安全嗎？速度夠快嗎？承接能力夠強嗎？今天的市場結構會被新技術顛覆嗎？目前，我們正在研究的應用包括：人工智能技術在上市後監管中的應用，區塊鏈技術在結算環節的應用以及

部份測試數據存儲的雲計算應用。當然，這些探索還處於初期階段，離最終落實尚有時日。

這些創新有的推行得快，有的推行得慢。在推進這些創新時，不時有朋友問我，你們的創新步伐是不是太慢了？我們也經常問自己，我們創新的步子邁得夠大夠快嗎？也有的朋友認為我們走得太快太急了。

要回答這個問題，必須回到香港交易所的特殊定位。首先，我們是香港金融市場的引領者，因為金融市場對整個城市的經濟發展舉足輕重，我們有深深的危機感和使命感，我們必須帶領市場各界朝正確的方向邁進，永遠努力不被邊緣化。與此同時，作為市場的營運者和監管者，我們必須兼顧各類市場參與者的利益，既要照顧到發行人和中介機構的需要，也要保護好投資者利益。因為這個特殊的定位，我們的創新跟一般公司的創新有所不同，既要積極進取、爭分奪秒，又要通盤考慮、循序漸進。

打個簡單的比喻：不知大家有沒有關注過狼羣的生活，狼是一種羣居動物，尤其是在食物匱乏、冰天雪地的冬天，狼通常會以羣體為單位活動。精明強悍的頭狼走在最前面，牠決定着狼羣狩獵、防禦和遷徙的大方向，必須勇往直前；牠的身後緊跟着狼羣的普通成員，牠們有的強壯、有的瘦弱，有些走得快，有些走得慢；走在最後面的是時刻保持警惕的護衛狼，牠要保護弱小，確保狼羣沒有掉隊者，護衛狼決定着狼羣前行的節奏和速度。由於狼羣的成員特別團結，儘管單隻狼的戰鬥力有限，但是依靠緊密合作，牠們可以打敗比自己體型大很多的獅子和老虎。

如果把整個市場比作一個狼羣的話，交易所的角色既像是這個狼羣的頭狼，也像是狼的護衛狼。我們有責任引領整個市場朝着最有前景的方向發展，我們必須靈活應變、與時並進。作為中央市場的營運者，我們同時也是市場的守護者，我們要守護每一個狼羣成員的安全，因此，當我們進行創新時，我們必須通盤考慮，統籌兼顧，照顧到發行人、中介機構、機構投資者、散戶投資者等市場各方的利益和訴求。

因此，在推進一些關係到市場重大發展方向或根本利益的創新時，我們

一旦找準了方向，必須一鼓作氣克服困難，全力推進，例如我們的互聯互通機制。而在推進有些具有重大爭議的改革時，我們因為要兼顧市場各方的步伐和利益，必須循序漸進、在取得市場的理解和共識後方能推進，例如上市機制方面的改革。

也就是説，狼羣在生死存亡的選擇面前必須不顧一切、奮力拼搏，但在長途遷徙的過程中，由於要照顧狼羣中所有成員的需要，狼羣的步伐也許不夠快，也許會因此錯過一片食物和水源豐富的森林，但是為了守護整個狼羣的安全，這是必須付出的代價。我們今天所處的市場環境如同大自然一樣風雲變幻，陰晴不定，根據市場環境，適時切換創新的節奏，也是我們的重要職責。

朋友們，身逢這樣的新時代，既是我們的幸運，也賦予了我們更多推動社會進步的使命，讓我們共同努力，敢想敢拼，成就香港更美好的明天！

2017 年 12 月 8 日

第二章
股票市場與市場監管

一年一度的美國期貨業協會國際衍生產品會議。

10 | 清水還是渾水？
—— 博卡行雜思

今年 3 月份，我應邀到佛羅里達州博卡拉頓市（Boca Raton）參加一年一度的美國期貨業協會國際衍生產品會議。在過去幾年裏，這個會議的國際影響力與日俱增，至今年已經成為了雲集業內所有重量級人物的盛事。博卡的陽光與海灘固然起到不小的作用，但我相信更重要的原因是，隨着金融危機之後各國陸續出台新的法律法規，行業面貌正在經歷劇變，無論是監管者還是交易所、參與者都不得不重新審視：衍生品行業將何去何從？

不出意料，今年大會主題正正就是國際資本市場的新法規、新監管。說起來，這還都是西方國家經歷金融危機洗禮後的「未了餘波」，旨在提升金融產品交易及結算的透明度、降低系統風險、加強投資者保護。須知道，直到今天，這類產品的交易結算仍多在透明度很低的場外市場（OTC）進行，加上西方市場多數採用「經紀／自營商」（broker-dealer）的傳統模式，監管機構及交易所對經紀背後的客戶信息所知甚少，因此其監管就顯得鞭長莫及了。因此，新法規主要集中在強制將相對較複雜的金融產品從場外市場放回場內市場與中央結算所，以及如何保障投資者權益上。

不過，今年有一件事出乎我意料，那就是幾乎所有的交易所都一面倒地把注意力放在亞洲，特別是中國。今年會議上，我獲邀參與了大會每年都很搶眼的國際交易所行政總裁戰略對話，其他嘉賓包括芝加哥商品交易所、歐洲期交所、洲際期貨交易所、納斯達克、紐約交易所及新加坡交易所等一眾交易所總裁。我發現，幾乎所有全球同業都將中國視為他們未來戰略規劃的重要一環，唯恐在這個「東方熱潮」中落後於人。這再一次提醒了我，香港交易所不是唯

一一個尋求中國機遇的交易所；而與其他交易所相比，雖然我們在把握中國機遇方面的位置角色確實獨一無二，但我們並不能想當然，掉以輕心。

然而，國際交易所及投資者爭相進軍中國，他們除了要面對中國資本管制這道「牆」外，還面臨截然不同的市場架構及監管環境。前者眾所周知，但大家對後者卻知之甚少。甚至，由於中國封閉的資本管制，有人可能會誤以為中國的市場結構仍處於落後階段。誰若真那麼想，只怕都要大跌眼鏡了。

事實上，有別於西方國家，中國已經發展了一套完全透明的一戶一碼的交易及結算模式 —— 就連每個投資者持有多少股份、付了多少按金等等，監管機構及交易所都可以實時知道。中國六家交易所共有 1 億 2,000 萬個直接連接託管銀行及結算所的投資者戶口。經紀／自營商要在不同客戶賬戶之間調動資金的空間很小，更不用說要挪用資金或進行其他不法行為。這全是中國上個年代初針對經紀／自營商舞弊問題嚴重而大刀闊斧改革市場結構的成果。所以到了今天，一切都變得透明、清澈。若說西方市場是一潭渾水，那中國市場就幾近蒸餾水矣。

那麼，是中國市場對呢，還是西方市場對呢？

古語有云：「水至清則無魚。」中國市場的問題就在於水太清了。投資者的資金固然安全，但金融機構的角色大大削弱，淪為執行代理，失去了創新動力；投資者對於市場產生可觀回報的能力也信心不大；監管機構成了市場的最終負責人，明顯不利市場長遠發展。所以，不能說中國完全對了。

反觀西方市場，監管寬鬆，市場猶如一池渾水，箇中貓膩甚多，養肥了許多以此為食的「池底魚」。有人說，西方市場是到了過份創新的程度。本世紀初以來，隨着水的渾濁有增無減，監管者、參與者漸漸看不清更掌控不了市場演變的方向和速度，最後導致有史以來最大的金融危機之一。所以，也不能說西方市場走對了。

我們香港正好處於東西拉鋸之間，這是挑戰也是機遇：一方面，要接通內地及國際市場這兩池大水，就需要投入大量人力物力來修建「渠道」；但另一方面，一旦大功告成，香港就能成為東西互通的必經之路。香港有着得天獨厚

的定位和優勢，這是歷史給香港的機遇。

我們深知面前的機會千載難逢，所以，為了香港交易所，也為了香港，我們都責無旁貸，決意加強自身能力。讓開市時間與內地市場同步是我們向這個目標邁出的第一步。投資數據中心等基礎設施為今後市場發展奠定基礎。此外，建立穩健的風險管理系統是另一個努力方向，去年完成的香港交易所風險管控體系綜合改革更是重中之重 —— 我們不僅要穩健管理今天的風險，亦要未雨綢繆，為明天可能遇見的風險作好準備。最後，我們要着手構想一個可兼取雙方之長的市場互聯互通方案。

正如在博卡所見所聞，全球目光都在看着中國；而我深信中國亦同樣期待着走向世界。香港身處這重大轉變的中心，明天的成功就取決於今天我們能否為東西兩方找出最合適的解決方案。香港人從來開拓進取、勤奮上進，我對香港的明天充滿信心。

2013 年 4 月 2 日

11 | 香港市場的另一歷史機遇 —— 寫在 H 股上市 20 週年

今年恰逢 H 股在港上市 20 週年，我們很高興舉辦了一系列活動來慶祝這個重要的里程碑事件。從青島啤酒來港上市成為首家 H 股公司算起，20 年已經過去，到今天我們仍然享受着 H 股機制的豐碩成果。古人云：「以史為鏡，可以知興替。」我想藉此機會在這裏分享一下我對香港資本市場的過去、現在及將來的一些思考。

在今天的人們看來，H 股上市甚至是內地公司在港上市都已是順理成章的事情。畢竟，自 2007 年起內地公司（包括 H 股）的成交量就一直佔到整個市場成交量的近三分之二以上，已經成為香港資本市場的主力軍。如果沒有 H 股，很難想像今天的香港可以躋身全球領先金融市場之列。H 股上市也為國際投資者創造了分享中國經濟增長的機遇，繼而吸引世界各地持續不斷的國際資金和大量金融才俊匯聚香江。不過，現在恐怕已經很少人知道 20 年前 H 股上市機制從無到有所遭遇的種種艱辛。

讓我們一起來回憶一下當時的情形。1993 年，中國仍然處於計劃經濟轉型的開始，市場上的非國有企業屈指可數，對於中國內地企業而言，現代企業管治仍是一個全新的概念；內地資本市場剛起步，完全不能滿足市場需求，更談不上證券市場監管和執法。簡言之，支撐資本市場運行的許多基本硬件軟件還不存在。因此，當有人提出讓中國公司來港上市這個大膽構思時，自然招來了多方質疑。打個形象的比喻，當時的中國就像一條方形的管道，而香港則是一條圓形的管道，實在看不出怎樣可以把它們連在一起。

然而，雙方都看到這背後潛在的巨大機遇不容錯過。中國內地當時正在

H 股上市 20 週年誌慶。

加快改革開放，大小企業（特別是國營企業）無不渴求海外資金。而在文化和地理上都與內地相近的香港，一個成熟的資本市場已經形成，投資者又熟悉和信任香港的市場體制，可以説，香港已經具備迎接這一歷史機遇的天時地利人和。不過，方管到底怎樣才能接上圓管？這真是一個難題。在此重要關頭，香港充分發揮其創新進取精神，在適當時候制定出適當的解決方案 —— 香港與內地攜手共建 H 股機制，在當時的香港《上市規則》中特別加設一章只適用於中國企業的上市規則條文。H 股機制，就是當年將方管和圓管連起來的「轉接器」。

　　在當時這個解決方案不無風險，也引起了不少爭議。其中一個主要爭議，是內地公司究竟應該遵守標準較高的香港主板《上市規則》，還是應該另闢一個標準較低的板塊上市。當時兩地市場的領導層最終達成共識：既然內地公司的目標是要進軍國際市場，那麼一開始就應當遵守最高標準的規則。今天回頭來看，當時作出這個正確決定着實需要遠見和勇氣。正因為這一正確的決定，此後內地公司赴港上市的個案源源不斷，帶來了香港市場的繁榮興旺；而另一方面，香港亦幫助內地經濟成功轉型，逐步邁向市場經濟。

今天，許多大型國企已經在香港上市，內地公司來港上市也已經比較成熟。有些人甚至開始擔心 H 股這股甘泉終會乾涸。對此我卻保持樂觀，原因有二。第一，中國證監會不久前放寬了內地公司來港上市的審批條件，加上 B 股轉 H 股的成功個案，H 股公司的上市來源依然十分豐富。第二，未來 H 股公司上市股份的全流通將大大推動市場的繁榮。目前，H 股公司的大部份股份均由政府部門持有，並不在市場流通。這些股份總值數以萬億，我們預計這些股份將來會逐步上市交易流通。到了那一天，我們市場的規模及流通量都將大幅提升。

其實，除了上述這些由現有 H 股機制帶來的增長機遇之外，香港亦正面向另外一大歷史機遇，這一機遇有望為香港帶來下一個繁榮的十年。不過，要把握這一機遇，可能意味着我們將不得不作出一些艱難的抉擇，情況一如 20 年前。

想認清這個新機遇，不妨先來看看自 1993 年以來的這 20 年中間發生了哪些變化，又有哪些東西不曾改變。

哪些東西不曾改變？內地經濟面向國際市場開放的意願和需求仍然很大。隨着國內貿易和經濟與全球接軌，內地對資本市場及金融業開放的需求與日俱增。同時，由於「一國兩制」，香港在金融市場、體制和與國際市場的互通等方面仍然保持着一定的優勢。簡而言之，香港對中國內地來說仍具有十分獨特的價值。

哪些東西經已改變？中國已經由一個「渴求資本」的國家變為「資本充裕」的國家，即從前是資本輸入國，現在已開始向資本輸出國轉變。香港未來的角色不再僅僅局限於為內地發行人提供國際資本，更具現實意義的已開始包括為內地投資者提供機會以接觸國際發行人及國際金融產品，同時也為國際投資者提供渠道，投資內地上市公司及金融產品。

那麼，香港今天是否已準備迎接這一歷史性的轉折了嗎？我們能夠與時俱進、準備好所需的基礎設施、制度和人才嗎？我本人對此信心十足。香港人既善於把握先機、銳意創新，又有着堅苦卓絕、同舟共濟的精神。我深信在這些

優秀品質的推動下，香港必能繼續突破自己，再創輝煌。

　　20 年前，香港上一代的金融人勇於創新並取得成功。今天輪到我們接棒了。讓我們攜手努力，為香港和內地市場的互聯互通共建新的機制，為下一個20 年創造新的繁榮。未來幾個月，我會在網誌中陸續與各位分享我在這方面的一些看法。

2013 年 9 月 2 日

12 | 投資者保障雜談

近幾週來，關於香港投資者保障、股份架構和股東投票權的討論十分熱鬧，各種聲音不絕於耳。我一直仔細傾聽着這些不同的聲音：有些聲音格外響亮，也有些較小的聲音不容忽視。每當我嘗試靜下心來思考投資者保障問題的時候，這些聲音總是縈繞在我耳邊，揮之不去。

一天晚上，我輾轉反側，耳邊響起了這些爭論聲，久久不絕。恍惚間，這些聲音的爭論一一展開……

第一個説話的是聲勢豪邁的傳統先生，他非常滿意香港現有的市場體制，完全不覺得有必要改變。「香港的體制長期以來運作非常順暢，為什麼現在要改變？這裏的市場之所以這麼成功，就是因為我們的投資者保障機制出了名的好。香港的《上市規則》十分清晰，誰想來香港上市都一視同仁。我們一直是走在世界前列的金融中心，近年還一再榮登首次公開招股集資額排行榜的榜首。對於我們來説，吸引發行人來上市集資完全不成問題，我們也不曾為任何公司妄開先例。好端端的為什麼要改變？」傳統先生搖着頭重重地坐了下來。

這時候，創新先生忍不住發話了。他是個髮型前衛的年輕小伙子，激情洋溢，語速極快。「傳統先生，你算了吧。多層股份架構有什麼問題？世界上大部份交易所都允許這樣做，只有香港墨守成規、不肯接受。看看那些在美國上市的科技公司，最大的幾家公司比如 Google 和 Facebook，都是以特別投票權來維護創辦人的地位。人們投資這些公司，就是因為相信公司創辦人獨特的眼光、業績紀錄和聲譽！創辦人關心公司的長期發展和利益，比起那些單靠短期套利賺錢的對沖基金和自以為是卻根本不懂如何經營創新科技公司的併購狙擊

手們好多了吧！你看看蘋果公司！喬布斯不就是在『完美』的企業管治程序下被踢出局，險些令蘋果破產的嗎？最終還不是靠把喬布斯請回來主持大局，才再創地球上的科技神話！」

創新先生滔滔不絕，冷不防聲音穩重的披露先生插嘴進來。「冷靜一點，創新先生。現在問題的關鍵不是創新的創辦人和進取的投資者相比孰優孰劣，而是訊息披露。監管機構只需要定下良好體制確保訊息披露準確並懲罰違規者。別忘了，若有公司以這樣的股權結構上市，考慮到手上並不平等的投票權，投資者願意為它們付出的價格自然也會打折扣。至於公司創辦人若要求有特殊投票權而投資者願意為這種架構的公司付出什麼樣的價格，就由市場來定吧。這種體制在美國和其他地方都很成功，既不損害公司價值，也沒有影響投資者利益。香港市場是時候與時俱進了。」

「不過，有一點要提醒大家，」披露先生繼續說，「美國的多層股權制之所以運行良好，是因為他們以披露為主的市場機制，與身經百戰經驗老到的機構投資者和一究到底的集體訴訟文化組合在一起，這些全都發揮着重要的制約作用，可抗衡同股不同權帶來的負面影響。如果香港要學習的話，必須有足夠的配套組合，既賦予創辦人足夠的動力，又確保他們誠實可信。如果你問我的意見的話，我認為循序漸進的改變要好過全盤複製美國的制度。」

「等一等。」又傳來了一個聲音。「你們人人都在說保障投資者，不如我們先問問投資者，看看他們到底想要什麼吧！」「好主意」，大家異口同聲地說。他們先問大基金先生。「我完全不關心一家公司到底在香港還是在紐約上市，因為在哪裏我都可以投資呀。我只關心那家公司是否一家好公司。我不喜歡同股不同權，但如果公司非要以這樣的架構上市，我也知道怎麼給它估值。」然而，另外一邊的小散戶女士卻感到很為難：「我不能投資美國股市啊，所以，如果有一家優秀的公司，請不要奪走我的投資機會啊！但話說回來，我真的不喜歡公司同股不同權，這對我們不公平。我希望監管機構可以幫忙照顧我的權益。」

接着我又聽到另一個聲音，是務實女士。「喂，各位各位，讓我們談點實

際的吧！我們香港人一直都以開拓、務實而聞名。我們曾經大膽引入 H 股和紅籌公司並大獲成功，我們也適時把握了小型民企來港上市的機遇。這一次，讓我們敞開雙臂迎接新經濟公司吧！如果香港錯過了中國下一輪上市大浪潮，我們大家都會輸掉！不只交易所和證監會將損失交易費及徵費、政府損失印花稅，經紀亦將失去數以億元計的佣金，投資者們更將損失投資這個時代發展最快、最有潛力公司的巨大機遇！香港怎麼可以錯過這些！」

慢著，有人實在聽不下去了！原來是道德先生，他很生氣大家竟然在赤裸裸地討論金錢。「你這是什麼意思，你…你……！」他大聲說道。「這是再簡單不過的事 —— 一股一票就是了，毋須再討論！你憑什麼聲稱創辦人可以享受特殊待遇！別忘記創辦人也有老去的那天 —— 當他頭腦不清自私自利時，你還願意給他機會獨攬大權，無止境地榨取公司的利益嗎？為了贏得一兩家大型公司，你就要出賣香港精神？那我們辛辛苦苦建立的聲譽何存？香港為什麼要學習美國？看看華爾街那幫人打著金融創新的幌子鬧出多大的全球金融危機。香港的體制就是這樣，不喜歡的人大可以捲好鋪蓋走人。」「還有一件事，」道德先生又繼續追問道：「香港交易所為什麼會考慮這個方案？是不是中國政府要求的？這要查一下。」

我感到氣氛開始緊張起來，大家坐立不安，但誰也不敢公然反對道德先生，因為……道德先生永遠是對的。可是，本來一直在自顧自聽音樂的未來小姐，此刻卻摘下耳機，向道德先生說道：「咱們就事論事，不要搞人身攻擊嘛。世界在變、中國在變，香港也應該要變啊。十年前香港錯過了科技革新的機會。展望未來，中國將湧現一大批代表新經濟的公司，尤其是在互聯網領域，它們可能會徹底改變中國未來十年的經濟面貌。這可是香港將中國故事和新經濟融合在一起、真正掌握全球領導力的好機會啊。」此時，未來小姐直直望向道德先生，說道：「你當然無所謂，你已經名成利就，但想想我們這一代香港人啊。」

道德先生心有不甘，反駁道：「但你們為了未來，難道就非要給創辦人特權嗎？」

「如果賦予創辦人特權是能吸引這些代表未來的新經濟公司來港上市的唯一之途,那就給他們好了。」未來小姐答道:「你沒有權力剝奪我們的未來。別忘了,你們今天投資的大公司明天都可能被這些新經濟公司徹底取代,到那時,我們年輕人怎麼辦?……」

未來小姐顯然煩躁起來。至此,夢中的我已聽得直冒冷汗……

「好啦,好啦,大家不要那麼激動。」我聽見一個熟悉的聲音在叫大家冷靜。謝天謝地,程序先生來了,真是人如其名,程序先生從來都是那麼深思熟慮。

程序先生繼續發表他的意見:「整件事不關誰對誰錯,也不是說特別股權結構對市場是好是壞,更不是說公司創辦人和進取的對沖基金究竟誰可以創造價值或破壞價值。每件事大家都可以證明有好壞兩面、甚至多方面。整件事不關香港到底應該擁抱明天還是活在往昔。大家都希望擁抱明天。」

這時,只見人人都坐了下來,聆聽程序先生講話。「這件事關係到審慎程序。」程序先生說道。「香港的《上市規則》非常清晰,如果要修訂條文,必須按照審慎程序進行。如果為了迎合新來者而朝令夕改,我們的公信力便蕩然無存。那麼什麼是審慎程序?就是說,如果公司要求的改變『有限、適度而且平衡』,又能根據香港現行整體上市機制的條文規定或精神合理地處理,豁免或者批准都可以斟酌。這亦是上市委員會及證監會一直以來的工作之一。另外,我們亦應考慮所行使的酌情權能否歸納為一項先例。這一點十分重要,因為香港是法治之都,監管者需要為將來尋求類似待遇的其他上市申請人劃定一條清晰的法規界線,並仔細闡述劃定這一界線的原因。」

程序先生續說:「如果要求超出了《上市規則》所許可的有限酌情範圍,那就要經過適當的公眾諮詢之後才可以修訂規則及政策,確保所作的改變經得起時間的考驗。這是香港的優良傳統,必須堅持。」

唉,夢中的我不禁在問,那答案到底是什麼?「為什麼不找答案先生請教一下?」有人建議。「對,好主意!」大家異口同聲。

我拼命想聽清答案先生的答案,但一下竟醒了過來!

現實中，哪裏會有什麼答案先生來給我們拿主意？我們只能依靠集體的智慧自己作決定。這裏最需要的，是客觀看待事情，不被負面情緒牽動，不受指摘影響，也不被個別公司或個案的具體情形而影響判斷。歸根究底，我們需要作出最適合香港、最有利於香港的決定，而不是最安全最容易的決定。

此刻，我已經完全從夢中醒來並回到辦公室。我迫不及待把這些聲音寫在這篇網誌裏，正要完稿之際，卻清楚聽到耳邊有另外一個聲音對我說：「小加呀，人們已經在批評交易所在這個問題上有既得利益不宜參與討論；儘管你自己認為這種説法沒有根據，但是此刻保持沉默置身事外不是更好嗎？」我思索許久但還是決定繼續這篇網誌，主要原因有三：

首先，沒錯，我是香港交易所的集團行政總裁，促進和保障交易所股東的利益是我職責的一部份。可是，正如我們的章程所規定，當「公眾利益」與港交所股東利益之間發生衝突時，我們永遠要把公眾利益放在第一位。正是出於「公眾利益」考慮，我才決定參與這一重要的討論。

第二，有關個別公司上市或政策改變的決策並不取決於我或香港交易所董事會。它們均由上市委員會和證監會審議決定，我只是眾多聲音中的小小一員。上市委員會的其他 27 名委員都是香港金融界的精英才俊，為了香港的利益，他們無私奉獻出自己寶貴的時間、智慧和豐富的實戰經驗。這一決策過程和證監會的監督安排正是為了保障香港的最佳利益。

最後我想説，我無意利用我的網誌去改變任何人的想法，或宣揚任何立場。我只是希望大家能在這一涉及公眾利益的重要議題上進行誠懇、公開、平衡客觀及尊重各方的討論。不論您是個人投資者，還是大機構負責人，只要胸懷坦蕩，只要抱着為香港最佳利益考慮的心態真誠表達意見，就不應為參與這場討論而感到羞怯、害怕或者愧疚，因為每個人的意見，對我們都同樣寶貴！

2013 年 9 月 25 日

13 「夢談」之後，路在何方？
——股權結構八問八答

自上次我在博客中「夢談」投資者保障以來，市場上出現了更多關於上市公司股權結構的討論，這是好事。不過，我們不能只停留在回味夢中的聲音，在現實中更需要勇於直面問題，共同擔當起「答案先生」的角色。今天，我想在此嘗試回答市場熱議的一些問題，分享一下我對於投資者保障與股權結構的看法。為了避免不必要的誤會，以下僅代表我個人的觀點。

一問：關於股權結構與投資者保障的討論似乎已經告一段落，你為什麼又舊事重提？

答：在前一陣子的激烈論戰中，各方都暢快淋漓地表達了自己的意見。有一些朋友獲得了精神勝利的愉悅，感覺很爽；也有一些朋友感到失望與惋惜。但問題是，大家都在自說自話，並不一定有認真傾聽和分析對方的發言，也沒有足夠的努力在這麼多不同的聲音中尋求共識。

面對香港金融業究竟應該如何迎接新經濟帶來的歷史機遇這一重大問題，我們仍沒有答案。在下一波新經濟浪潮中，中國創新型公司將佔據相當大的比重。對於香港而言，丟掉一兩家上市公司可能不是什麼大事，但丟掉整整一代創新型科技公司就是一件大事，而未經認真論證和諮詢就錯失了這一代新經濟公司更是一大遺憾。

在我看來，這個問題關乎香港的公眾利益，並且已經迫在眉睫，不容逃避。這需要我們有承擔、有勇氣去進一步尋找答案，否則，就白白浪費了一個為香港市場規劃未來的重大機遇。因此，我決定在此率先說出我的拙見，希望

拋磚引玉，引發更多有識之士對於這一問題理性和智慧的探討，為香港找到一個最好的答案。

二問：創新型公司與傳統公司有什麼不同，為什麼它們值得投資者在公司治理機制上給予新的思考？

答：創新型公司與傳統公司最大的不同在於，它取得成功的關鍵不是靠資本、資產或政策，而是靠創始人獨特的夢想和遠見。回顧這些創新型公司的成長史，我們不難發現，每一個偉大的商業計劃最初都起源於創始人一個偉大的夢想。蘋果公司的成功起源於喬布斯發明一台改變世界的個人電腦的夢想，Facebook的成功源於朱克伯格希望以互聯網改變人們交流方式的夢想，谷歌的成功源於佩奇和布林想要通過鏈接把整個互聯網下載下來的夢想。這些創始人的偉大夢想和創意成就了創新型公司，也成為了它們最重要的核心資產。毫無疑問，對於這類公司而言，創始人應該比任何人更珍惜他們自己的「孩子」、更在意公司的長遠健康發展，也恰恰因此，眾多投資者鍾情於這樣的公司。

創新型公司還有一個重要的共同特點，就是它們的創始人創業時都沒什麼錢，必須向天使投資人、創投、私募基金等融資來實現自己的夢想，這就使得他們在公司中的股權不斷被稀釋；一旦公司上市，他們的股權將進一步下降、作為公司發展方向掌舵人的地位將面臨威脅。在公司的長期利益和短期利益發生衝突時，他們甚至可能會被輕易地逐出董事會。

為了鼓勵創新，為了保護這些創新型公司的核心與持久競爭力，國際領先的市場和很多機構投資者在這方面已經有了新的思考與平衡，他們認為給予創始人一定的空間與機會掌舵，有利於公司的長遠發展，也是保護公眾投資者利益的一項重要內容。

三問：給予創新型公司創始人一定的控制權與保護公共股東利益是不是一對不可調和的矛盾？

答：在一個好的制度設計下，它們並非不可調和。制度設計的關鍵在於創始人的控制權大小必須與市場的制衡和糾錯機制相匹配，以減少創始人因錯誤

決策或濫權對公司和其他股東帶來的損失。偉大的創始人是可以創造出偉大的公司，但權力不受約束的創始人也可以讓偉大的公司轟然倒下。因此，制衡與糾錯機制必不可少，一個市場中制衡與糾錯機制越強大，給創始人的控制權就可以越大，反之亦然。

四問：在維持現狀與雙層股權這兩個極端之間，是否還有其他的可能性？

答：對於這一問題，市場意見紛紜，提出的建議也很多，從最簡單的堅持同股同權到最極端的雙層股權都有。

最簡單的可能是維持現狀，不給予創始人對於公司控制權任何形式的特殊權力，但這不無代價。若要選擇這個選項，香港可以保持傳統公司治理機制的純潔與簡單，可以輕易佔領道德高地，但是也可能意味着香港主動放棄了一大批引領經濟潮流的創新型公司，從而失去我們市場未來的核心競爭力。

而與維持現狀相對，另一個極端是允許上市公司發行附有不同投票權的雙類或多類股票（即創始人所持股票的投票權高於普通公眾股票的投票權）。這類制度在美國及歐洲很多海外市場運行多年，Facebook 和谷歌等大型 IT 公司均採用這種多層股權結構上市。

不過，香港如果要引入這一制度恐怕將會引發爭議。支持者認為香港應向以披露為主的成熟市場大步進發，讓市場和投資者自由決定，而反對者則認為這是香港在倒退，因為香港和海外市場區別巨大，香港中小投資者無法與強勢的大股東有效抗衡。

這兩個極端之間其實有很多不同的可能性，但最具代表性的分水嶺在於是否給予創始人多數董事提名權。

分水嶺的一邊可能是允許創始人或團隊有權提名董事會中的少數（例如 7 席中的 3 席、9 席中的 4 席等等），並對高管之任命有一定的影響力。支持者認為這種安排不會對現有同股同權制度造成任何實質改變，同時可以在制度上使創始人對公司保持重要的影響力，不用顧慮隨時會被強勢股東聯合踢出董事會。這一安排贏得共識的關鍵在於如何確立創始股東對高管任命（特別是行政

總裁）的影響力，這需要監管者設計出精巧的制度安排，既保障創始人及團隊掌舵公司的穩定性，又不對同股同權的基本原則產生實質性衝擊。

分水嶺的另一邊是讓創始人或團隊可以提名董事會多數董事，但股東大會可以否決創始人的提名；除此之外，所有股份同股同權。支持者認為這樣的機制可以使創始人通過對多數董事的提名，實現對公司一定的控制，但反對者認為這可以使創始人以很低的成本實現對董事會乃至整個公司的有效控制。

有可能讓正反兩方達成共識的關鍵是這一提名制度的糾錯能力與有效期限。如果創始人的提名屢次被股東否決仍能繼續提名，那這種控制權就可能已造成實際的同股不同權；如果這種提名權在股東大會否決一至兩次後即永久消失，這就會使創始人極其認真嚴肅考慮提名以求得股東支持。同時，當其他股東與創始人在根本利益上有重大衝突時，其他股東可以通過一、兩次否決就收回這一特權，這樣的安排可以大幅降低該制度可能被濫用而引發的爭議。

五問：如果市場達成共識要對現有制度做出一些改變，我們應該如何確保程序公義？

答：如果選擇維持現狀，我希望是經過仔細論證和綜合考慮後作出的主動選擇，而不是因為屈於壓力、懼怕爭議或者懶於作為的後果，因為這關乎香港的未來。

如果我們考慮修訂上市政策及規則，則應該根據修訂幅度的大小選擇相應的審慎程序來推進。輕微的改動也許只需監管機構行使酌情權；而稍大的變化則需要事先向業內人士進行一些「軟諮詢」（Soft Consultation）使決策更周全；更大的改革則必須經過全面市場諮詢，有些甚至需要經過立法程序。

當然，現實中需要討論的情況可能比這些更複雜，需要具體情況仔細分析。簡言之，無論做出任何選擇，我們都必須經過審慎客觀的程序，體現法制尊嚴和程序正義。

六問：現在熱議的「合夥人制度」是不是一種可行的上市方式？

答：老實說，我不明白這個問題與我們討論的上市公司股權治理機制有什

麼必然的邏輯關係。

傳統意義上的「合夥制」與公司制是兩種完全不同的公司治理機制，很難想像如何將它們揉在一起：

- 合夥制是一人一票、合夥人之間通過合同相互制約；而公司制則是一股一票、股東之間通過公司章程、公司法等「標準契約」來定義權利與義務；

- 合夥人這個集體是由合夥人之間的合同約束，誰進誰出由合夥人達成共識而決定，由此來體現合夥人公司的價值傳承等等；而公司制下股權依出資比例而定，股東之間的關係是依靠「標準契約」來規範，股東通過在市場上買賣自由進出。可以說，前者是人治，後者是法治。

上市公司只能是採用以股權為基礎的公司治理機制，監管者不會也無法在上市公司制度層面將這不可相容的「水」和「油」揉在一起。

在上市公司治理機制下，監管者只關注股東、董事和管理層這三類人羣之間的權力與責任關係，是否屬合夥人與此無關。當然，部份股東、董事或管理層可以自行組織合夥人公司或其他團體來維護共同追求的某種特定價值觀和管理理念，但這並不是上市公司監管者的關注點。當這樣的組織形式對上市公司運作產生影響時，監管者會要求適當披露。

七問：如果市場同意要為創新型公司來港上市做出一些規則修訂，我們應對適用於什麼申請人做出怎樣的限制？

答：假如市場同意給予某些創新型公司的創始股東一些特殊權利，這些權利也應該僅適用於有限的情況。例如：

- 這家公司必須是代表新經濟的創新型公司，因為這是整個討論的出發點，這一制度「例外」並不是為其他傳統公司設計的；當然，亦有需要對何為「新經濟」、「創新型公司」下一個更準確的定義；

- 獲得此類有限權利的必須是創始人或創始團隊，因為這也是討論的出發點，這一制度「例外」不應該被隨意轉讓或繼承；

- 創始人必須是股東並持有一定股權，因為討論的基礎是股東的權利，要保證創始人與股東利益的整體和長期一致性；一旦創始人或創始團隊手中的股份降到一定水平下，這一制度「例外」也應自動失效。

此外，還可以考慮施加最小市值或流通量等條件，以確保這些公司中有相當數量的成熟機構投資者來監督這些特殊權利不會被濫用。類似這樣的限制方案還可以有很多，但總而言之，不是所有公司、所有人都能享受特殊權利。

這裏值得一提的是，「合夥人制度」本身是否可以是一個條件呢？如前所述，「合夥人制度」是公司自身激勵人才、留住人才和追求特定價值觀的管理制度，市場監管者毋須評論其優劣，但它不應與公司的股權制度混為一談。如果「合夥人」符合開列的條件，例如他們是創新公司創始人或團隊，並且是持有一定股份的股東，那麼就可以被考慮，否則就不行，這與申請人是否採用「合夥人制度」並無必然關係。

八問：作為香港交易所集團的行政總裁，你上次發表的言論已經招致一些非議，認為你有為香港交易所謀私利之嫌，甚至有人認為你在為個別公司上市開道，此次你再發網誌不怕引火燒身？

答：我不害怕，因為發表這篇網誌之目的正是為了香港的公眾利益，這一公眾利益遠遠超出了某一家公司是否來香港上市。

何謂公眾利益？在我看來，它首先包括崇揚法治精神、捍衛程序正義、維護市場的公平、公正、公開與秩序；同時，公眾利益也應該包括發展市場、確保香港作為國際金融中心的長期核心競爭力。一個心懷公眾利益的市場營運者和監管者，必須綜合考慮其職責與目標，充分聽取市場各方意見，並找出最有效的方案，最大程度實現全市場的共贏。

同時，我也相信，一場有智慧的討論會聚焦於問題的實質和觀點的論據，而不會拘泥於討論者的身份和地位，更不會以揣測討論者意圖來逃避這個影響香港長期核心競爭力的問題。

最後，我想說，以上回答僅代表我的個人意見，我並非以上市委員會委員的身份在此發言，既不代表香港交易所董事會的意見，更不代表上市委員會的

意見。在這個問題上是否進行公開諮詢、如何諮詢、何時諮詢完全取決於上市委員會及香港證監會的決策與指導。

　　我之所以願意在這裏袒露心扉，是因為我相信香港是一個理性社會，能夠開展有智慧、有擔當的討論，希望我的直率表達能夠呼喚更多有識之士為這一重大問題獻計獻策。我期待各位一起加入這場討論，香港的公眾利益需要您！

2013 年 10 月 24 日

14 釐清「斷路器」疑雲

　　隨着馬年的開市鑼聲敲響，香港的市場又變得熱鬧起來，我和香港交易所的同事們也對新一年的工作滿懷憧憬。

　　在剛剛過去的蛇年，我們收穫頗豐：2013年我們的市場復甦勢頭強勁，首次公開招股集資額名列全球第二，交易所買賣基金及多種衍生產品合約成交量均創下歷史新高；集團旗下的子公司倫敦金屬交易所（LME）多項產品交易量亦刷新歷史紀錄。不過，正如我以前所説，任何時候我們都不能自滿。因為我們身處一個競爭激烈、瞬息萬變的行業，要在這個行業中脱穎而出，我們必須時刻全神貫注，確保我們的市場富有競爭力，同時維持審慎的風險管理水平和投資者保障標準。

　　這幾個星期以來，我被多次問及有關市場的微觀結構的看法，尤其是「斷路器」這個問題備受關注，各方人士紛紛發表意見，討論十分熱烈，這在我看來是一件好事。既然如此，在這裏我也想説一點自己的看法。

　　首先，什麼是「斷路器」？所謂「斷路器」指的是一個機制，是指價格在極短時間內突然出現極端波幅以至超過某一預設幅度從而觸發有關證券或證券組別、甚至整個市場的交易暫停的一個機制。設立該機制的目的是給予市場一個冷靜思考的機會，避免由於程式錯誤或烏龍指令等非基本面因素造成的價格波動引發恐慌性反應。在不同的「斷路器」機制安排下，在短暫停頓期間，交易可以、又或不可以在一定限制條件下繼續；暫停結束後市場隨即恢復正常交易，「斷路器」的觸發條件也將被重新設置。

　　以前，我們的市場也不時討論過「斷路器」機制，當時得出的結論是沒有

必要。有人認為「斷路器」機制不適合我們的市場、形同強制干預市場運行，如今舊事重提只是浪費時間。然而，時移勢遷，過去所作的決定未必適用於現在的情況。香港交易所必須維護市場的公平有序，事實上這也是我們的法定職責，因此我們必須時刻保持警覺，密切關注各種市場變化會否令我們需要引入新機制才能繼續維持市場秩序。

在正常情況下，投資者在市場中博弈尋求平衡價格，毋須市場運營者的干預。但是近年來，電腦應用的深入推廣幾乎改變了每個市場的交易方式，我們的市場自然也不例外。現在的交易較十年前更方便快捷，而且經常通過電腦程式自動完成，但這也就可能引發錯價盤出現（即使是小概率事件），造成過度反應，危及市場秩序。固然，這些過度反應最終都會在市場力量下消失，但可能當中要經歷相當的動盪，甚至嚴重影響市場信心。近年來，這樣的事故在海外市場屢有發生，迫使我們不得不思考：類似的事故未來會不會在香港上演？

一些市場參與者曾向我們表示擔憂，不知香港市場是否有適當的措施防範人為及機器錯誤所引發的混亂。事實上，國際證券事務監察委員會組織（國際證監會組織）已要求全球各個市場就此進行檢視，證監會今年初實施的新電子交易規例亦正是朝着同一方向推進。我想，現在也許是時候重新討論香港市場是否需要「斷路器」機制了。

這就是今天我們再次探討「斷路器」機制的原因。在此，我也想澄清市場上流傳關於「斷路器」機制的一些誤解。

誤解之一，是認為香港交易所推出「斷路器」機制是為了與內地交易所接軌。其實，探討「斷路器」機制之目的並非為了與內地接軌。我們討論「斷路器」機制的目標有二：第一，考慮香港市場是否需要引入「斷路器」機制；第二，如果需要，就要制定一個能夠適合香港市場獨特情況、滿足香港市場獨特需求的機制。

誤解之二，是以為「斷路器」機制就必定像內地及部份其他亞洲市場一樣設定嚴格的每日價格限制，這其實未必。在許多設有「斷路器」機制的海外市場，觸發交易暫停的價格限制並非固定的，而是隨市場變動而動態調節，使價

格發現的過程盡可能貼近常態。最關鍵的問題是，如果要在香港推出「斷路器」機制，我們就必須找到最適合香港的模式。

誤解之三，就是以為「斷路器」機制等同停牌。這也是不一定的。許多交易所採用的「斷路器」模式提供了短暫的「冷靜期」（一般為數分鐘），期間可在符合若干條件的情況下繼續交易，例如可以限制的價格執行交易盤等。

在我看來，適合香港市場的「斷路器」機制必須滿足以下條件：

(1) 可降低由非基本面因素（例如錯誤程式）所引發的市場驟然極端波動風險；

(2) 有助於維護市場秩序；

(3) 具有足夠的靈活性，容許由基本面因素推動的價格變動；

(4) 操作直接，方便市場人士理解及執行。

要滿足這些條件並不容易，因此，在得出結論前我們必須就此展開廣泛深入的探討。

有關這方面的討論現才剛剛開始。我們的市場究竟是否需要「斷路器」機制？如果需要，又應該是哪種類型的「斷路器」？對於這些問題，香港交易所並未得出任何結論。只有在充分諮詢市場並綜合考量各方意見之後，我們才會作出決定。在這個問題上，大家可能會各有己見，不過，如果沒有通過公眾諮詢這個平台集思廣益，如果沒有仔細的聆聽、辯論和思考所有意見及建議，我們就不可能為市場做出最佳決定。所以，我希望大家多一點耐心，在積極發表意見的同時能夠保持開放態度，聆聽他人意見。

當然，研究「斷路器」機制絕非我們今年唯一一項戰略議程。除了開展各項新業務計劃以外，我們也在努力研究鞏固既有業務的方案。

我們深知，任何新措施的推出都不無挑戰，因為它們關係到各方市場人士的不同利益。因此，我們會仔細考慮各方參與者的意見。日後如果決定推行新舉措，我們一定會全面諮詢市場；若建議措施獲得市場支持，我們一定會給予市場充分時間作準備和調整才付諸實施。我們也會盡力闡釋新措施，務求交易所參與者和投資者都能充分理解。

未來我們就這些新措施進行諮詢時，希望所有市場人士都能踴躍發言，為我們獻計獻策。只要我們一起努力，我們一定可以進一步提升香港的競爭力，在 2014 年更上一層樓。

祝大家馬年身體健康！萬事如意！生意興隆！馬到功成！

2014 年 2 月 13 日

15 | 香港需要您的 聲音和智慧！

去年大概這個時候，我曾經寫過兩篇網誌，淺談不同股份架構下的投資者保障，拋磚引玉，希望有一天香港社會能在這一涉及公眾利益的重要議題上進行坦誠和富有建設性的討論。

如今，這一天終於到來了！上週五，香港交易所在網站上公佈了《不同投票權架構概念文件》，就香港是否應該允許同股不同權上市架構提出了一些概念性議題，啟動了更為系統的市場諮詢，希望能夠聽到更多有智慧的意見。

近年來，由於收到不少有關同股不同權上市架構的查詢，上市委員會內部一直有討論這個問題，去年由於某家公司的緣故，更多市場人士參與了這一討論。值得高興的是，這場討論並沒有因為某一家公司的離去而擱淺。相反，討論的深度和廣度不斷升級，早已超越了最初的狹隘議題。所以，才有了今天我們看到的這份概念文件。

這份文件不同於香港交易所以往發出的許多諮詢文件，它並未針對《上市規則》的修改提出具體改革方案，而只是就一些概念性議題先行進行探討。它的主要議題只有兩個：一是香港聯交所是否應該允許公司採用同股不同權架構上市？二是如果允許，應該在什麼條件下允許？

換言之，在這場關乎香港市場未來競爭力的重要討論中，我們現在剛剛邁出了第一步，就是傾聽市場到底有多少聲音認為現行的同股同權原則可以有靈活和變通的空間與必要，如果這樣的聲音夠多夠大，我們才能邁出第二步——去討論應該如何變通的問題，如果市場對這一改革缺乏足夠的共識，我們也許就沒有必要邁出第二步。

　　雖然不涉及《上市規則》的具體改革方案，但是這份諮詢文件全面梳理了香港市場目前所處的現狀以及其他主要市場的做法，並附有一些實證研究結果，為公眾理性、深入參與這場大討論提供了十分豐富的背景資料，值得仔細一讀。

　　拿到這份文件，也許有人會批評我們的效率太慢，「都已經過去十幾個月了，怎麼才看到一份概念文件？」老實說，在這個競爭激烈、分秒必爭的年代，誰不想一日千里呢？可是，在香港這樣一個崇尚法治，追求程序公義的社會裏，效率、公平與公正，到底如何兼顧與權衡，永遠都是擺在政府和監管者面前的一道難題，需要智慧，也需要耐心。尤其是這樣一個關係到香港市場未來競爭力的重大諮詢，更需要客觀、充分反映市場各方的不同意見。當效率與程序公義產生矛盾時，我們很可能錯過大好事，但也可能因此避免鑄成大錯。

　　據我所知，在過去的十多個月中，上市委員會，香港證監會和香港交易所上市科為了起草這份諮詢文件付出了大量辛勤的汗水，也經歷了無數輪熱烈的討論。他們如此仔細縝密地準備這份諮詢文件，正是為了遵循程序公義的精神，為香港尋求最大化的公眾利益。在此，我要由衷地為他們的嚴謹公正點讚，我更要為香港的程序公義和法治精神鼓掌！

　　在這份文件中，香港交易所完全保持中立，沒有任何預設立場，只為提供一個理性辯論的舞台，讓市場各方都能就此議題暢所欲言，為香港的明天集思廣益。過去 20 年，香港憑藉銳意進取的創新精神，成長為一個多元化的國際金融中心。世界日新月異，變化是唯一不變的主題。未來 20 年，香港市場能否持續保持競爭優勢再創輝煌，取決於我們能否時刻保持警醒，能否在堅持法治精神的前提下與時俱進，能否理性聽取各方的聲音求同存異。

　　朋友們，香港的明天需要您的聲音，請為香港市場的未來踴躍發聲！無論您是一位經紀人，一名律師，會計師，一位上市公司老闆，一名投資者，還是一名普通市民，您的聲音對於這場諮詢都同樣重要！

2014 年 8 月 31 日

16 | 中國特色、國際慣例、市場結構內外觀

　　A股市場巨幅震盪稍見緩解，我原本不想在這個時候對A股市場評頭論足、事後諸葛亮。然而，6月底我在上海陸家嘴論壇就內地A股市場發言的部份小標題，近日多番被引述評論，當中確有好些誤會。本來，參與陸家嘴論壇的多是金融業內人士，他們在現場已聽到發言的整體及其上文下理，我也毋須多加背景，但我當天有些用詞可能不夠清晰，引起了部份公眾誤解，我希望有系統地再説明清楚。

　　我當日主要談的，其實是嘗試客觀分析A股獨有的市場結構及其對動盪中的內地市場可能造成的影響，並非評價哪個市場制度誰好誰壞，也並非要鼓動投資者應該到哪個市場。我認為，內地在獨特的歷史路徑下形成的「穿透式」帳戶管理與中央託管制度，可算是全球僅有：對監管者而言，它是「最透明」、「最扁平」的場內市場結構；若論到對投資者託管資產的保障而言，它也是「最安全」的；事實上，內地近年確是鮮有投資者的資金或股票被中介機構挪用的事件發生。但是，在這制度下，內地市場內卻缺失了像香港這樣的國際市場中的多層次、多元化、專業化的機構投資者，也欠缺了各自以專業優勢、業務特點判斷市場的中介機構，導致內地市場由散戶主導價格形成機制，往往容易導致強烈的羊群效應，形成單邊市場趨向。市場同質化往往容易形成單邊追漲殺跌，在市場形成極大動盪。

　　就內地而言，經歷救市的市場現在其實更需要開放、更需要國際化。對香港市場而言，在發生動盪之後則更應認識到自己市場的機會與挑戰、更應主動

地發揮香港獨特的優勢，準備好在內地市場恢復元氣與再次崛起時，為香港金融市場發展闖出更廣闊的一片天地。

我曾在多個場合，包括一些國際性行業組織年會及香港證券業界舉辦的講座，都從「穿透式」制度視角談過內地與香港市場結構的異同。既然現在坊間不少人也對這課題感興趣，我在這裏補充一下我的具體看法。

一、A 股市場和香港等國際資本市場有什麼重要區別？

最重要的區別之一，在於國際市場是由券商、交易所、及不同類型機構投資者形成多層次的市場，而目前內地市場則是「扁平」、「穿透式」的以散戶為主的市場。

所謂「扁平」的市場，是指 A 股市場目前還可以說是散戶主導的單元市場，交易金額約 90% 都是散戶的交易。而在擁有多元、多層次架構的成熟市場，投資者的組成幾乎是倒過來的：以香港市場為例，機構投資者佔比大約是六、七成，散戶佔比三成；在美國市場，機構投資者交易佔比長期超過 70%。

所謂「穿透式」市場，則是指內地獨有的「一戶一碼」制度。在這市場結構下，所有投資者（包括散戶）的交易、結算戶頭都集中開在交易所、結算公司及託管公司的系統內，作為中間層的券商通常不像國際同行一樣可以管理投資者的股票、錢財及保證金，投資者的財物都在集中的統一系統中託管。

從場內監管層面而言，這種市場結構容許監管當局一眼望穿底，每一戶口在交易什麼一目了然，參與者在股票帳戶和資金帳戶層面的違規操作空間有限，市場內藏污納垢的空間不多，理論上這種帳戶制度在監管方面是直接高效的。反觀在多層次的國際市場中，監管機構與交易所只有透過券商或大型機構才能查處終端投資者，對監管者而言可謂相對缺乏透明度。若單從這角度看，內地這一在獨特的歷史路徑下形成的穿透式帳戶制度可算是對監管者「最透明」的場內市場。

二、為什麼內地會形成「扁平」、「穿透式」的市場結構？

　　檢視國際成熟市場的發展歷史，國際市場包括香港都是從下至上，分階段長期發展起來的。投資者從富人開始起步，催化了中介機構（券商）的誕生。當時每個券商猶如一個小型交易所，各券商「跑馬圈地」，為自己的客戶互相買賣產品並提供配套服務。此後，券商集合成立交易所，二級市場機構逐步形成。在投資者方面，全球自七十年代後，成熟市場中的大批中產階級崛起，開始湧入資本市場，但儘管如此，直接進入市場的散戶仍為少數，迅速發展起來的保險及退休基金等機構投資者成為券商的主要客戶。在國際資本市場上，基本是上市公司、中介機構與機構投資者之間的三元博弈，鮮見散戶直接入場。

　　內地的市場架構雖然最初亦是參照海外多層次市場經驗，但我們不要忘記內地市場只有短短 25 年的發展歷史，證監會、交易所、券商及散戶股民幾乎是同時誕生、一起長大，期間券商們甚至可以說曾有一段時期被稱為「野蠻生長」快速擴張的 10 年。急速成長期間難免出現「壞事」，例如券商挪用客戶保證金炒作股票，莊家操縱股市，這令到內地監管機構痛下決心，清理整頓市場中介機構，即 2004 年起歷時數年的券商綜合治理，並因應散戶對於政府的訴求，建立了獨特的市場監管和風險管理手段。在此次治理後，結果就是市場中間層被事實上「拿掉」，A 股變成了現今的獨特的、單元、扁平、穿透式市場。

三、因着市場結構不同，A 股市場和國際市場的監管理念和手法
　　有什麼不同？

　　這兩種市場結構反映了不同的監管理念和手法。在成熟資本市場，前提假設通常是市場參與者都是「好人」，有「壞人」出現欺負投資者時，或投資機構、中介機構違規時，監管機構便會事後執法，嚴肅處置，即監管邏輯是暴露出問題後，加強監管，事後改正。監管當局主要功能是一個「裁判員」，監管裁決「大人們」之間的博弈與遊戲，不會傾向保護市場中任何一方的利益。

　　在這樣的市場中，難免不時會出現一些「壞人」，譬如 2011 年在美國宣告破產的明富環球（MF Global）便曾大規模挪用客戶資金。再者，多層次的市場結構不容監管當局一眼望穿底，這也是美國監管機構當年未能及時發現大量有毒資產在中介機構層面積累的原因之一，最終導致雷曼倒閉、次貸危機，引發系統性的問題。然而，即便是經過 2008 年金融海嘯之後，國際市場監管機構總結教訓，也只是決定要全方位地監控金融機構的風險集中程度，保持資本充足率，加強場外市場監管，以確保金融機構更健康和市場更穩定，並未因此改動分層次的市場結構。

　　反觀在內地市場，監管機構切身了解自身處於一個「新興加轉軌」的市場，新興市場意味着市場參與各方的經驗還需要積累，轉軌則意味着內地市場是從計劃經濟體制起步轉型而來，機構投資者的發展大大滯後於廣大散戶湧入市場的節奏，這就逼使內地的監管者不得不在制度上設有大量保護散戶的措施，期望盡可能通過各種制度設計事先防範、甚至爭取取締壞人壞事，特別是「以大欺小」，結果不免加入了太多「家長」情懷，改變了國際市場上常見的三元博弈平衡，使缺乏經驗的散戶投資者都產生較為強烈的依賴，缺乏防範投資風險的獨立意識。

四、那麼，「穿透式」市場「最安全」這話從何説起？

　　這話主要是針對內地獨特的「穿透式」市場結構而言，由於個人投資者的錢、券、物均在系統中統一中央管控監測，許多被禁止行為都已事前在軟、硬系統之中被限制，中介機構如券商幾乎不再可能偷竊、佔用或挪用客戶資產，散戶也不會輕易因疏忽承擔過大違規犯法的風險。由是，至少在投資者的資金帳戶層面，中國的市場便難出現因大型機構倒閉而對投資者資產安全造成威脅。

　　不過話又説回來，在這市場中散戶在帳戶層面的財產雖然是安全了，但隨着機構在市場的參與度相對減少，大規模的市場風險卻可能相應上升了。在國際市場中，客戶的錢財物置於中介機構的託管下，中介機構有強大的客戶資源

激勵它們去創新、去服務、根據不同客戶羣的風險取向合理配置投資及管理風險。在市場動盪中，這些機構以各自的專業優勢、業務特點、與理性判斷博弈市場，不同機構觀點各異，市場上不容易產生過強的單邊效應。這情況就如紐約時代廣場上每年等待除夕倒數的人羣被員警用隔離帶分片分區管理，他們之間雖可流動，但不能成批同時向一個方向快速亂竄，從而有助於防範發生「踩踏事件」。

相反，在扁平、穿透式的內地市場下，由於市場的主體是單元同質化的散戶羣體，對市場方向的判斷缺乏機構的制衡力量，很容易導致強烈的羊羣效應，容易形成單邊市場趨向，在市場動盪時，擁擠踩踏的風險往往大增；這情況就猶如當人羣在廣場上都向着同一方向熱舞，就有可能出現「踩踏事件」。

近年，內地大力發展基金業，在着手改變散戶主導的投資者結構方面也取得了明顯進展。然而，這些新晉的機構投資者特別是開放式基金的資金來源同樣還是散戶為主，因此基金的投資決策會受大量個人投資者因市場變化而進行份額申購、贖回的影響。同時，由於基金的力量不夠大，在散戶活躍佔主導的市場中，出於業績的考慮，基金的行為會出現趨同於散戶而出現「散戶化」。這些散戶的投資習慣仍然影響、甚至可以說是「挾持」着基金經理的投資理念。反觀在成熟市場，基金投資者多為養老基金、保險資金等各種長線價值投資基金，在動盪時期他們往往成為穩定市場的力量。

五、為什麼又說「扁平」市場「最民主」？

「民主」這詞在此也許不是一個最恰當的比喻，但當天說的「最民主」指的是投資者參與市場價格形成的過程。在內地，股票投資是覆蓋面最為廣泛的金融投資管道，目前滬深股票帳戶已經超過二億個。散戶的訂單通過券商通道直接放到市場上競價，直接參與價格形成過程。在國際市場，參與市場價格形成過程的主要是機構投資者與大券商，他們經過專業謹慎、理性的判斷作出投資決策，由他們代表眾多散戶決定證券的市場價格。

　　換句話說，內地股市是世界股市中平民百姓參與最直接、最普及、普羅大眾最關注、對民生影響最深遠和「萬眾皆股」的市場。正因如此，散戶的需求遂成為監管政策制定的最重要驅動力量。無論是在上市公司一級市場融資、大股東減持、漲跌停板、融資融券、期貨市場門檻、及證券資產託管等一系列制度的設計都是假設散戶沒有經驗、需要保護、需要幫助。這些制度安排往往在牛市的形成與發展中很有效，但在市場震盪時卻容易扭曲市場功能，有時效果甚至適得其反。

六、既然 A 股市場結構那麼「透明」、「安全」與「扁平」，市場怎麼還會出現這麼大的震盪？

　　正如上所述，內地在特定的歷史條件下形成的一個單元、扁平、完全穿透式的市場架構，優點很明顯，但缺點也同樣突出，就是缺乏多層次、多樣化機構投資者的制衡與對沖。市場同質化往往容易形成單向追漲殺跌，在市場形成極大動盪。再者，恰恰是因為內地市場結構的高度透明與扁平，基於機構監管的不同分工體制，在場外配資活動所產生的巨大風險反而可能容易被忽視了。

　　從市場上目前已披露的資訊可見，直接催生 A 股市場這次股市大幅波動的似乎是大規模場外程式化配資，這些場外配資具有一些鮮明特點：首先是規模巨大，來源充沛，但透明度有限；其次是扁平的互聯網連結大大降低了股民獲得配資的門檻與成本；再者，這些場外配資的槓桿率其實有不少已經大大超出了合理風險管控邊際，而這些融資又未有如場內受到監管者監管的金融機構的兩融業務（融資融券）般嚴格管控投資者的資格與限制。結果，當市場波動時，配資系統為管控自身資金風險而設的高度敏感程式交易啟動，更擴大了市場的震盪。

　　對世界任何資本市場而言，場外程式化配資是一個完全陌生的「新鮮事物」，特別是在互聯網金融發展模式具有海量小額特性的今天，它對這次股災到底起了什麼作用還有待進一步研究，但迄今市場對它已有兩點共識：一是它

基本落在了一個監管「盲區」；二是它加大了市場的震盪幅度。

七、既然內地與國際市場諸多不同，內地與香港市場還應該互聯互通嗎？

內地市場近期的動盪殃及港股，不少朋友因而對內地與香港市場互聯互通計劃產生懷疑，這很容易理解。不過，香港是完全開放的經濟城市及國際金融中心，全球任何地方發生大事如美國息口變動以至歐債危機，香港市場都難免震盪，更何況是來自毗鄰的內地市場？事實上，從二十幾年前 H 股來港上市起香港市場就已經與內地市場不可分割地聯繫在一起，今天的港股市場超過 60% 的市值與接近 70% 的股票交易量均為內地企業，其他上市公司的業務不少也與內地息息相關，大量來自內地的投資者也已在香港市場活躍多年，這意味着把香港屏蔽起來、與內地市場隔斷並不現實。

再者，香港的開放以及港股通使更多國際與內地投資者來港，長遠而言，這會令香港成為亞洲時段真正的世界金融之都。當然，任何事情都有成本與風險，我們要做的應是如何在把握機會之餘、審慎管控好風險。面對挑戰不逃避，這才是香港精神。

對於這次 A 股股市大幅波動的起因、救市的成效、以及對於未來市場發展的長遠影響，內地的監管機構和市場人士相信必會有更深入、更精準、更全面的反思。惟可以肯定的是，內地資本市場雙向開放與人民幣國際化的發展路徑不會因此逆轉，內地市場和國際市場從監管到市場運作雙向而行也是大勢所趨。對於內地市場而言，A 股市場的結構問題若得不到改善，其市場開放和國際化進程必然受阻，而若內地要改變過於同質化的投資者羣體，則需要考慮引入多樣化、風險收益偏好不同的投資者羣。滬股通等互聯互通計劃便是其中一個重要「快捷方式」，能夠說明 A 股引入境外成熟且多樣化的機構和個人投資者，從金融安全和市場穩定角度實現安全可控的開放。

八、內地市場的動盪為香港市場帶來什麼啟示？

對於香港而言，內地與國際資本市場結構現存的巨大差異，卻也正正突顯了香港市場獨特的價值與優勢。

就以近期股市大震盪為例，雖然香港沒有內地市場硬件上的穿透與透明，但市場整體運作公平有序，市場訊息保持靈通。這大概是由於香港市場身處國際資本前沿，曾受到多次金融危機的衝擊，監管者、中介機構都各司其職及早防範好風險。歷練的監管者、專業的中介機構以至經歷市場大幅波動洗禮的投資者對股災及市場風險司空見慣，能夠在市場動盪之中沉着應對、處變不驚。合規守法的文化深入香港市場人心，加上嚴謹的監管制度，香港市場過去多年即使在環球股災中也不曾出現系統性市場危機。

這些優勢令香港絕對勝任作為內地資本市場雙向開放的橋頭堡。當內地試點資本市場開放之時，香港作為世界最成熟的金融市場之一，不單可為內地提供國際資本市場運作方式的試驗田，為國際投資者提供便利且低成本參與內地資本市場的機會，同時也把香港金融市場的未來發展引領到新里程；除卻鞏固了當前香港股票市場的優勢和提升流動性，更有助強化香港作為境外首選人民幣離岸中心的地位，且同時促進香港建設其他資產類別的交易中心，繼續推進互聯互通計劃完全符合香港發展的長遠利益。

結語

香港金融市場過去的成長與內地的發展密不可分，內地資本市場開放及人民幣國際化未來亦會為香港市場帶來巨大機遇。今天 A 股市場面臨着嚴峻考驗，雖然在短期內突顯了香港市場的穩定性與相對優越性，但這不應是我們自我陶醉的時候，因為從長遠來看，一個強大的、更國際化的中國內地市場才是保證香港長期繁榮穩定的基礎，才能使香港在全球其他金融中心之中保持其相對優勢。

今天的香港，恰恰處在中國市場改革開放、融入世界的大潮之核心，甚至是風口浪尖上。這就要求我們不僅對自己的市場充滿信心，更要求我們精準理解和掌握內地和國際市場的精髓，尤其是兩地市場的差異，這才能讓我們充分認識包括香港等國際市場的優勢與挑戰何在。這樣，我們的市場才可以保持靈活高效地與全球各種市場制度銜接自如，才能讓我們更有智慧、有膽識、自信地去迎接市場未來發展的新挑戰。

2015 年 7 月 16 日

17 | 關於三板和新股通的初步設想

2016 年，我們將開始實施新的 3 年戰略規劃。《戰略規劃 2016-2018》裏提到很多新的設想，一些設想已經比較成熟，可以很快付諸實施，一些設想尚處初步醞釀階段，需要向市場集思廣益後進行完善。

最近不少朋友對我們提出的三板和新股通設想頗感興趣，也有很多疑問，相信今天的業績發佈會上可能也有很多記者朋友就此提問。不少朋友很想知道這些設想的出發點是什麼？必要性在哪裏？有無實現之可能？今天，我想從一個市場運營者和參與者的視角，梳理一下我個人關於這些問題的思考，希望拋磚引玉、集思廣益，與大家共同尋找前行之路。需要指出的是，這些問題的最終決策須監管機構廣泛聽取多方意見後統籌安排。

關於三板

股票上市業務是香港交易所的核心傳統業務，也是香港作為國際金融中心的一大競爭優勢。經過過去幾十年的快速發展，香港市場已經憑藉其質素和國際化程度躋身全球領先金融市場之列。但作為股票市場營運者和前線監管者，我們不敢也不應沉湎於過去的成績，不斷提高市場質素和競爭力是我們義不容辭的責任，我們時常在思考：香港的市場有哪些需要改進的地方，一個最理想的上市市場應該是怎樣的？

在我看來，一個最理想的市場，應該讓好公司很容易上市融資，也能迅速

把爛公司清理出局，這樣才能優勝劣汰，吐故納新，維持市場質素，有效發揮股票市場資源配置的功能。

那麼問題來了，怎麼能讓一個市場好公司多而爛公司少呢？相信很多人會說，提高准入門檻啊，要麼將客觀財務運營指標大幅提高，要麼實行更嚴格的實質審核。

提高准入門檻的方法確實有助於擋住部份壞公司，也肯定會把一部份潛在的好公司關在門外。但是，再嚴格的上市標準也無法保證上市時的好公司在上市後不會變成壞公司，更無法阻止這些壞公司利用各種財技滿足上市交易的標準成為受人詬病的「老千股」或「殭屍股」，事實上，近期出現的一些「問題股」在當年上市時也是完全符合上市標準的。因此，單靠提高准入標準無法防範壞公司。如果僅盯住上市門檻或寄望上市審查而不加強退市機制，就容易產生有些好公司進不來、不少壞公司出不去的現象。

對於清除爛公司更加有效的方法應該是加強上市後監管，加大執法力度，特別是強化退市機制，防止大家所謂的「殭屍公司」或「殼公司」長期滯留市場、浪費資源、拖累市場信譽。一旦擁有了有效的退市機制和充分的投資者保護措施，反而可以制定更加寬鬆的准入門檻、讓更多有潛力、有活力的創業型公司有機會准入我們的市場。也就是說，一個理想市場的「進」「出」機制必須匹配，寬鬆的「准入」機制必須配備嚴格的「退出」機制。

毋庸諱言，今天的市場還沒有達到這種理想狀態，儘管大部份在港上市的公司都很優秀，但我們的市場也存在一些問題，譬如一些新興的好公司還不能來上市，同時一些問題公司仍然滯留市場。那麼，我們應該如何改革才能夠邁向理想的方向呢？

僅從分析的角度而言，改革的方式似乎有兩種：一種是直接改革現有市場尤其是創業板，全面修訂《上市規則》，進一步放鬆准入門檻，引入更有活力的新興公司，同時從嚴制定和執行上市後的監管和退市標準。這樣的改革最直接、最有效，但是它勢必會影響大量既得利益（現有「殼股」生態圈中的參與者與投資者的利益）。在香港這樣一個注重程序公義和私人產權保護的法治

社會，這樣有爭議的改革必須在諮詢公眾後取得廣泛市場共識的前提下才能啟動。因此，這樣的改革是一項系統性工程，牽一髮而動全身，需要勇氣與智慧，也可能需要較長時間才能實施。香港市場此前的歷史教訓告訴我們，推行這樣的大改革不可不考慮其可能帶來的市場衝擊。

如果存量改革短期難見成效，另一個可能就是做增量改革，譬如新設一個市場（三板），然後用新市場的新氣象來帶動和倒逼現有市場（主板和創業板）改革，這才是三板市場的由來。在三板這個全新的市場板塊上，監管者可以考慮全新的上市規則：一方面設立比較寬鬆的准入門檻，不把潛在的好公司關在門外；另一方面強化投資者保護措施和退市標準，加快爛公司強制退市過程。如果擔心這樣的新板對於投資者來說風險太大，也可以在成立初期為這個三板設立一定的投資者適當性門檻，先允許一些風險承受力較強的專業投資者自願准入。等到這個新的板塊發展到一定階段，股票市場優勝劣汰的投資文化已經深入人心，這個新板塊的新氣象就可以起到示範作用，帶動存量市場的改革。

之所以想到三板這樣的概念，就是因為這是一個全新的市場，不存在既得利益，啟動的阻力可能會較小，短期內可能比較容易見效、且對現有市場應該不會有很大的衝擊。創業板可以與三板並行營運，逐步改革提升。總而言之，三板只是一個存量改革無法啟動情況下的一個潛在備選方案（Plan B），具體細節還有待研究。這個設想總的原則是易進易出，歡迎好公司、清走爛公司。

存量改革與增量改革兩種方式各有利弊，需要大家以開放的態度、創新的思維鼓起勇氣去開拓。要麼我們改革現有市場，一勞永逸地解決問題；要麼我們盡早在新增市場啟動改革，以時間換空間，闖出一片新天地。究竟應該選擇哪一種方式進行改革？或者說在什麼樣的時機才可以進行哪一種改革？這些問題關係到香港市場的長期競爭力和發展，非常值得市場各界深入探討、在監管機構的統籌下盡快做出抉擇。我最不希望看到的是迴避這些問題不作為，那樣不僅會令我們的市場停滯不前，甚至很可能令我們的市場陷入「想要的好公司來不了、不想要的爛公司送不走」的尷尬境地。

關於新股通

新股通是我們正在研究的一種滬港通延伸方式,就是將現有的股票通模式從二級市場延伸到一級市場,讓兩地市場的投資者都可以在未來申購對方市場發行的新股(包括 IPO 和新股增發),具體可以分為「內投外」(允許內地投資者認購香港新股)和「外投內」(允許國際投資者從香港認購內地新股)。

如果得以實現,新股通將為兩地市場創造巨大的共贏。對於香港而言,新股通的「內投外」可以豐富香港市場的流量,吸引海外名企大規模來港上市;對內地來說,「內投外」可使內地投資者能在滬港通體系下廣泛投資大型國際公司,實現中國國民財富的全球權益配置。前一階段內地金融市場流行一個詞,叫做「資產荒」,描述的是投資者有大量的資產配置的需求,但是找不到優質的資產。「內投外」可以幫助內地投資者解決這個「資產荒」的問題。反過來看,「外投內」對香港而言,它將進一步提升香港作為國際投資者投資中國內地資產首選通道的競爭力。對於內地而言,「外投內」則可以幫助內地公司引入更多國際機構投資者,改善投資者結構,完善新股發行定價機制。

有不少朋友問我,香港與內地實行完全不同的上市審核機制,要實現新股通是否面臨很大的監管障礙,真有實現的可能嗎?在我看來,無論一件事情多難多複雜,只要能為雙方創造共贏,就有實現的可能,就值得我們用積極開放的心態去探索。讓我們來分析一下新股通在兩地監管問題上可能有什麼樣的問題需要解決。

新股通的「內投外」即允許內地投資者購買香港新股,這在香港應該不存在法律障礙,但需要中國證監會的許可。我認為這並不是不可逾越的監管障礙。目前的滬港通監管規則中已經包含了香港已上市公司對內地投資者增發新股的條款,其實認購公司增發的股票和首次上市發行的股票並沒有本質區別,如果未來內地監管機構能夠再向前一步,將股票通模式從允許認購增發新股到允許認購 IPO 新股,新股通的南向之旅就暢通了。

　　當然，我們的合作夥伴上海證券交易所與深圳證券交易所也都有着自己的國際板夢想，希望吸引國際公司直接在內地上市。如果這一願景能在中短期內實現，我們提出的新股通「內投外」就不一定特別受內地交易所的歡迎，沒有內地監管機構和內地交易所的支持，新股通「內投外」就不可能實現。但如果內地國際板這一願景在中短期內不容易實現，我們提出的基於滬港通模式的新股通就可以為內地提供一種新的「國際板」路徑，如同現有的滬港通模式一樣，新股通可以實行監管合作、執法互助和交易所利益共享，充分體現互惠共贏的合作原則。一個完善的資本市場肩負着雙重使命，一大使命是為上市公司提供便捷的融資渠道，另一大使命則是為投資者提供豐富多元的投資機會。新股通「內投外」的可以利用香港獨特的制度優勢幫助中國國民財富實現有效、安全的全球權益配置，我認為應值得內地監管者通盤考慮。

　　而新股通的「外投內」（允許香港投資者和國際投資者認購內地新股）在內地並無法律障礙，因為這些公司在內地上市、受內地監管，在香港如果僅對專業投資者開放也無重大監管顧慮。內地一直希望吸引海外機構投資者，允許國際投資者認購 A 股市場的新股符合內地資本市場對外開放的大方向。香港目前的監管法規允許在香港以外市場上市的公司向香港的專業投資者發售新股，只是出於投資者保護的考慮不允許它們在香港對普通散戶定向發售。因此，如果新股通「外投內」僅對專業投資者開放，相信香港監管機構的顧慮會比較低。如果新股通的「外投內」要對香港的散戶投資者開放，必須在香港的監管法規調整後才能實現。

　　當然，新股通還只是我們提出的一個初步設想，具體怎麼操作還有待我們在兩地監管機構指引下深入溝通和探討。

　　總而言之，我們樂於傾聽市場的聲音，歡迎各界朋友就三板和新股通暢所欲言，與我們一起在監管機構的領導下促進香港資本市場的健康發展。

　　為了讓大家更容易理解新股通，請參閱以下列表：

新股通	香港監管	內地監管
「內投外」 （內地投資者認購香港新股）	公司在香港上市，受香港證監會監管，無香港法律障礙	需中國證監會許可。但已有配股先例，有可能找到共贏方案
「外投內」 （香港投資者和國際投資者認購內地新股）	無監管顧慮，若： • 只允許對專業機構投資者開放 • 暫不允許對香港散戶開放	公司在內地上市，受內地法規規管，因此無大的法律障礙

2016 年 3 月 2 日

18 | 關於「老千股」

近期，港股通的交易量持續放大，內地投資者對港股市場的投資熱情和關注度也越來越高。在大家對於港股市場價值的熱烈討論中，我也聽到了一些關於「老千股」的申訴（在這裏，「老千股」主要是指大股東不以做好上市公司業務來盈利，而主要通過玩弄財技和配股、供股與合股等融資方式損害小股東利益）。到底什麼是「老千股」、誰是「老千股」、怎麼管制「老千股」，可能涉及很多技術及規則執行細節，不是一兩篇文章能夠講得清楚的。我今天只想換一個角度來談談我對這問題的看法。

一、在資本市場，我們都希望好人很多、壞人很少，好人做事很容易、壞人做事特別難。那到底誰是好人、誰是壞人呢？

關於這個問題，1,000 個人可能會有 1,000 個不同的答案。而對於監管機構來說，這個問題其實很簡單：遵守規則的人就是好人，違反規則的人就是壞人。除此標準外，監管機構既不能有自己的感情好惡，也不能施加自己的價值判斷，唯一用來判別壞人的標準就是規則。在香港的法制框架內，即便是壞人，也需經過公義的程序才能被定罪，但這往往需要時間。

二、既然知道有壞人會幹壞事，為什麼不採取有效的措施事前防範，事後懲罰豈不是亡羊補牢？

財務造假、內幕交易、操縱市場等等是全球各大市場監管機構都嚴厲打擊的行為，對於幹這些壞事的人，大家都是深惡痛絕的。如果能夠事先看出誰是壞人，相信沒有一家監管機構不願意挺身而出，遺憾的是，監管機構並沒有孫悟空的火眼金睛，無法事先一眼識別出誰是好人誰是妖孽。

於是，對於如何防止壞人幹壞事，不同市場採取了不同的監管哲學，一種監管哲學是假定所有人都有作惡的動機與可能，對各種融資行為都採取嚴苛的審批制度，通過事前盤查來篩除壞人；一種監管哲學是假定絕大多數人都是遵紀守法的好人、從事正常的市場交易，監管者儘量不干涉市場自由，主要通過事中監察和事後追責來懲罰壞人。

前者的好處是可以將不少壞事扼殺在搖籃中，讓壞人無法作惡。對於一個散戶為主的市場來說，這樣的監管理念可能是不二的選擇。而對於一個機構投資者佔主導地位的市場，現實情況是，再嚴厲的事前審查也無法完全杜絕所有違規（除非關閉整個市場），而過於嚴苛的審批必然會妨礙好人的自由、窒息正常的市場活動；而且，如果審查過程中賦予監管者過多的自由裁決空間，容易滋生腐敗與尋租。

香港市場採取的後一種監管哲學，其好處是保障了好人的自由，提高了資本市場的效率，當然代價是不能把壞人在事件未發生前、或立刻就擋在門外。這樣的監管哲學不是推卸管制的責任，而是說管壞人的着眼點主要是強制披露責任、確保股東審批程序和加強事後違規檢控，通過懲罰震懾違規。當然這樣的監管理念會與香港市場以機構投資者為主的這一市場結構息息相關。

在監管層面上，香港證監會和香港交易所分別肩負不同的職責。作為上市公司的前線監管機構，香港交易所根據《上市規則》監管上市公司和董事的合規情況（例如是否及時進行信息披露、是否按規定召開股東大會等），但儘量不干預上市公司股東的決策自由，且無權監管投資者行為。

香港證監會作為獨立的法定機構，根據《證券及期貨條例》全面監管整個市場，因此，有關收購、股份回購及私有化等上市公司事宜和內幕交易、操縱市場等違法行為，由於可能涉及投資者的行為，均由證監會負責事中監管和事後檢控。

三、為什麼香港允許公司那麼容易地配股、供股、送股與合股？

在我看來，所有這些再融資制度都是市場的中性工具，允許它們的初衷當然是為了幫助好公司更容易、更方便、成本更低、效率更高地融資與發展，這也是許多內地企業選擇來香港上市的一大主要原因。

再融資制度可為有經營困難的公司解困，當然他們拯救失敗會給參與融資的股東帶來損失。與此同時，也有一些壞人會利用這些融資手段直接侵害小股東利益。就像廚房裏的菜刀一樣，好人用它是為了用來切菜切肉，但也難免有一些壞人會拿它來殺人或搶劫。不過，如果因為有幾個壞人拿菜刀殺了人，我們就把所有的菜刀都收起來，肯定會給大家做飯帶來極大的不便。因此，如何在不影響大眾生活的前提下有效降低有人拿菜刀行兇的威脅是香港監管者不斷努力尋求的平衡，而最重要的監管責任應落在事後強而有力的追責以阻嚇其他潛在壞人。

四、既然香港市場的監管不能把壞人事先擋在門外，普通散戶投資者（尤其是缺乏投資經驗的內地投資者）該如何保護自己，躲避「老千股」？

投資者首先必須嚴肅對待自己的投資責任，既然你是自己投資盈利或損失的最終承受者，你就應該是這筆投資安全性的第一責任人。一定要買自己了解的股票，不要輕信小道消息，遠離誘惑。如果真的不熟悉市場，那就去人多、亮堂的地方，那裏壞人較難藏身。壞人大多喜歡躲在偏街小巷人少的角落，資

本市場上也是一樣。因此，缺乏經驗、人生地不熟的散戶投資者應該儘量走寬敞亮堂的大道，投資那些人人皆知、往績良好、信息披露充分透明、風險較低的公司。膽大勇猛或者想以小博大的投資者如果選擇走偏街小巷，那就一定也要格外小心，提高風險意識。

其實，儘管香港的再融資機制靈活方便，但大股東使用時都必須嚴格按照《上市規則》召開股東大會得到股東授權或同意。很多情況下，只有得到小股東單獨同意後才可發行。上市公司就再融資計劃發通告到股東大會中間往往都有相當一段規定的股東通知期。因此，選擇投資這些公司的投資者一定要仔細跟蹤公司公告，及時了解公司的再融資計劃，積極參與股東大會投票，並時刻查察投資，及時作出適當投資決定。

總而言之，要想避開老千股，儘量走大道，避免串小巷，提高警惕，謹慎決策。出事後再喊警察，往往為時已晚，再好的警察，也無法替你的損失買單。

正是考慮到了內地投資者可能對於香港市場規則了解不足，兩地監管機構在劃定港股通的合資格股票範圍時採取了循序漸進的原則，滬港通先選擇了一些市值大、流動性好的股票作為試點，然後才在深港通中加入了市值較大的中小盤股，並繼續保持了投資者准入門檻。初次出海投資的內地投資者如同初學游泳的人，應該先去淺水區練習，然後才去深海遨遊。

2016 年 9 月 11 日

19 | 淺談香港證券市場的監管分工

　　自我加入香港交易所工作以來，經常有朋友問我香港交易所和香港證監會的監管分工到底有何不同。最近，香港證監會加強執法力度、早期介入一些上市申請的審批，再次引發不少傳媒朋友的關注和查詢。

　　梳理下來，大家似乎主要關心以下兩個問題：

(1) 香港證監會早期介入上市審批是不是意味着香港交易所上市監管體制失靈或工作不力？

(2) 香港證監會與香港交易所若「並行審批」會不會讓市場受到重疊監管而感到無所適從呢？

　　對於從事金融行業的朋友而言，相關法規和條例已經對香港交易所和香港證監會的分工和定位有詳盡的解釋，自然無需我再累述。但對於一些圈外的朋友而言，要弄明白我們的分工有何差異也許並不容易，在此，我想用一個虛擬城市的交通警察與刑警的互動來打個比方，方便大家對這個問題有一個不一定完全準確、但簡單明瞭的理解。

　　在這個虛擬城市中，交通警察與刑警共同的目標是維護社會治安、保障人民安全。但在達成這一共同目標的過程之中，他們的任務和使用的工具存在一定的差異。香港證監會的角色就像是這個虛擬城市中的刑警，而香港交易所的角色則更像是交警。

　　交警的首要任務是維護交通秩序、保持道路順暢與安全。他們的日常工作包括組織駕照考試／發放運營執照（類似上市審批）、安排車輛安全與排放年檢（類似上市後的合規監管）、建設與運營交通指示和監控系統（類似市場監

察）、攔截和處罰違例駕駛人員或違例營運車輛。對於違例者，交警可處以警告、扣分直至吊銷駕照或執照的一系列懲罰，其權力來源於交通管理條例，就好比香港交易所的監管權力來源於《上市規則》。總的來說，交警只有民事執行權，一般不具備直接刑事執行權。

刑警的主要任務是維護社會的整體治安，他們的日常工作包括跟蹤監察、搜查取證、拘捕和審查疑犯、提交刑事檢控等等。簡而言之，他們的主要工作重點是抓壞人，懲治犯罪。交警負責的交通安全領域只是刑警工作的一個組成部份，刑警的任務也包括監督交警的工作，在特殊情況下，也可直接干預交警的個案處理，包括牌照許可及發放。刑警的權力來源是刑法，就如同香港證監會的權力來源於《證券及期貨條例》。

那我們來分析一下大家可能關心的問題：

一、 刑警開始介入一些交警的工作是不是意味着交警辦事不力或交通管制的條例失靈？

如前所述，交警與刑警的任務、職責、工作重點是不同的。與其肩負的不同任務相對應，交警與刑警配備的工具也不同。交警僅配備哨子、指揮棒、酒精探測器和路障設置等，一般不配槍。在日常交通管制上，這些工具是有效的，但碰到重大案件時沒有什麼威懾力。而刑警則配備各類槍枝、手銬等多種大火力武器，必要時還可以進行搜查與拘捕。這些武器可威懾罪犯，並非為管理普通民眾而配備的。

同樣的道理，交警與刑警在執勤中的側重點也有所不同。交警執勤的側重點是交通的安全與順暢，雖然必須懲罰違規駕駛或運營，但一般不會為了個別闖紅燈或超速行為而在高峰時刻或路段輕易追車攔截，造成大面積交通堵塞。而刑警處理的多是大案要案，如果不能儘快將罪犯抓獲，往往會對公眾和社會治安造成嚴重危害，這種時候該封路就得封路，該追車就得追車，不能放縱違法犯罪。因此，刑警的執法如果給普通民眾帶來一些不方便，大家一般也都可

以理解。交警與刑警的有效配合與合作，可以使警力資源效益最大化：交警在維護交通順暢的前提下，不能忽視追究違規行為的責任；刑警在抓壞人的同時也應儘量減少給大眾出行帶來的不便。

二、更嚴格的警力實施是不是可以減少或者杜絕違例與犯罪？

是，但又不是。

交警加緊巡邏確實有助於發現和降低違例。如果交警在巡邏中發現一些違例行為有上升趨勢，自然應該通過增加巡邏、抽查和設立更多超速限制、或禁止停車區域的方法來加強執法。為了嚴懲違例和犯罪，交警還應該及時將更多案例移交刑警處理。

但是，交警可以通過收緊事前審批或發牌（類似上市審批）禁止所有「壞人」上路嗎？這恐怕非常困難。

大家試想一下，如果這些人考駕照的時候還沒有暴露出壞人的「潛質」，而且已經成功的通過了所有路考，交警就必須依法發放牌照。當然，如果交警發現了某一類型的司機在路上違例的趨勢偏高，也可以考慮有針對性的修改考試內容或通過標準，更有效地把一些潛在的「壞司機」攔在門外。但即使如此，也不可能保證把所有的壞人都擋在外面。而且，駕駛考試內容或通過標準的改變不能輕易做出，做之前應該諮詢社會，以免影響到大多數的好人。與此同時，已拿到執照的人也不會因為標準變了而需要重新申請。一個法治社會不能追溯性地去用新法追究舊事。

在這裏，交警與刑警在執法許可權上有比較大的區別。交通管理規則一般很詳盡，駕駛考試內容、發牌門檻/程序、違規處罰標準等方面都有清晰具體的指引。交警的裁決許可權很小，不能憑主觀判斷或價值標準來隨意決策，否則會出現不公平、不一致、甚至導致尋租腐敗。

相對於交警，刑警的執法權力主要來源於刑法。保障公共利益與公眾安全是刑警執法的最高原則。刑警擁有更高的執法力度和空間。比如，一個人已

經通過了交警的筆試和路考，但如果刑警有充分理由相信這個人上路會影響公眾利益，刑警可以命令交警拒發駕照。再比如，一個司機接受酒精測試時未超標，也沒有違反其他交通規則，交警一般只能放行，即使交警認為這個司機可疑，也很難隨意拘留他，因為交通法規沒有賦予交警這樣的執法許可權。在這種情況下，交警應該將情況上報給刑警，只要刑警有充分理由，例如懷疑此人可能危害公眾利益，就可以訊問甚至拘留這個司機。總之，加強警力和執法能夠威懾一部份壞人，讓他們不敢上路。

三、 公眾應該如何理解與面對交警與刑警的「並行監管」呢？

要回答這個問題，我們首先必須認識到交警與刑警權力之間的重大區別。交警的權力只能在一個有限的小尺子（類似《上市規則》）下運作，他的工具是小哨子加小警棍；而刑警的權力則來源於一個大尺子（類似《證券及期貨條例》），他的工具是大警笛和大頭槍。鑒於這樣的不對等關係，刑警有權隨時隨地干預交警的工作，也可以直接越過交警執法。

當然，為便於公眾適應和接受，及基於有效運用警力資源的考慮，交警與刑警達成了一定的默契與理解；在常態的交通管制上，交警用小尺子、小哨子、小警棍在前線監管，刑警拿着大尺子、大警笛、大頭槍在二線督戰，有權隨時干預和督導。在這種諒解下，刑警不會隨便響着警笛在大街上亮槍，也不會越俎代庖拿去交警的小哨子、小警棍直接指揮交通。這種分工安排明確，各有側重點，也便於公眾清晰理解。

具體到駕照發放上（類似上市審批），交警和刑警也有較清晰的分工。交警負責筆試（類似資格審查 eligibility test）和路考（類似合適性審查 suitability test），而刑警則負責更嚴厲有關公眾利益的背景複查。一般而言，筆試與路考在先，背景調查在後。當然，刑警有權隨時隨地干預筆試與路考，也可以「前置性」地開展深度背景調查，直接在筆試或路考前截查考生，甚至不允許考生進一步參加考試，但刑警不會也不應直接進入考場監考筆試或路考、也不會逼

考生在交警處考完後再到刑警的考場再考一遍。

　　必須指出的是，「刑法」的地位遠高於一般交通管理條例，因此，如果刑警認為路況惡化嚴重，需要臨時改變與交警的執法分工和安排，刑警根據法律有更廣泛的執法權力，他們完全有權做出改變，甚至直接代替交警上路執法，或直接進入考場監考。但無論如何改變，刑警和交警的通力合作都十分重要，並應與公眾充分溝通，否則容易引起公眾困擾。

　　用交警與刑警的比喻來形容香港交易所與香港證監會的關係並不一定完全準確，但我希望用這樣淺白的描述為大家了解我們的不同分工提供一個新的視角。

　　回到香港現實的今天，我想補充一點，香港交易所雖然是一家上市公司，但不是普通的商業機構，《證券及期貨條例》規定，香港交易所必須將公眾利益放在商業利益之上，為保證落實這一原則，香港交易所一半的非執行董事（包括董事會主席）必須由特區政府任命，行政總裁的任命也必須得到香港證監會的批准。而且，從商業角度看，上市公司的質素關係到香港交易所的最大利益，只有優秀的上市公司越來越多，才能吸引更多投資者，為交易所帶來更多的交易量，作為市場的營運者，香港交易所沒有理由放鬆上市審批，為了一點蠅頭小利損害香港市場的長期利益。

　　此外，儘管目前市場上出現了一些低素質的上市公司，對香港市場的聲譽造成了不良影響，但它們無論在數量、市值還是交易量上，都只佔整個市場的極小部份。俗話說，林子大了，什麼鳥都會有。我們在努力持續改進上市公司質素的同時，也應客觀和理性地看待市場上出現的一些問題，避免非理性誇大負面影響、以偏概全。

　　簡而言之，在市場發展與嚴格監管之間，永遠需要保持一定的合理平衡。這是非常考驗監管者智慧和勇氣的永恆難題，我們日常工作的很大一部份，正是思考和學習如何解答這道難題，儘管永遠找不到完美的答案，但是希望能夠在大家的幫助下，不斷完善我們的監管工作，爭取交出一份令人滿意的答卷。

2017 年 7 月 23 日

20 | 關於完善香港上市機制的一些思考

今年 6 月，香港交易所就完善香港上市機制刊發了兩份建議方案，徵求公眾意見。文件的諮詢期大約還有兩週就要結束了，我誠摯地希望大家能夠抽出寶貴時間回應這次諮詢，在限期前給我們更多真知灼見，共同推動香港市場的進步！

這次的諮詢，特別是有關創新板的設立引發了市場的熱議，不少朋友都在不同場合提出了不少中肯的意見。在此，我想與大家分享一下近期朋友們與我們討論得比較多的幾個問題及其所帶來的思考，並進一步闡釋我們建議的初衷，希望拋磚引玉，集思廣益，以下所寫絕非香港交易所就這次諮詢的既定立場或結論。

一、 設立創新板可能是香港上市制度近 20 年來最大的改動，如此大動干戈到底為什麼？

答案其實很簡單。香港作為一個國際金融中心，必須要 Know Your Client，即「認識你的客戶」或知道誰是你的客戶。二十多年前，通過 H 股的引入，我們有了兩個客戶：中國的貨主（上市公司）與世界的錢主（投資者）；通過互聯互通，今後 20 年我們又要增加兩個新客戶：中國的錢主與世界的貨主。在世界主要金融市場中，真正有可能擁有四個客戶（中國貨和錢／世界錢和貨）的國際金融中心也只有香港。擁有四個客戶的香港就可以保證另一個 20 年的繁榮嗎？還不夠。

今天的世界（包括中國）都在向新經濟、新科技大步邁進。我們身處一個創新頻現、創業家精神爆發的大時代，幾乎每一天，都有新經濟公司湧現，在改變人類生活的同時，它們也創造了激動人心的投資機會。只有能夠聚集新經濟上市公司和懂得新經濟投資者的國際金融市場，才有可能在新時代佔據新的領導地位。但是，香港目前的市場制度還只能聚集傳統的貨主與傳統的錢主。新經濟和新科技對我們目前市場制度的適應性提出了直接的挑戰，但也為我們改善市場結構、提升國際競爭力提供了天賜良機。

如果把市場發展看成是一場 400 米的接力賽，香港市場由於存在發行人來源地過於集中、傳統行業佔比過高的結構性問題，可以說，我們在第一棒上並沒有領先，但幸運的是，還沒到不可逆轉的地步。只要奮起直追，跑好第二、第三棒，我們仍有希望贏在第四棒。不要忘了，香港是唯一真正有可能擁有中國貨和錢及世界錢和貨的國際金融中心。

二、 當年設立創業板（GEM）也是基於類似的初衷，但推出後並不成功，你憑什麼認為創新板就能成功呢？

創業板成功與否，仁者見仁、智者見智。但即便結論是不成功的又如何？難道說一次失敗，就永遠失敗？失敗是成功之母，我們不能因為過去的一次嘗試失敗就放棄所有的嘗試。如果嘗試的話，我們至少有成功的可能性。如果試都不試的話，我們永遠不可能成功。同時，這一次我們希望在推出創新板的同時，也能對創業板同步進行改良，有利於提升創業板的素質。

三、 香港沒有科技創新發展的歷史，沒有培育科技企業的強大生態系統，創新板能成功嗎？

如前所述，在發展新經濟領域，香港在接力賽的第一棒並沒有領先，但由於香港的獨特定位，我們絕對有跑贏的基礎條件。在港上市的企業中，已經有

具有全球領導地位的新經濟巨人，我們沒有理由悲觀，我們必須奮起直追。

也正因此，我們計劃在 2018 年推出一個全新的平台：香港交易所私募市場 HKEX Private Market。它將使用區塊鏈技術為早期創業公司及其投資者提供一個股票登記、轉讓和資訊披露的共用服務平台。它將是一個不受《證券及期貨條例》監管的場外市場，作為創業企業的孵化器，它將為這些企業及其投資者進入資本市場提供「學前培訓」。相信這個平台將有助於香港創造支持新經濟發展的生態環境。

四、 到底什麼公司算新經濟公司？你們怎麼判斷一個公司到底是不是新經濟公司？

定義新經濟公司的一個基本原則是看這家公司的主要發展驅動力是不是創新（包括技術創新與商業模式創新等），也就是看驅動公司發展的是傳統意義上的有形資本（資金、固定資產），還是新時代的無形資產（專利、技術、模式創新等），更通俗的說，看一個公司的發展主要是**靠錢（資本）**還是**靠人（創造力）**。新經濟公司可能來自生物技術、醫療保健技術、資訊技術服務、軟件、電子商務等新興輕資產行業，也可能源於傳統行業的商業模式的創新。而且，新經濟公司的定義可能會隨時間演變，因此，我們在建議文件中徵詢市場意見，希望能為新經濟公司作出原則性的定義。

我相信，根據大家討論之後確定的原則性定義，應該能夠對大部份的擬上市公司作出準確判斷，對於少數公司的判斷可能會存在分歧，需要進一步辨析。這就跟上市審批中的合適性（Suitability）審查一樣，大部份個案比較容易達成共識，但有小部份個案存在眾說紛紜的空間。與合適性審查不同的是，是否新經濟並不是某個公司能否上市的決定因素，只是決定這個公司是否應在創新板上市的因素。在香港現有的《上市規則》中，很多規則都是按原則性定義來判斷的。儘管原則性定義可能會帶來一些主觀判斷的空間，甚至存在誤判的可能性，但我們應該敢於嘗試、允許試錯，惟有如此，我們才能不斷學習，不

斷進步。不能因為擔心有犯錯的可能，就畏首畏尾、裹足不前。

五、 何不只接受已經在美國上市的不同投票權架構公司來香港做第二上市，這樣上市機制的改動要小得多，也容易得多？

在諮詢文件中，建議設立的創新主板會允許已經在美國上市的不同投票權架構公司來香港做第二上市。但是，我們沒有理由僅僅容許已在美國上市的不同投票權公司來港上市；這樣的安排既不合理，更不自信。事實上，過去多年的發展已經證明我們的市場機制是行之有效的，我們應該對香港自己的上市審批能力有自信，而不是依賴美國監管機構的審批作背書。無論是 H 股 20 年前來港，還是今天的互聯互通，我們香港都是第一個吃螃蟹的人，今天怎麼能説只有別人吃過的螃蟹我們才敢吃呢？

六、 為什麼不直接改革主板來容許採用不同投票權架構的公司來港上市？

直接改革主板絕對是可以考慮的方案之一，但我們為什麼會建議在主板之外另起爐灶設立一個創新板呢？眾所周知，主板是香港市場的主體板塊，《上市規則》和監管條例詳盡繁多，許多規則根據多年的實踐約定俗成，如果要對主板《上市規則》進行如此大的改革，勢必需要進行更加深度的市場諮詢和討論，也勢必引發更多市場爭議。

打個通俗的比喻，假設我們一個大家庭住在一個傳統的大房子裏，孩子們長大了，希望對我們的傳統廚房更新改造，加入新一代的智慧廚電。但要在現有的廚房中更新換代，必然會大費周章，水電氣一併改造工程浩大，會給住在房子裏的人帶來很多不便。所以我們建議了另外一個裝修方案，在主廚房邊上加建一個新廚房，這樣在建設過程中不僅不會影響家人使用主廚房做飯，而且，一旦新廚房建好後，兩個廚房同時投入使用，還可以解決全家眾口難調的

問題，在條件成熟、大家有共識時，隨時還可以考慮把兩個廚房之間的牆拆掉，打通兩個廚房，合二為一。

總之，我們提出設立創新板建議的初衷是為了在給市場引入活力的同時儘量減少對主板市場的影響。當然，我們對所有的改革方案都持開放態度，希望大家集思廣益，能夠討論出完善上市機制的最佳方案。

七、 既然創新初板是高風險板塊，為什麼還採用低門檻、輕審查上市條件呢？這不是違反常理嗎？創新初板將來會不會演變成一個充斥着「殼股」的市場呢？

這個問題聽起來很有道理。所有人都希望監管者讓好公司很容易就能上市，同時用高門檻和嚴格的審查把壞公司擋在門外，然後讓所有投資者都能參與投資好公司。大家必須認識到，這樣的理想既不現實，更不可持續。我們今天在創業板上經歷的不少挑戰均來自這樣不切實際的管理理念。公司上市時與其說有好壞之分，不如說有高風險和低風險之分。尤其是新興公司，它們中間只有少數可以成為「金鳳凰」成功，很多都會失敗，沒人能事先準確地看出哪一隻是「金鳳凰」。我們要想參與「金鳳凰」的成功，就得讓所有有潛力化身鳳凰的公司容易上市，也要有接受失敗的準備。

之所以建議創新初板採用相對寬鬆的上市條件及監管機制，是因為這些「準鳳凰」尚處於初創階段，它們沒有資源聘請昂貴的中介機構來通過嚴格的上市審批和合規要求。如果對它們設定十分嚴格的上市門檻，它們就上市無望。但這些公司的性質，決定了它們的股票投資風險較高，所以我們初步建議僅開放給專業投資者。

創新初板會變成一個「殼股」市場嗎？答案是不會。首先，正因為上市相對容易，上市地位本身就不會變成有價值的「殼」供買賣或操縱。同時，如果大家同意推出創新板，我們也應該考慮在後續的規則細化中設計更嚴格的持續合規責任（包括維持最低市值、最低成交量等等），容易進，也容易退，不達

標者將被強制退市，創新板初板絕不會淪為一些朋友所擔心的只進不出的「殭屍股」市場。

八、只允許專業投資者參與可能飛出「金鳳凰」的創新初板對於廣大散戶公平嗎？

回答這個問題並不容易，核心是如何平衡散戶參與投資的權利與其承受風險的能力，全球各大市場的監管選擇並不相同。在高度發達和成熟的美國市場，基本原則是只要披露充分，投資者（不區分機構與散戶）可以自由選擇，買者自負，監管者不為散戶設定更多的保護措施。而以散戶為主的內地市場則以保護「弱勢群體」的原則對上市公司上市設置了嚴格的門檻和審核標準。香港既是高度發達的國際市場，也有很強的中國特色，必須找到最佳的平衡。這次在創新初板的設計中也是如此。

我們要思考的問題是：我們的市場願意接受哪種程度的投資風險？縱觀創新板諮詢文件的建議，其實如同下圖中的這條光譜，提供了具有不同風險、上市條件和監管標準的多種板塊供大家討論：在這條「光譜」最左端的是香港交易所計劃明年初推出的私募市場，是一個不受《證券及期貨條例》監管的場外

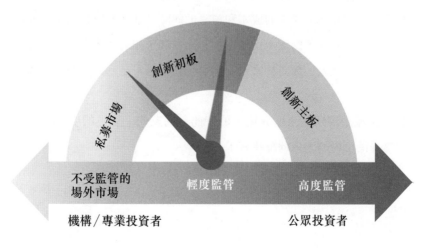

市場，風險最高；中間的是輕度監管的創新初板，最右端的是與主板一樣監管要求的創新主板，對應最高的上市門檻和最嚴格的監管要求，還有較低的投資風險。

大家可以看到，越往光譜的左邊走，融資門檻越低，投資風險越高。最左端的私募市場提供非證券監管下的股權私募功能，在這裏每 100 家公司裏可能只成就幾家偉大的公司，大部份的企業都有可能以失敗而告終。這樣的高風險肯定讓我們感到不安，因此，創新初板的門檻必須右移，但到底移到何處更合適，主要取決於大家有沒有信心讓散戶參與，如果創新初板不讓散戶參與，就可以儘量往左移，如果讓散戶參與，就必須儘量向右移。

光譜上的每一個區間，都有不同的利弊，諮詢文件中的建議只是一個討論的起點，具體如何選擇需要廣泛聽取市場的意見，方能找到最有利於香港市場發展、最能保持香港國際競爭力的方案。

以上只是最近一些朋友們常與我討論的問題，我相信大家可能還會有其他一些問題，囿於篇幅，無法一一盡述。

我們身處一個日新月異的時代，我們必須及時調整自己的定位和步伐，與時俱進。過去幾年，我們一直在思考：如何在堅守香港市場核心價值的前提下，不斷鞏固香港的國際競爭力？如何在不降低投資者保障的同時給大家提供更加豐富的投資機會？誠然，這些問題都沒有簡單的答案。任何改革都面臨成本和風險，我們在顧慮改革的成本和風險的同時，卻往往容易忽視不改革的成本與風險，因為不改革的成本與風險經常要在若干年之後才會顯現，付出代價者很可能是下一代人。面對挑戰，我們這一代人是不是應該更勇敢地擔當起我們的責任呢？

現在距離諮詢期結束還有兩週的時間，朋友們，香港的明天需要你的聲音，請為香港市場的未來踴躍建言！

2017 年 8 月 1 日

21 | 新經濟、新時代，
香港歡迎您！

　　今天，香港朝着上市機制改革邁出了重要的一步！經過幾個月的公開諮詢，香港市場已經就完善《上市規則》、擁抱新經濟達成了積極的共識，探明了市場改革的大方向，我們剛剛在網站上公佈了這次諮詢的結論。

　　我想對每一位參與了此次諮詢的投資者、發行人、中介機構和監管機構代表表示由衷的感謝，感謝你們真誠地發出了自己的聲音，感謝你們用集體智慧推動了香港市場的進步！我相信很多的朋友都會為香港點一個大大的讚——香港，我真的為你的開放和務實而感到驕傲與自豪！

　　這不是一次容易的改革。作為香港市場近二十多年來最重大的一次上市改革，它從一開始就面對前所未有的爭議，尤其是是否接納不同投票權架構、未盈利公司上市等議題，在全世界任何一個市場都是不容易抉擇的監管選擇題。四年前，我們只能以夢境為由嘗試拋磚引玉，希望啟發市場就這些關係香港公眾利益的議題開展討論，但根本不敢奢望我們的市場能夠在短期內達成共識、啟動改革。

　　幸運的是，在這四年中，新科技和新經濟已經成為驅動世界經濟發展的新浪潮，它們在深刻改變人類，特別是中國人的生活。新經濟在推動社會進步的同時，也創造了激動人心的投資機會，獲得了全球投資者的擁抱與追逐。幸運的是，香港是一個充滿內省和開拓精神的城市，雖然我們錯過了一兩個大的IPO，但是大家開始認真地思考香港應該如何與時俱進、如何鞏固自己獨特的國際金融中心的優勢了，大家開誠佈公的討論也越來越有建設性與可行性。幸運的是，我們的政府和監管機構是富有遠見和擔當的，不僅沒有因為困難而卻

步，而且一直積極領導與推動這場諮詢和改革。

正因為有這麼多人奉獻真知灼見，我們對市場未來發展的構想才能越來越清晰。正因為有這麼多香港人的認真和堅持，我們當年的夢想才能變成現實：在遵循程序公義、保護投資者利益的前提下，市場各方達成了共識，真的「做出了最適合香港、最有利於香港的決定，而不是最安全最容易的決定」！

根據諮詢中達成的共識，我們將在主板《上市規則》中新增兩個章節，列出有關尚未有收入的生物科技公司和採用「不同投票權架構」的新經濟公司來港上市的規則框架，我們也將修改第二上市的相關規則，方便更多已在主要國際市場上市的公司來港進行第二上市。總之，改革後的主板將能聚集更多類型的上市公司，尤其是高成長創新公司。與此同時，我們也提升了主板及創業板上市門檻，並對創業板重新定位，希望創業板能夠繼續吸引更多優質的中小型公司。

需要指出的是，這次的諮詢不是一場非黑即白的是非題，更不是一場必須分出誰勝誰負的選舉。諮詢只是用公開公平公正的方法讓市場各方各抒己見，讓監管機構聽到所有的聲音，讓我們通過集體的智慧求同存異。現在，市場已找到最大公約數，為市場改革邁出了重要的一步，當然，在最大公約數之下，市場各界對於不同問題的看法永遠不可能也無必要完全統一起來。這正是香港程式正義和法治精神的體現。

也許很多朋友會問：「不同投票權架構」真的是一種更好的公司治理架構嗎？吸引新經濟公司一定要靠接納「不同投票權架構」嗎？投資者能承受無收入公司的風險嗎？其實，我們無意改變任何投資者對於這類多元化公司的既定喜好，我們只是想把上市的大門再開得大一點，給投資者和市場的選擇再多一些，因為不想把非常有發展前景的新經濟公司關在門外。如果您對「不同投票權架構」或無收入公司始終心存疑慮，您絕對應該避開這類公司，修改後的《上市規則》也不會影響您的利益；但是，如果您非常看好一家公司的前景，並且不介意它的多元化結構，修改後的《上市規則》將為您提供更多投資機會。

不少朋友關心新規則中有關投資者保護的問題，我想大家需要分清楚兩個

問題：一是如何保護小股東的利益不受控股股東侵害；二是控股股東是如何取得控股地位的。我必須強調，在新的規則中，現有規則對於第一個問題所涉及的投資者保護措施絲毫沒有因為引入「不同投票權」架構而改變或者減少。《上市規則》中新增的章節只是改變了以前只允許通過資本投入來獲取控股地位的規定，現在允許新經濟公司通過合同（即公司章程）形式來獲取控股股東地位。為了應對未來有可能出現的新問題、新風險，新的章節還將對這一權利設定更具針對性的特殊投資者保護措施。

明年第一季度，我們計劃推進《上市規則》的細則諮詢，歡迎大家屆時繼續發表意見，與我們一起制訂出全面、完善的細則規定，實施最有利於香港市場發展、最能保持香港國際競爭力的方案，尤其是一定要設定好相應的投資者保護措施。我們的市場不僅要讓更多更新的發行人走進來，還要讓我們投資者的利益得到更好保護。

股票市場如同一個百貨商場，只有貨架上的貨（上市公司）更加豐富、更加符合顧客（投資者）的需求、更加安全，才能吸引四面八方的顧客，才能人氣旺、有活力。

我堅信，這次改革完成後，我們將可以同時為上市公司和投資者提供更加豐富的選擇，從而讓我們的市場變得更加多元化、更富有活力，讓香港這個國際金融中心更有全球競爭力。很多新經濟公司都對我們的上市機制改革表現出了濃厚的興趣，近期，我們已經收到了越來越多新經濟公司有關來港上市的查詢。

朋友們，主板《上市規則》中新增的兩個篇章和新改的第二上市規則將只是一個開始，我期待着和你們一起共同書寫香港市場發展最華麗的新篇章。

2017 年 12 月 15 日

22 ｜ 獨角獸與王老五

—— 在 2018 中國（深圳）IT 領袖峰會上的發言稿

導讀：以「王老五尋親」的比喻簡單介紹了美國、內地和香港三個市場在上市機制和監管上的差異。由於所處的發展階段和市場環境不同，三個市場的監管考量和上市機制註定不同，各有千秋。獨角獸和新經濟的到來，引發了香港和內地上市機制的改革。內地正在討論的 CDR 試點，可能會是一個重大的創新，但也會面臨着不容易的抉擇，到底是用獨角獸的回歸來引發現有發行邏輯的化學改變，還是沿用現有的發行邏輯來約束獨角獸，都將考驗監管者的智慧。香港與內地兩個市場一向優勢互補，合作大於競爭，大家共同服務於中國的經濟發展，兩地市場正在進行的上市改革可以創造共贏。

我想打個通俗的比喻，從美國、內地和香港三個市場維度來談談獨角獸上市這件事，咱們不妨把它叫做「王老五尋親記」，主要人物有新郎王老五、新娘、岳父和岳母。上市公司就是王老五，投資者好比是新娘，監管者是岳父，岳母就是政府、媒體、專家和眾多吃瓜羣眾。

在美國、內地和香港這三個不同的市場，「王老五尋親記」分別上演着不同的劇情：

美國的股票市場經歷了百多年的發展，已經比較成熟，投資者也比較成熟，以機構投資者為主，可以說美國的新娘（投資者）多是擁有高學歷、高收

入、高閱歷的「三高人士」，人家追求自由戀愛，不喜歡包辦婚姻，人家想投什麼就投什麼，投好了就賺錢，投不好也不怨天尤人。所以美國的岳父（監管者）比較輕鬆，不用替新娘操心挑選新郎，他對新郎只有一條要求：不能撒謊，如果你有不良愛好，你得說清楚，你抽不抽煙、喝不喝酒，都得說清楚，只要不騙人就可以和新娘自由戀愛。當然，美國的新娘可不是好欺負的，如果新郎欺負新娘的話，新娘請來的離婚律師（集體訴訟）很可能會讓他一無所有。

內地市場王老五人數眾多，公司上市融資的需求很大。而新娘以散戶為主，都非常單純、熱情，但是也特別容易輕信和情緒化。所以岳母（政府、媒體、專家）認為內地的新娘們天生是弱勢羣體，要求作為岳父的監管者特別心疼散戶，也特別有保護意識，不得不千方百計幫新娘們甄選德才兼備，還要做好防範，不讓新郎欺負新娘。因此，內地的岳父工作特別辛苦，特別操心。而新郎也必須排長隊慢慢接受審查，有些到了 30 歲還是王老五。

而香港算是處於中間地帶，既不像美國那麼完全婚戀自由，也不像內地有眾多審查。王老五和新娘的戀愛還是比較自由的，只要符合婚姻條列的要求（符合香港《上市規則》），岳父一般都不會干預誰娶誰、何時娶。香港這個地方的岳父（監管者）相對也比較輕鬆，但是也有不少活要幹，因為畢竟還是有散戶。

香港的岳母總的來説還是尊重市場，認同投資者應該盈虧自負，但也不時對岳父指指點點。

這三個市場本來各有各的邏輯，各有各的優勢，互不影響，王老五們根據自己的條件，可以選擇去三個不同的市場相親。

不過，獨角獸來了，寧靜被打破了。這些昔日的王老五都在國外找到了富有的女朋友（PE/VC），註冊在國外，用自己的聰明才智在中國這片沃土上取得了巨大的成功，成為獨角獸了。近幾年來，這些鑽石王老五紛紛開始操辦婚事、進入資本市場了，絕大多數特別優秀的鑽石王老五都到美國去了，這下香港坐不住了，內地也坐不住了，大家都卯足勁推進上市改革，想吸引這些獨角獸。這些獨角獸出走的原因主要有四個：

首先是結婚的年齡限制，也就是上市的市值要求。三個地方對市值的要求不同，美國基本不管，香港和內地都對上市設定了一些市值的門檻。

第二個學歷，也就是盈收要求。美國不要求學歷，只要你説清楚，中專畢業或小學畢業都可以到資本市場娶妻，但是不能撒謊，有人願意要你就要你。香港有學歷限制，香港的學歷限制必須是大學畢業（必須有營業收入）。內地要求新郎必須是研究生，大學學歷還不夠，還要讀研，讀研是什麼？就是要有利潤。

第三個是人品，也就是各種潛在風險。美國不關心人品，但是新郎不能掩

蓋，必須如實披露。香港還是要求沒有不良行為的記錄，最好不抽煙、不喝酒。內地則要求德才兼備，你必須要德才兼備，德育、智育、體育都要好，這麼多姑娘都等着，王老五得把水準提高了才能進來。誰來評判新郎的人品呢？當然是岳父（監管者）。

　　更核心的是第四個問題，婚前協議，也就是特殊投票權機制。咱們內地和香港都是講究夫妻平等、同股同權，但是現在很多新經濟公司為了保證創始人對公司不喪失話語權都引入了特殊投票權機制，也就是在結婚前，新郎和新娘簽一個婚前協議，規定各自的權利，很可能不是按照出資比例來約訂婚後的話語權的。美國是允許這種婚前協議的，但是香港和內地目前都不認可這種婚前協議。

　　這四個東西，在美國都可以，但在香港都不行，四五年前我們希望留住一個巨大的獨角獸，但經過了一番的艱苦努力還是沒有成功。所以香港正在進行上市機制改革，希望能把以上的問題解決了，簡單來說，我們把年齡限制（市值要求）做了調整，學歷要求只對生物科技類公司進行了調整（允許沒有收入的生物科技公司來上市），也接受了婚前協議（特殊投票權架構），對於品行的要求依然保留。同時，允許已在美國和英國上市的企業以 HDR 的形式來香港第二上市。

　　香港通過這次改革實屬不易，特別是在對於特殊投票權架構的安排上，儘管我們本意上希望對美國目前盛行的非常寬鬆的特殊投票權制度加上一定的限

制，但最後仍然基本採用了美國模式，以保證改革後的新政對於獨角獸有足夠的吸引力。

在我看來，內地推進上市改革也會遇到類似的挑戰，特別是面對眾口難調的岳母和預期很高的新娘，內地的岳父將極為辛苦，壓力也特別大。

最近內地正在研究通過 CDR 吸引獨角獸回國，這是一個重大的創新。如果推出 CDR 試點的話，就相當於是在內地婚姻法（《證券法》）沒有改之前推出了一個專門針對外籍華人在中國的婚姻暫行條例，允許他們進入中國資本市場了。

當然，因為婚姻法沒有改，所以推出 CDR 也必然面臨着一些重要抉擇。譬如，在引入特殊投票權時是選擇像香港一樣基本引入美國模式，還是引入必要的中國元素來限制獨角獸的某些投票權？如果限制了，獨角獸還會來嗎？如果不限制，岳母和新娘會有意見嗎？再譬如，怎麼監管獨角獸通過 CDR 上市後的再融資、並購，是用 A 股目前的邏輯來監管和約束還是接受這些公司主上市地的監管政策？還有怎麼選擇獨角獸，是繼續靠岳父審查遴選機制，還是依靠市場來選擇？諸如此類的等等問題都可能都是考驗監管者智慧的難題。

內地、香港和美國市場有其不同的邏輯。美國市場崇尚「自由戀愛」，對市值、盈收、投票權都沒有限制，最重視「不能說謊」。與香港一樣，內地市場現有的上市法規還無法相容這些獨角獸的特殊，而社會各界又對吸引新經濟有巨大預期，監管機構壓力山大。大規模引進對市場存量影響大，要全面改革內地上市制度和法規並非一朝一夕之事。不想過大影響存量市場，改革就只能從有限試點下手。

香港處於兩者之間，既足夠國際化，又了解中國的國情。在中國經濟發展的進程中，香港市場一直充當着連接中國內地和海外的轉換器作用，並且與內地資本市場優勢互補。雖然中國證監會和香港證監會是不同的監管機構，但關係親密，一國兩制的香港可以利用自身的優勢為內地的市場改革提供充足的時間和空間。我相信兩地市場正在進行的上市改革可以創造共贏，共同服務於中國的實體經濟。

2018 年 3 月 26 日

23 | 明日起航的香港上市新規，祝你一帆風順！

　　明天將是個值得紀念的大日子！從明天起，修訂後的主板《上市規則》將正式生效並開始接受新經濟公司的上市申請，香港資本市場將以更加開放的懷抱來迎接創新型公司上市。

　　這次改革是香港市場近 25 年來最重大的一次上市機制改革，也是最具爭議的一次改革。過去 5 年，我們已對這次改革進行了廣泛深入的討論與思考。從五年前做的關於投資者保護的夢（見本章第 12 節〈投資者保障雜談〉）到今天新上市制度的啟航，是一次難忘的旅程。在新規即將啟航的前夜，我想將這些集體思考做一個階段性的整理。

這不是一次容易的改革，香港交易所為什麼要推行這次改革？

　　香港是一個不大的城市經濟體，我們必須與時俱進、靈活應變，我們必須超越我們的經濟體量、從全球大局來考慮問題，才能保持國際競爭力。我們的目標是連接中國內地市場和國際市場，要實現這一遠大目標，我們就必須密切專注世界經濟和全球市場的發展趨勢，尤其是美國市場上正在發生的事情。

　　過去幾十年，香港從一個區域性的金融市場華麗轉身，發展成為舉世矚目的國際金融中心，這是了不起的成就。但是，我們在一個非常重要的方面已經落後，我們沒有以正確的姿勢擁抱新經濟，特別是在容許有特別投票權架構的公司、未有營業收入的公司上市，以及第二上市等重要領域還不夠靈活和開放。在這方面，紐約已經比我們和很多其他市場更有競爭優勢，這也是美國市

場能在統領科技革命方面脫穎而出的一大重要因素。

從國際競爭趨勢看,香港的國際競爭力正面臨巨大的挑戰,如果我們的上市機制仍然一成不變,我們就會落後於其他主要國際市場,落後於新經濟時代的發展。因此,無論這次改革有多難,我們都必須成功,而且需要沿着這個方向繼續堅持走下去。

在推動改革的過程中你們遇到的最大挑戰是什麼?

幾年前當我們剛剛嘗試啟動這場改革的時候,新經濟的發展趨勢還不像今天這樣明朗,是否應該容許「同股不同權」在香港引發了強烈爭議,不少反對的聲音認為引入不同投票權架構一定會削弱投資者保護,如何說服反對者、打消市場的疑慮,凝聚市場共識,成為我們必須面對的最大挑戰。

其實,引入不同投票權架構一定會犧牲投資者保護是一個誤解。事實上,這次的改革絲毫沒有削弱我們目前的上市制度為小股東提供的保護。目前上市制度為小股東提供的保護措施根本沒有任何改變,改變的是我們怎麼看待控股股東能夠獲得其控制地位的方式。

過去,我們只認可一個股東通過為公司貢獻的金融資本來獲得控股股東的地位。股東的控制權需要與其出資額相當。改革後的《上市規則》讓人力資本(如智慧財產權、新商業模式、創始人的願景等)也被承認和接受,成為獲得控制權的一種方式。換句話說,我們並沒有改變小股東如何受到保護。我們只是打開大門,允許控股股東通過不同方式來獲得控制權。

在無營收公司申請上市方面,為什麼只對生物科技行業情有獨鍾?

首先,這是生物科技公司的獨有特點決定的。一般來說,無營業收入的公司都處於高風險的早期發展階段,投資者很難判斷公司發展前景。而生物科技公司產品的研發、製造和上市過程都受到國家醫藥監管當局的嚴格監管,它們

新上市制度 2018 年 4 月底生效。

每一階段的發展都有清晰明確的監管標準和尺度，這一特點使得生物科技公司可以在資本市場上提供清晰具體的披露，供投資者來判斷投資風險。

此外，在中國，我們深深感受到生物科技行業大發展的春天已經到來，天時、地利、人和都具備了。為什麼這麼說呢？

天時：今天，我們有幸身處在一個科技大爆炸的年代，由於生命科學、人工智能、大數據等等科技的突破和相互賦能，生物科技將產生巨大的突破，有可能改變我們的生活和人類的命運。

地利：中國是全球人口最多的國家，目前已經是世界第二大經濟體，但是很快將面臨嚴重的人口老齡化的問題，隨着越來越多的中國人走向富裕，大家會越來越關注健康管理和生命的質量，都希望能夠活得長、活得好。生物科技在中國大有可為，這是和我們每個人都息息相關的一個重大產業。

人和：中國的監管機構已經認識到了人民的需求，銳意改革藥品審批機制，與國際監管機構合作互認，加快創新藥的研發審批程序。

剩下的就是資本支持了。作為中國的國際金融中心，香港應該在這一關係

人類命運的重大變革中擔當重任，發揮我們資本市場的優勢，為生物科技行業雪中送炭。

香港市場這次改革歷經五年才得以完成，而內地 CDR 試點在幾個月內即將出爐，這是否説明香港的制度已變得低效與僵硬？

我們大家都希望改革能高速高效地完成。但是，香港市場是法治社會，崇尚程序正義與規則穩定。為保證市場規則的嚴肅性、持久性與可預見性，任何對市場有重大影響的改革都必須進行廣泛深入的市場諮詢，讓市場所有利益羣體都有發聲的權利和機會。只有這樣，改革後的規則才能得到市場各界最廣泛的接受，才能經得起市場和歷史的檢驗。開放的香港資本市場高度國際化、市場化、專業化，所有參與者都可以隨時選擇不投資香港，我們規則的改變也因此必須高度尊重法治和程序正義。而規則一經改變，就必須保持相對的穩定與一致，不能朝令夕改。總而言之，五年時間可能是長了些，但我們積累的經驗與教訓也許可以使未來的改革更快、更有效率。我們始終相信，正確的改革經得起歷史的檢驗，這次穩健完成的改革更適合於香港。

上市改革後的香港市場與美國和內地相比有何異同呢？

改革後的香港市場已經像美國市場一樣，基本擁抱了新經濟公司，在不同投票權架構和收入門檻方面已經不再存在制度上的重大障礙和劣勢。與此同時，香港更接近內地市場，在文化、語言和交易習慣上「更接中國地氣」，而互聯互通有使在港上市的公司可以直接引入內地投資者。這種「兩全其美」的制度設計讓香港市場比起美國市場具有巨大的吸引力和優勢。

與內地市場相比，香港市場更加國際化、市場化，監管基本以披露為本，在上市審批、發行結構、價格與時間安排上完全按市場原則監管，更加靈活和自由。

簡而言之，香港處於美國與內地市場兩者之間，既足夠國際化，又瞭解中國的國情。上市改革後的香港，比美國更溫暖，更像家。比內地更開放、更市場化、更國際化，既有家的溫暖，又有看世界的自由。

A股市場正準備推出的CDR（存託憑證）試點，是針對香港上市改革推出的競爭之舉嗎？對香港市場有什麼影響？

內地推出CDR試點，主要是為了讓更多內地投資者分享新經濟帶來的投資機遇。這是勇敢的一步，也是內地資本市場發展到一定階段之後的必然之選。對於新經濟產業這麼重要的經濟引擎，內地市場進行發行體制改革來歡迎這些公司上市應該只是早晚的事情，我不認為他們是為了跟香港競爭而推這個試點，如果這麼想的話，是把內地監管機構的格局想得太小了。

具體來看，CDR目前已公佈的規則還不相容無收入公司，除非細則有變，CDR試點對香港推出的生物科技板塊沒有什麼影響。CDR推出的影響主要會反映在沒有盈利的或採用不同投票權架構的境外註冊新經濟公司上市方面。這裏包括兩類公司，一是已在國際市場上市的巨型中國民企；二是仍未上市的「獨角獸」。對於前者而言，一旦被選中參與CDR試點，它們就先行完成CDR上市。如果CDR不能與國際市場自由流通，這些企業未來也一定會考慮在香港上市，因為這樣可以部分解決內地CDR市場與國際市場的價格傳導。而對未上市的「外籍」中國獨角獸來說，無論是否能被選中CDR，都一定會首選新政下的香港市場做正股IPO。因為CDR是正股的預托證券，而正股正常上市，不僅有助於CDR價格發現，而且為發行人在現有境內外制度框架下提供更有效、更靈活、更豐富的融資管道。因此，「獨角獸」的最優方案是CDR與正股同時上市。

從宏觀發展角度看，CDR試點對香港的影響大致會有兩種可能。一種可能，CDR是個家數、規模和頻率相對有限的試點，那就只是一種象徵意義上的突破，暫不能承載大量新經濟公司的融資需求。如果是這樣，對香港市場的

實際影響暫時有限。

另一種可能是利用 CDR 的契機全面改革 A 股市場現有發行體制和監管邏輯，對新經濟全面開放內地資本市場，取消發審機制，允許公司根據市場情況隨時發行上市，大大簡化上市後的再融資、大股東減持及海內外並購等企業行動的監管。這種可能性到底有多大，現在還難以判斷。如果真是如此發展的話，改革後的香港上市新政在允許採用不同投票權架構方面就不一定有很大的相對優勢，但內地市場如此大規模的改革開放一旦啟動，將會對資本市場產生長期和深遠的影響，這樣的變化一定會為香港帶來巨大的外溢優勢，反而更加帶動香港市場的發展。

打個比方，在全球資本市場這個大班級裏有很多同學，內地市場和香港市場是兩位同桌的同學，考試科目是國際化、市場化的改革開放。如果內地市場改革開放緩慢，那麼它在考試中可能只能得到 55 分，同桌的香港可能有 60-70 分，我們算是有相對優勢。如果內地市場開始大變革了，那內地市場的體量、優勢和潛力就會得到充分釋放，它的成績很可能是 95 分。但這樣的大變革必然會同時為香港帶來巨大的外溢效應，香港也可以提升到 85 分，比起全班其他同學，雙方的絕對優勢都提高了，這對於兩者來說都是好事。一種是高水準上的相對劣勢，一種是低水準上的相對優勢，如果從中選擇的話，我當然更喜歡前者。因此，我認為香港只要充滿自信，與時俱進，就可以永遠處於不敗之地。

新的《上市規則》實行後，你有什麼顧慮嗎？

我的顧慮主要有兩方面：

一是如何界定可以採用不同投票權架構上市的新經濟公司，到底怎樣定義新經濟？這是一門藝術，而不是科學，對於絕大部分的上市申請，根據我們公佈的相關指引信就可以清楚地給出是或否的答案，但是對於少數的申請個案，恐怕存在仁者見仁、智者見智的空間。在這方面，我們要盡量找尋正確的答案，但也要給監管者一定的「容錯」空間，監管者會審慎檢視新規的運作情況

及個案、虛心聆聽市場回饋，致力與各界共同完善制度。

二是如何控制沒有營收的生物科技上市公司的投資風險。這類公司屬於高風險高收益的投資板塊，如果研發成功，公司的股價可以一飛沖天，如果失敗，公司的股價可以一文不值，因此只適合比較有經驗的成熟投資者，我們會給這類上市公司的股份名稱添加特別的標記「B」，向投資者提示風險。

我們在為此類公司設定上市門檻時特別諮詢了很多業內專家的意見，盡量在控制風險和豐富投資機遇之間尋找平衡點，設定一個比較合適的上市門檻，並根據行業特點為此類上市公司設定了特殊的資訊披露要求，讓發行人將所有潛在的投資風險披露給投資者，方便投資者能夠據此作出正確的投資決策。我們還特別成立了一個生物科技顧問委員會，邀請了一些生物科技投資界的專家為我們的上市科提供意見。

此外，我們也對生物科技企業的除牌程序制定了更嚴格的規定，以防止炒殼。

有些投資者也許會問，你們為什麼不仔細審核，只選擇那些將會成功的生物科技公司來上市呢？老實説，我們沒有這樣的 DNA，我們不認為我們比市場更聰明，不可能越俎代庖為投資者做出決策。我們只能着眼於制定更明確更清晰的披露標準，讓投資者能在充分知情的情況下做出自己的投資決定。因此，我想提醒投資者，在買賣股票之前一定要仔細研究上市公司的基本面，為自己的投資負責。

生物科技行業是一個新興的產業，對於生物科技上市公司的監管，我們還是一個新兵，我們針對生物科技上市公司的上市機制還有不斷優化和進步的空間。我們一定虛心傾聽市場和專家的意見，不斷地學習和改進，希望市場能夠給予我們一定的時間和試錯的空間，也希望大家繼續給我們多提寶貴意見，讓我們共同把香港打造成為生物科技公司的創新搖籃。

作為香港市場的營運者和監管者，我們既要滿足好上市公司（資金需求方）的融資需求，又要保護好投資者（資金供給方）的利益，我們永遠行走在學習平衡之術的路上：如何在提供投資機遇和控制投資風險之間保持平衡，如何在

上市機制的靈活性和投資者保護之間保持平衡，如何在堅持原則和保持競爭力之間保持平衡，如何在程序正義和積極作為之間保持平衡……這些是我們一直思索並嘗試不斷改進的平衡練習題。

幸運的是，我們在改革的路上一直都有你們的理解和支持。

朋友們，新的規則明天就要啟航了，讓我們共同祝福它能乘風破浪，為香港市場開啟更美好的明天！

2018 年 4 月 29 日

24 | 如何迎接生物科技新秀

香港上市新規已經生效幾週了，毫無疑問，香港資本市場將迎來生物科技發展的新時代。啟程之際，我們對未來既充滿了信心與期待，也不免對旅程中可能出現的風雨有一些忐忑和緊張。未來的旅程注定不會平坦，我們必須做好充分的思想準備。在此，我想與大家分享我對一些問題的思考。

修改後的香港上市新規第一次向連收入都沒有的生物科技公司敞開了大門，你們為什麼邁出了這麼大膽的一步，而且這個門為何只對生物科技公司開放？

一般來説，沒有收入的公司都處於高風險的早期發展階段，投資者很難有可靠的標尺來判斷公司發展前景和投資風險，所以我們以前的《上市規則》不接受沒有營業收入的公司來上市。但是，生物科技公司比較特別，因為它們的產品研發、製造和銷售過程都受到國家醫藥監管當局的嚴格監管，它們每一階段的發展都有清晰明確的監管標準和尺度，這一特點使得生物科技公司可以在資本市場上提供財務指標以外的清晰具體的披露，供投資者來判斷投資風險。

而且，一個服務實體經濟的資本市場，不能光想着怎麼在企業富貴時錦上添花，更應該考慮如何為推動社會進步的行業雪中送炭。生物科技行業的發展關係着人類的命運，作為中國的國際金融中心，香港應該在這一重大科技革命中擔當重任，發揮我們資本市場的優勢，為生物科技行業雪中送炭。

雪中送炭很高尚，但是風險也很大，因為生活在嚴冬中的人更容易凍死。與其他行業相比，生物科技行業有哪些特殊的投資風險？

首先，生物科技公司的產品研發週期長，一種新藥從研發到最後獲批在市場銷售，短則三、四年，長則達十年。

第二，由於生物科技公司的產品事關公眾健康，受到政府嚴格監管，在獲批生產前幾乎不可能有營業收入，因此它們的產品研發需要大量的資金投入。

第三，生物科技公司的產品研發失敗風險高，例如，一些研發中的創新藥物即使通過了一期和二期臨床測試，在三期臨床測試中失敗的也比比皆是。即使通過了所有臨床測試，監管當局也有可能因為其他因素不批准該產品上市。

除了上述上市公司本身的風險之外，投資生物科技板塊可能還面臨兩大市場風險：

一是股價波動風險。生物科技板塊不同於一般的行業，它們上市後的表現容易兩極分化，要麼大喜，要麼大悲，由於新產品的研發成敗與是否通過審批決定着生物科技公司的生死，任何有關產品研發與監管審批進程的信息都容易給股價帶來劇烈波動。

二是內幕交易的風險。由於高度的專業性和信息嚴重不對稱，生物科技板塊內幕交易的風險會明顯高於其他板塊。一方面，有關產品研發的信息對於股價的刺激明顯高於其他行業，內幕交易的潛在回報高；另一方面，生物科技行業的產品審批受到產品安全性和療效、社會需求、醫改政策、監管取向等多重因素的影響，並非所有信息都是透明的，由於公開信息比較有限，產品獲批的可預見性非常低，內幕交易的空間顯著高於其他行業。

由於生物科技行業的這些特徵，國際市場上涉足生物科技公司的投資者多是擁有較高分析能力和抗風險能力的機構投資者，散戶鮮有參與。

既然風險這麼高，你們為什麼不把上市門檻定高，把風險高的公司都擋在門外？

我們沒有按照這個邏輯來思考。最高的門檻就是把門關起來，或者根本不開門。既然已經決定打開門，就必須努力在發展和審慎之間尋求合理的平衡。門檻主要設定在最小市值和臨床測試進度方面：如果門檻設的太低，可能會把過高風險的公司放進來，不利於市場的長期健康發展；如果門檻設的太高，又無法滿足生物科技行業的合理融資需求，而且可能把很多急需資金的好公司擋在門外，讓投資者錯失良機。

我們在設定《上市規則》時特別諮詢了很多業內專家的意見，盡量在控制風險和豐富投融資機遇之間尋找合理平衡點。我們最終選擇設定了 15 億港元市值和已通過一期臨床測試、即將進入二期臨床測試這兩大主要上市門檻，並根據行業特點為此類上市公司設定了特殊的訊息披露要求，以幫助投資者判斷投資風險。

如果門檻不是很高的話，交易所是不是應該嚴格把關，只放好公司進來，把濫竽充數的公司擋在門外？

交易所的確會嚴格把關，但把關的焦點是在訊息披露上，而不是實質性前端審查。上市門檻是客觀標準，公司一旦達標，交易所就不能通過任意改變門檻或用其他主觀判斷來對個案公司進行取捨審查，只能嚴格要求公司詳盡準確披露所有投資者應該知道的信息。

必須指出的是，作為市場的營運者和監管者，我們永遠不可能比市場更聰明，無論監管者怎麼用心審核，都沒有辦法將所有的壞公司擋在市場之外，更不可能為投資者做出決策。

如果交易所沒有市場聰明，那交易所組織的生物科技諮詢小組是行業的專家，他們是不是有火眼金睛能夠幫投資者把關呢？

答案是否定的，生物科技諮詢小組不是法定機構，只是聯交所和香港證監會的顧問，完全不參與上市申請個案的審批。為保證公平公正和避免利益衝突，諮詢小組的運作會有嚴格的防火牆機制。小組的主要工作功能是幫助聯交所上市科和上市委員會了解生物科技中的專業知識和經驗，制定詳盡和專業的披露指引，並在實踐中幫助核查招股書中的專業披露內容。換句話說，生物科技諮詢小組成員是被動的顧問，不是主動的審查者，寄望生物科技諮詢小組把關審核不僅不現實，更是對其功能的嚴重誤解。

既然生物科技板塊風險這麼大，監管者又不能通過前端實質審查來確保質量，交易所是不是應該主動控制上市數量與節奏，避免過熱炒作？

的確，我們好不容易才凝聚市場共識修改《上市規則》，邁出了迎接生物

科技新秀的第一步。最希望看到的是市場穩步發展、細水長流，最不願意看到的就是公司一擁而上、紮堆上市，投資者不加思索爆炒這些生物科技新秀的股票，透支上市公司的盈利前景，最後一旦有什麼消息刺破股價泡沫，斷崖式的股價暴跌會令投資者損失慘重，更令市場對生物科技望而生畏，只餘一地雞毛。這樣的情景在美國、歐洲和台灣都曾經上演過，暴漲暴跌不僅背離了我們想讓投資者分享生物科技成長紅利的善良初衷，也會對香港來之不易的生物科技孵化環境帶來傷害。

正因為此，我非常理解大家為什麼希望交易所主動管理入市規模與節奏。但是，香港一向信奉法治精神、程序正義和市場至上的監管邏輯，我們不認為監管者應該替市場做主，人為地干預正常市場運作機制。我們的任務是確定清晰明瞭的規則，聚焦真實適時披露，事後嚴格執法懲治。我們不可能有科學合理的方法調整市場供需，即使我們能夠公平合理地管理公司入市節奏，也無法保證生物科技板塊一定成功或能夠幫助投資者規避風險。即使這樣的「流量管理」在短期內也許能夠人為地保持市場供需平衡，它一定會為我們市場的長期健康發展帶來其他不良後果。無論是上市公司還是投資者，在生物科技板塊的發展歷程中都必須經歷市場的磨練與洗禮，沒有捷徑可行。無論這些上市的生物科技新秀最終是鳳凰涅槃，還是一敗塗地，都是投資者必須面對和接受的現實與風險。如果想讓香港的生物科技板塊成長為參天大樹，就不能把它們放在呵護下的溫室裏。因此，儘管「流量管理」這個想法的出發點很好，但它不符合香港市場的基因和基本邏輯，我們也不會採用。

那投資者應該怎麼做呢？

投資者在買賣生物科技公司的股票之前一定要仔細研究上市公司的產品研發、臨床實驗結果和監管程序等細節，為自己的投資負責。而且，鑒於生物科技行業的專業性和複雜性，投資這一板塊需要了解足夠的專業知識和行業背景，這樣的投資機遇只屬於擁有豐富投資經驗和一定風險承受能力的成熟投資者（主

要為機構投資者），生物科技公司並不適合所有的投資者，尤其是中小散戶。

要防止這個板塊過早過熱引發崩盤，我想提醒投資者，尤其是對生物科技不熟悉的中小散戶投資者，一定要小心、小心，再小心！冷靜，冷靜，再冷靜！

如果你看不懂生物科技公司的招股說明書，請千萬不要投資；如果你對生物科技行業的政策與發展不甚了解，請千萬不要投資；如果你的心臟不能承受股價一天漲跌 20% 甚至更高的波動，請千萬不要投資。如果你不能接受投資的股票價值有可能歸零的風險，請千萬不要投資。

那非專業的散戶投資者可以怎樣參與生物科技板塊的投資呢？

從國際市場的經驗來看，香港生物科技板塊可能還需要幾年才能全面成熟。我們的市場才剛剛起步，從生物科技行業分析師等專業人才的儲備、專業投資機構的積聚、再到整個生態環境的成熟，都需要時間沉澱。萬丈高樓平地起，基礎一定要打好，最初上市的一批生物科技公司的市場表現將對市場長遠發展產生重要影響。

我建議對於生物科技了解有限的散戶投資者應該先觀察再試水，在機構投資者與上市公司充分互動和博弈、形成穩定的價格趨勢之後再逐步嘗試。此外，鑒於生物科技板塊的特殊性，個股投資風險很高，如果能用一籃子個股組合的方式來投資可以分散風險，東方不亮西方亮，投資者可以在數十間生物科技公司上市後，通過投資相關生物科技指數基金來分享這一板塊的整體機遇。

總之，生物科技板塊的啟航來之不易，需要我們共同珍惜它、守護它。因此，接下來我們希望能夠與業界和媒體攜手，多做一些生物科技的投資者教育工作，幫助投資者走近生物科技行業。同時，我們也歡迎市場各界多提寶貴建議，幫助我們在工作中不斷完善對生物科技類上市公司的披露要求和監管。我相信，在我們的共同努力下，假以時日，香港一定可以發展成為一個孕育生物科技創新的搖籃。

讓科技改變生活，讓科技改善健康，讓科技延續生命。有了生物科技板塊，我敢大膽夢想，讓我們一起活到 120 歲！

2018 年 5 月 23 日

第三章

互聯互通與共同市場

2014 年 11 月 17 日，滬港通開通。

25 ｜ 中國資本市場雙向開放的首班車 ──「滬港通」概覽

　　自 4 月 10 日李克強總理在博鰲宣佈「滬港通」項目後，我們已正式全面啟動了「滬港通」項目的市場溝通與前期準備工作，包括在網站上刊登了全面的實施綱要與各類指引、資料手冊。我相信大部份對「滬港通」感興趣的朋友已經通過不同途徑對「滬港通」有了一個基本的了解。所以，我寫這篇日誌不是為了重複這些信息，而是希望借助這篇日誌讓未來有可能關注「滬港通」的新朋友更簡便地了解「滬港通」的基本結構以及它設計背後的理念與意義。

　　我想在下文分三個方面談談「滬港通」：

- 什麼是「滬港通」？
- 它的設計有哪些特點？
- 「滬港通」意味着什麼？

一、什麼是「滬港通」？

　　「滬港通」是在內地與香港兩地證券市場之間建立一個交易與結算的互聯互通機制。在這個機制安排下，兩地的投資者可以通過委托本地證券商，經本地交易所與結算所買賣、交收對方市場上市的股票。

　　在「滬港通」機制下，香港投資者通過香港的證券商直接買賣特定範圍的 A 股（即上證 180、上證 380 的成分股、以及同時於聯交所及上交所上市的 A+H 股），其訂單會依次經過香港的證券商、香港交易所、香港交易所在內地設立的子公司，最後到達上交所完成訂單撮合。

　　在「滬港通」機制下，香港投資者與香港的證券商進行結算、交收。而香港的證券商則與香港結算公司進行結算交收。最後，香港結算代表整體香港與國際投資者統一與中登公司完成結算交收。也就是說，對於香港投資者而言，買賣 A 股的過程與買賣港股無異，並不需要在內地經過任何額外的手續。

　　對於內地投資者也是同樣的道理，他們可以通過內地券商經上交所直接在內地購買試點範圍內的港股（即恒生綜合大型股指數、恒生綜合中型股指數的成分股，以及同時於聯交所及上交所上市的 A+H 股），並通過中登公司在內地完成結算交收。

　　下圖是「滬港通」機制的一個概覽：

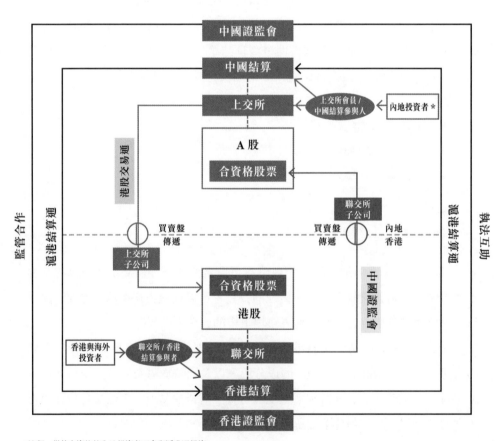

*註釋：僅符合資格的內地投資者可參與滬港通投資。

二、「滬港通」的設計有哪些特點？

「滬港通」是內地循序漸進開放資本市場的一大創新，旨在以最小的制度成本，換取最大的市場成效。即兩地市場盡可能地保留沿用自身市場的法律、法規、交易習慣，同時盡可能地降低對投資者的限制因素，讓市場力量發揮決定性作用，實現最大幅度的中國資本市場的雙向開放。「滬港通」的設計遵循謹慎的原則，充分考慮各方因素和風險。「滬港通」的特點可以總結為以下幾點：

(1) 交易總量過境，實現最大價格發現

投資者買賣對方市場股票的本地訂單將會被直接傳遞到對方交易所平台，與當地買賣訂單一併進行總撮合。「滬港通」股票的交易價是在結合兩地買賣盤總量後，總量對盤形成的價格，不會分割流動性，以保證達致最佳的價格發現。

(2) 結算淨額過境，實現最小跨境流動

股票結算與資金交收將通過本地結算所先在本地進行對減，然後再以淨額方式與對方結算所作最終結算。兩地結算所互為結算參與者，分別代表兩地參與者實行淨結算。淨額交收可以避免資金大進大出，控制資金跨境流動風險，減低對銀行資金池造成的波動。

(3) 人民幣境外換匯，實現全程回流

香港投資者買賣 A 股，是用離岸人民幣支付，再由香港結算統一向中國結算支付人民幣。

反之，內地投資者買入港幣股，將支付人民幣，再由中國結算把人民幣帶到香港、在香港兌換成港幣、最後交付給香港結算進行結算；賣出港股時，資金也將最終以人民幣「原路返回」。

所以，所有的跨境資金流動都是以人民幣進行。這樣做可以最大程度降低對在岸人民幣市場及在岸外匯儲備造成匯率影響。借助香港這個離岸人民幣中心，完成雙向人民幣換匯，實現投資人民幣的完整環流，大大加快人民幣國際

化進程的步伐。

（4）結算交收全程封閉，實現風險全面監控

流入「滬港通」系統內的資金只能用來買賣規定範圍內的 A 股或港股，一經賣出，套現資金只能在本地結算系統沿原路返回，而不會以其他資產形式留存在對方市場。這樣做能夠有效防範洗錢活動，控制熱錢無序流動風險。

而且，在「滬港通」下，所有的交易行為都在交易所、結算公司的系統內進行，誰在買賣、買賣什麼、以什麼價位建倉、出倉等都是一目了然，均有清晰記錄，能夠有效監控市場。

（5）本地原則為本，主場規則優先

正如同在不同地區駕車都要遵守當地交通規則一樣，「滬港通」交易須分別遵守兩地市場適用規則。這個原則體現在上市公司、證券商、交易與結算三個層面。

第一，雙方投資者投資對方市場標的股票時，受對方上市公司證券監管者保護，標的公司的披露與投資者保護責任受標的公司上市地監管者約束。

第二，證券商仍然遵循發牌地的有關規則、法律。

第三，交易與結算以標的公司所在市場的交易、結算規則與習慣不變為本。與此同時，出發地投資者還須遵循出發地市場的某些特殊的交易與結算規則。

整體而言，兩地監管機構基本沿用各自現有法律與監管規則，以最低制度成本實現最大市場化運作。

（6）結構高度對稱，利益高度一致

互聯互通的結構設計是高度對稱的——兩地交易所設立的子公司互為對方交易所參與者，結算所也互為對方的結算參與者。而且從監管層面上看，兩地監管者的擔憂與關注也是相互的、對稱的，所以兩地市場的權利與義務對等。正因為雙方的利益、訴求高度一致，所以在監管互助與投資者保護方面雙方更容易形成共識，有助於打破「各家自掃門前雪」的國際監管困局。

（7）收入平均分享，實現互利共贏

兩地交易所與結算所互相平分跨境訂單為其帶來的收入，實現互助互利、共創雙贏。「滬港通」不是零和遊戲、不存在此消彼長；不存在誰多誰少，它是兩地證券市場坦誠合作的最佳模式。

（8）初期實行額度管控，確保平穩運行

「滬港通」是一個長期的、結構性的計劃，試點初期設置額度管控，是為了平穩有序、風險可控地推出「滬港通」。在額度控制上，實行「實時監控」和「先來先用」原則，交易所不負責分配額度，完全按市場化原則運行，不存在尋租空間。

在「滬港通」推出初期，分別設每日額度和總額度：南向港股通的每日額度為 105 億人民幣、總額度為 2,500 億人民幣；北向滬股通每日額度為 130 億人民幣、總額度為 3,000 億人民幣。需指出的是，所有額度均是買賣對減後的淨額度。也就是説，「滬港通」一旦進入買賣常態後所能支持的交易量實際上會遠遠大於額度數量。

在運行順利後，我相信兩地監管機構、交易所將參考額度使用程度、運行情況、以及市場影響等因素綜合考慮後作適當調整。

（9）試點模塊化結構，未來可靈活擴容、擴量、擴市

「滬港通」的設計已預建可延伸性。一旦成功運行一段時間後，在監管當局批准下，我相信它的模式可以靈活地拓展至更多的投資標的、更高的投資規模，也可以滿足其他地域、市場和資產類別對跨市場、跨監管體系互聯互通的要求。

三、「滬港通」意味着什麼

「滬港通」是中國內地深化經濟體制改革、穩步推進資本市場雙向開放的重要嘗試。經過三十多年改革開放的探索，內地從開放貿易、開放企業融資和併購逐步走到了今天的資本市場雙向開放，可謂水到渠成。但是，這種制度改

革的歷史性、複雜性、不可逆性也使得決策層在進行制度設計的時候特別注重風險控制。

「滬港通」正是在這樣一個歷史背景下誕生的開創性嘗試。它綜合了目前所有開放措施的主要功能，首次在不改變本地制度規則與市場交易習慣的原則下，建立了一個雙向的、全方位的、封閉運行的、可擴容的、風險可控的市場開放結構，為制度與規則的逐步改革贏得寶貴的時間與空間。同時，由於跨境資金流動全部是人民幣，所以「滬港通」可以成為人民幣國際化新的加速器，可謂一舉多得。

對於香港而言，「滬港通」的意義同樣重大。香港過去 30 年的繁榮發展與內地開放息息相關，這一次也不例外。當內地試點資本市場開放之時，作為重要國際金融中心與世界最成熟的金融市場之一，香港具有獨特的優勢成為內地資本市場雙向開放的橋頭堡。同時，「滬港通」完全切合香港未來發展路徑的設想：鞏固當前香港股票市場的優勢和提升流動性、建設人民幣離岸中心、建設其他資產類別的交易中心和互聯互通等……可以預見，「滬港通」將成為香港資本市場發展新的里程碑和轉折點。

至於我們應如何看待「滬港通」，我的看法是：

- 「滬港通」不是一個簡單的新政策、新舉措或新產品：「滬港通」代表了中國資本市場雙向開放大格局中的一個新思路、新方法、新結構，具有很大的想像空間。

- 「滬港通」不是一個 A 股短期救市方案，「滬港通」宣佈後對 A 股與港股都產生了短期的和有限的利好，如果這樣的利好能夠持續，大家自然會皆大歡喜。但「滬港通」推出更主要的考量則是改變市場機制，引入新的市場活力與理念，形成一個長期的制度安排，逐步實現中國資本市場的雙向開放。「滬港通」的戰略意義體現在制度的穩步持續與結構上的互聯互通。開始的靜悄悄沒關係，長久穩定的市場活力才算最終勝利。

- 「滬港通」也不是一個中央送給香港的大禮包：我個人非常不喜歡將內

地與香港之間的關係描述成支持、送禮、贈予、扶持這一類非平等、非互利的關係。香港今天的繁榮得益於中國經濟的蓬勃發展，也歸功於香港人的勤勞與智慧，歸功於「一國兩制」。一國使內地對香港更有信任與信心，願意以香港為試點開始；兩制又使香港可以為國家發展提供一條至關重要的、新的、不可替代的發展路徑。「滬港通」的推出是「一國兩制」成功實踐的眾多代表作之一。

- 「滬港通」表面上看是兩地交易所之間股票互聯互通的商業性安排。但在本質上，它卻是中國資本市場雙向開放這一盤大棋中的一步好棋。在這個棋局中，交易所只是計劃的執行者，是一個兵、一個卒，具體實施則由兩地金融監管當局共同策劃、設計與協調。而支持這一棋局的更大的歷史背景則是中國新一代領導人開啟的讓市場發揮決定性作用的新一輪改革開放。

總而言之，「滬港通」是中國資本市場雙向開放列車的首發車，而香港就是其第一站。對於這趟列車來說，當前最重要的是平穩開出，安全抵達。就如當年的 H 股上市，經歷 10 年的時間涓涓細流方才匯成大海，徹底改變了香港資本市場的面貌，我相信「滬港通」對香港資本市場的影響也會一樣地深遠，而且它同樣需要時間來發展和完善。我相信，不用十年，甚至不用五年，這趟列車將會走得更快、行得更遠。

2014 年 5 月 8 日

26 | 滬港通答疑之一

　　近幾個月來，大家都在為滬港通計劃的準備工作爭分奪秒。令人欣慰的是，滬港通各項準備工作進展順利，業界各方反應也十分積極。我們計劃在 8 月和 9 月聯合上海交易所舉行數次滬港通市場演習，稍後我們將在香港交易所網站公佈獲准提供滬港通服務的券商名單。

　　在我們與國內外投資者的廣泛交流中，有投資者曾就滬港通的投資者保障及一些結構性的制約提出疑問，這些制約主要包括額度設置、標的股票限制、假期安排、交割前端監控以及監管責任分配等等。我希望在未來一兩個月就這些問題通過網誌形式與大家交流。

　　回答這些問題之前，我們必須從滬港通的歷史使命說起。我們都知道，中國資本市場的雙向開放是歷史發展之必然，但鑒於兩地市場結構現存的巨大差異，其實現途徑卻一定充滿艱辛與挑戰，不可能一蹴而就。如果我們坐等兩地市場最終逐步自然融合接軌，我們可能須等待五到十年。有沒有辦法在不根本改變兩地市場現行結構與規則的前提下提前實現高度市場化的互聯互通，成為滬港通設計理念的原始動力。

　　如果用水管來比喻的話，那麼香港代表的國際市場的管子可能是方形的，而內地市場的管道則是圓形的，兩邊閥門結構及水壓也是不同的。滬港通恰如一個連接轉換器，它的挑戰就是要在不改變兩邊「水管」形狀的前提下把兩邊的市場有效有機地連接起來，既要保障水流暢順，還要保證系統能夠充分承壓，做到水不淹、水不漏，閥門開關靈活自如。為了完成這一使命，滬港通在設計上既要高度市場化，也必須遵循謹慎的原則，充分考慮各方因素和風險，

因此，在試點初期不得不在一些制度安排上設定一些限制，比如額度、投資者門檻、有限投資標的等等。也就是説，儘管滬港通比既有其他開放方式具有前所未有的市場方便與自由度，但在推出初期，滬港通仍不能完全讓投資者像投資本地市場一樣靈活與熟悉。

我相信，在滬港通機制順利運行一段時間後，大家對於風險的認識會更加清晰，上述的限制肯定會越來越少，我們會與上海交易所及兩地監管機構一起，傾聽業界及投資者的聲音，不斷探索完善滬港通的制度安排。希望大家能夠多些耐心，能以鼓勵創新的心態來看待它，給我們不斷改善的空間和動力。

接下來，我想回答一些有關滬港通的問題。近幾個月來，我們收到了很多投資者朋友們發來的問題，相信未來還會收到一些新的問題，由於時間關係，無法馬上一一作答，頗感抱歉。我將在最近幾篇網誌中挑選一些最具代表性的問題（想必也是大家都很關心的），一併回答。

問：投資者進行跨市場投資，是否會受到他／她原來所在市場的監管保護？

答：滬港通的一大特點是「本地原則為本，主場規則優先」。所以，當投資者到了對方的市場，就應遵循對方市場的規章制度與交易習慣，同時也享受對方市場監管機構的投資者保護。例如，原則上説，中國證監會負責監管 A 股上市公司，但不能替內地投資者來監管港股上市公司；內地投資者投資港股時，其境外投資不再受中國證監會法例的直接保護，而要遵循香港市場的法規要求，其投資者權益也須通過香港證監會來主張；反之，香港投資者投資滬股通也是一樣。

但是，與投資者自己到對方市場開戶口投資不同，滬港通帶給了中港兩地投資者在本地市場、投資對方市場的便利。所以，兩地監管機構為了滬港通的順利開通，正在開展大量監管合作、執法互助的工作。這就好比一個人到 A 餐館吃飯，A 餐館就要對自己賣出的食物向其顧客全盤負責，但無須對隔壁 B 餐館的飯菜負責。但今天開通滬股通，就好比把 B 餐館的菜也一併擺在 A 餐館桌上，方便客人在這家餐館吃到隔壁餐館的菜。這時候，A 餐館對 B 餐館提供的這盆菜的責任就不能是零了，但也不可能是百分之一百。這就需要兩個

餐館協商責任分工、互相配合。因此，這次滬港通之所以能夠推出，中國證監會和香港證監會功不可沒，為了向兩地廣大投資者提供前所未有的投資便利，他們承擔了大量額外的監管責任，他們是真正有擔當、有遠見的監管者，值得我們尊重與敬佩。

需要強調的是，在跨境投資者保護問題上，最重要的其實是培養「買者自負」的態度和能力。監管者可以打擊違法行為、讓廣大投資者免受非法之徒帶來的損失，但卻不能讓投資者避免由於沒做好「功課」而造成的後果。而跨境監管的難度與挑戰更應促使參與滬港通的投資者主動學習，充分認識跨境投資的風險。所以說，在滬港通的監管保護問題上，我們要做到「買者責任自負、主場監管盡責、客場監管配合」，三者之間充分互動與平衡。

問：滬港通計劃為何要設定額度？未來是否會取消額度？

答：滬港通是在中國內地資本項目尚未完全開放的情況下推出的一項創新計劃，在試點初期設置額度管控（具體額度計算方法參見滬港通投資者資料手冊），是為了平穩有序、風險可控地推出這一計劃。在運行順利後，我相信兩地監管機構將在綜合考慮額度使用程度、運行情況、以及市場需求等因素後適當調整額度管理機制，甚至可能逐步取消額度管理。所以大家不必太過擔心額度夠不夠的問題。

問：滬港通當前的額度分配是否公平？

答：與 QFII 額度制度不同，滬港通計劃的額度不是按申請行政分配給個別的券商，而是按照「先到先得」的原則分配，確保公平與高效。額度的分配完全依照買單的時間先後排序，與買賣的訂單大小無關。為防止出現不公平「佔位」的情況，我們將嚴格監控，儘量縮小買盤與最近成交盤之間的價差，從而防止不成交的買盤擠佔額度；而且不允許修改訂單 —— 如果投資者需要修改已經發出的訂單，必須先取消該訂單，然後根據屆時額度餘額情況發出新訂單、重新排隊。

問：滬港通為何只能在兩地市場均交易的日子才開放，日後會否調整？

答：如此安排主要是因為兩地市場的交易日並不一致，考慮到交易結算所涉及的本地及跨境付款事宜以及券商和銀行界的營運安排，尤其是兩地券商一般都不會在假日期間營業，滬港通只在上交所及聯交所均為交易日、且結算安排均可有序運作的日子才開通。

有些投資者擔憂，在本地市場長假停市期間萬一遇到市場急劇動盪，投資者將無法入市拋售持有的對方市場股票，儘管這是非常合理的擔憂，但解決這一小概率事件的唯一方法就是讓業界完全改變本地市場假期安排，人為地去適應對方市場的開市時間。如果在一邊市場假期期間開放滬港通，業界的營運成本和負擔將會大幅增加，投資者及從業人員也需要改變原有的休假安排及風俗習慣，例如每年香港的聖誕假期和內地的春節和國慶長假。目前的安排是兩害相權取其輕的選擇。在滬港通開通後，兩地交易所會繼續與券商及銀行業界探討在假期提供跨境交易服務的可行性及市場需求。

問：投資者該如何管理假期風險？

答：由於上述交易日期安排，必然出現一邊市場休市而另外一邊市場仍在交易的情況。投資者需要提前了解滬港通的假期安排並根據自己的需要採取相應的風險管理措施。比如，聖誕、春節以及國慶假期間滬港通將關閉，但香港或上海有一邊市場則會正常交易，通過滬港通計劃持有對方市場股票的投資者如果擔心對方股市在此期間出現不可預料的波動，可能需要提前減倉規避風險，安心過節。

問：所謂「前端監控」是怎麼一回事？

答：「本地原則為本、主場規則優先」是滬港通的主要原則之一。根據內地法規，目前內地投資者只能出售其中國結算戶口內前一交易日收市時已經持有的滬股，換言之，上交所和券商的系統會對投資者的戶口進行交易前預先檢查，內地投資者不能對自己戶口內尚未到帳的股票發出賣盤訂單。而國際投資者的

通用做法是 T+2，即在交易日通過券商下單，成交後兩天結算時才將股票從託管銀行交付券商。大型基金一般不會將證券存托於券商。由於香港與內地市場的證券存託管系統不同，香港交易所無法對投資者的戶口進行持股量檢查，因此，香港交易所將對券商指定的中央結算系統股份戶口的持股紀錄進行類似檢查，以確保券商發出的賣盤訂單並沒有超過其持股量。

鑒於兩地市場的這一重要差別，國際投資者在賣出滬股時也必須改變其在國際市場上「先賣再結算」的交易習慣，遵循內地 T+0 的結算規則，在交易日早晨 7:30 前將欲售股票轉至券商端才能在本日賣出，從而確保香港結算與中國結算每日結算時不會出現無法交付的情況。

我們理解國際投資者會對這一交易模式感到不習慣，這也可能會對其初期參與交易有一定影響，但暫時也在所難免。在滬港通順利推出後，我們會繼續與市場保持密切溝通，探索能否在股票借貸規則或 IT 系統方面找到可行的改進方法。

最後，我想強調的是，滬港通是一項長期制度安排。它將對整個市場所有參與者開放，市場參與者可以根據自身情況，審慎選擇參與的時間與程度，不必急於一時。在滬港通推出之後，我們也會根據市場需求，不斷探索滬港通改進、升級和擴容的最優方案。滬港通可謂中國資本市場雙向開放的首班車，列車發出首日的上座率是否火爆，其實並非最重要的考量因素，最重要的是要確保列車平穩安全地行駛。

在未來一兩個月，我還會繼續通過網誌回答投資者朋友們有關滬港通的進一步提問。歡迎廣大投資者朋友們繼續關注滬港通的動態，讓我們一起迎接滬港通的到來。

2014 年 8 月 10 日

27 | 寫在 1117 號列車啟程前的心裏話

隨着稅收政策的明朗，滬港通的一切準備工作已經就緒，還有一天，中國資本市場雙向開放的首發車——1117 號列車就要從滬港同時啟程了。

作為這趟列車的築路者之一，我相信我和不少香港及內地的工作夥伴心中此刻百感交集，有激動，有期盼，也有緊張，有志忑……在我個人而言，當然，最多的還是感激，感激香港交易所上下同仁的辛勤努力，感激兩地監管機構的勇氣和擔當，感激合作夥伴——上交所和中國結算——與我們的緊密協作，也感激業界同仁的鼎力支持！

同時，我們也要感謝媒體朋友們一直以來對於滬港通的關注。最近這幾天，你們常常追問我，「你對滬港通開通首日的交易有什麼預期？」「到底是流入滬市的資金多還是流入港股的資金多？」

老實說，我對於滬港通首日的交易量沒有任何預期。滬港通是一項全新的嘗試，如同一條剛剛開通的高鐵路線，開通首日的上座率是高是低皆有可能。無論當天的交易是火爆還是冷清，我們都當以平常心來看待。反正路已經開通了，車天天都有的，您大可以根據自己的時間安排行程，從從容容地買票上車。

至於開通首日到底是滬股通交易量大還是港股通的交易量更大，也不是我所關注的。也許有人會擔心，北向的交易多了會不會帶走香港的錢，或者南向的交易多了會不會帶走上海的錢？千萬不要忘了，在經濟日益全球化的今天，任何市場的資金都是不停流動着的，而且這種流動一直處於加速中。經濟基本面決定了資金流動的方向，其他因素只能影響流動的速度。

近日來，我們也經常聽到朋友議論滬港通到底對香港較好一些，還是對內

地更好一些。我認為這都問錯了問題。正確的問題應該是,「有了滬港通是否比沒有滬港通更好?」無論對香港還是上海,答案都是肯定的。所以,與其糾結於對誰更好,我們更應該着眼於是否能夠雙贏。更何況,滬港通的真正意義並不在於滬港兩地存量上的重新分配,而是在於通過市場的融合與碰撞、把目前還未進入資本市場的巨額增量引入到兩地資本市場。

對於我們這樣的「修路者」而言,滬港通首日首週甚至首月的上座率完全不是我們的關注重點。我們最關注的指標是滬港通列車的安全指數和舒適指數。在設計和準備滬港通計劃的過程中,我們一直遵循謹慎的原則,盡量充分考慮各方因素和風險,盡可能地降低對投資者的限制因素,最終就是想讓大家的投資之旅可以安心、順暢。

不過,滬港通畢竟是一條全新的投資之路,兩地的法律、法規、市場結構、交易規則均存在明顯差異,縱然我們演習一千遍,開通的初期很難保證不出現這些或那些的插曲。萬一出現這樣的不足之處,敬請您體諒,我們一定會竭盡全力尋求最佳解決方案,盡快改善您的旅途體驗。

現在,一個全新的旅程即將開始,我衷心祝願搭乘滬港通列車的每一位乘客旅途愉快、心想事成!

2014 年 11 月 15 日

28 ｜ 滬港通答疑之二

經過大家的共同努力，滬港通已經順利運營一週了。對於像滬港通這樣的大型市場基礎設施項目，原本不應在短短一週內就急於進行評估，但近幾天我們的確收到了業界朋友們發來的許多問題，所以我決定還是在此做一個階段性的小結，回應市場的關注。

你對滬港通這一週的成交滿意嗎？

如果從運行的安全性和穩定性來打分的話，我覺得滬港通至少可以得個「良」，因為開通這一週整體運行安全平穩，沒有出現什麼問題。如果從交易量來看的話，沒有出現火爆的局面，可能低於不少市場人士之前的預期，但我個人並不擔心。因為滬港通是一座天天開放的大橋，而不是一場音樂會，無法用一週或者一個月的上座率來衡量它的成敗。相反，它的價值可能需要兩三年或者更長的時間才能得到驗證。老實說，我對於滬港通的短期成交量未敢有任何預期，因為在偉大的市場力量面前，我們都是謙卑而渺小的，我不知道橋兩邊的人到底會選擇什麼時候過橋、會往返多少次；但長期來看，我對這座大橋的價值是充滿信心的，因為我知道，有了橋，河兩岸的人就不會再因為人為原因分隔兩岸、不相來往。

你覺得北向滬股通為什麼會出現交易先熱後冷的情況？

據我所知，很多海外機構都對投資 A 股充滿興趣，但是由於滬港通的稅收政策公佈不久，很多大型投資機構還沒有來得及做好準備，也有一些長線基金因為不習慣現在的「前端監控」機制，希望等我們明年搭建好股票追蹤系統後再參與滬股通。在滬港通開通第一天，之所以出現額度迅速用盡的情況，可能是因為一些 QFII 機構在換倉，他們在內地賣出了原先持有的一些上海股票，又通過滬股通在香港把這些股票買了回來，希望把手中的 QFII 額度騰出來投資其他的內地資產。

總的來看，國際投資者會在未來一段時間內做好一系列準備，分期分批逐步參加滬股通，他們投資的主要動因是價值判斷與指數跟蹤。由於滬港通消息公佈已經七個月了，各種短期套利機會逐步消失，加上年底將至，滬股通短期內應該不會看到大量短炒資金流入，這並不是一件壞事。

港股通的成交為何如此清淡？

港股通的冷清開場的確讓包括我在內的很多市場人士有點始料未及，但細細思考後也覺得並不奇怪，可能主要與以下因素有關：

(1) 短期套利機會缺失： 與 2007 年「港股直通車」不一樣，兩地市場間巨大的估值差已經不復存在，不少 A+H 股票的 H 股甚至已貴過 A 股；

(2) 內地機構投資者還未來得及進場： 包括公募基金、保險公司在內的很多大型內地機構投資者還在等待監管機構公佈相關的投資指引；

(3) 50 萬人民幣資產的門檻： 在今天的 A 股市場，有相當大部份交易是由持有 50 萬以下資產的投資者進行的；這一門檻未來是否降低，還有待監管當局的綜合考慮；

(4) 港股通合資格股票無小盤股： 滬港通從大盤指數股開始試點是一個主動的政策選擇，主要是為了更有效地進行跨境監管和風險管理。未來

滬港通合資格股票肯定會逐步擴容，但不會因為交投的活躍度而輕率調整；

(5) 對香港市場的熟悉程度：對於內地投資者來說，香港是一個完全陌生的市場，市場上可供他們了解相關上市公司的研究報告還比較有限，他們熟悉香港市場和滬港通的規則都需要時間。

歸根結底，在短期內港股通的發展還有待投資者的熟悉，以及規則的完善和擴容。但長遠來看，港股通的潛力不可限量，因為港股通所吸引的不僅是證券市場的存量資金，而更重要的是現在尚未投資於證券市場的增量資金。這一點我將在下文展開闡述。

既然兩邊都有很多投資者還沒有完全做好準備，為什麼不等他們都準備好了再推出滬港通？

這是一個雞生蛋還是蛋生雞的問題。為方便機構投資者而搭建的股票追蹤系統只有在滬港通推出了之後才有市場需求，至關重要的稅收政策也只有在滬港通推出時間確定後才有可能正式公佈與生效。如果不知道滬港通哪一天開通的話，投資者恐怕很難做出「最後一公里」的投資、合規和技術安排。所以，想在滬港通推出前把一切配套設施都百分百準備好是不現實的。

滬港通就像是一個初生嬰兒，這個孩子已經懷胎十月該生了，儘管我們還未佈置好育嬰房，我們還是應該先把孩子生下來。我們完全可以等孩子出生了再買尿布，再買嬰兒床也不遲！而且，有些東西可以因應孩子量體裁衣，或許更可以少買錯、少走彎路。

香港交易所未來如何推廣和完善滬港通計劃？

正如前面所言，滬港通這個孩子已經出生了，我們現在的主要任務是好好照顧他、培育他。在滬股通方面，我們正在為機構投資者搭建滿足前端監控要

求的股票追蹤系統，預計明年 5 月可以推出。在港股通方面，我們將積極與內地監管機構溝通，推動相關政策出台，以方便更多內地機構投資者參與香港市場；我們也將為內地券商的投資者教育和推廣活動提供支持，幫助內地投資者了解香港市場。此外，我們也會認真傾聽市場的聲音，不斷探索滬港通升級和擴容的最優方案。

目前滬港通的市場反應會不會影響你們修建其他的互聯互通大橋？

對於這一點，我沒有完全的把握，有可能大家會先停一停、看一看，但也有可能是大家建橋的信心更足了。幾年之後再回看現在的市場反應，也許會發現，對於滬港通這樣大型的金融創新來説，現在的淡定其實是一件好事。在我們討論滬港通的方案設計時，監管機構最大的擔憂就是市場波動風險，擔心滬港通會不會像 2007 年港股直通車的消息那樣引發市場瘋漲，因為任何瘋漲的行情，最終都會以慘烈的下跌收場，到頭來最受傷的還是中小投資者。從本週滬港通的波瀾不驚來看，現在投資者的心態還是比較冷靜和理性的，這樣的市場氛圍其實更適合我們建新的大橋，至少監管機構不必擔心有太多人要搶着過橋引發踩踏事件了。而且，新建的橋與路可以形成一個更科學、更互補、更全面的國際化網路，加速中國資本市場的發展，提升其國際影響力

目前滬股通的成交遠遠超過了港股通，滬港通會不會「南水北調」抽走了香港的資金？滬港通是不是更有利於內地？

短期來看，滬股通的成交確實遠遠超過了港股通。但是，長期來看，港股通的成交量增長潛力一定會超出很多人的預期，因為相對於海外投資者來説，中國內地居民的海外資產配置需求更大。20 年前，中國需要引入大量的海外資本；未來 20 年，中國的資本需要走出去，而且資產配置多樣化的需求也會大大增加，大量財富會從銀行存款轉向證券和其他資產類別，這是大勢所趨。

我相信很多中國人會從港股通邁出投資海外的第一步。

　　而且，香港從來都是一個資金自由港，每天都有來自全球的資金流進或流出香港，這些資金的永遠都是追隨投資機會的，就算沒有滬股通，這些資金也可能流向其他海外市場。

　　至於滬港通現在到底給誰的好處更多，就像我在上一篇網誌所說，這不是我們應該糾結的問題。做生意的人都知道，雙贏是任何商業合作成功的前提，要想尋找合作機會，首先要想怎麼讓合作夥伴贏，然後才是自己贏，一味算計怎麼讓自己多贏的人注定很難找到合作夥伴。香港背靠一個巨大的本土市場，面對內地市場大步邁向國際化的巨大機遇，憑藉「一國兩制」給我們帶來的優勢，香港沒有不贏的道理。因此，我們都應充滿自信，放眼未來。

　　路遙方能知馬力。滬港通是一項長期制度安排，它的價值需要數年才能得到準確的檢驗。如同當年的 H 股上市，經歷二十多年的時間涓涓細流方才匯成大海，徹底改變了香港資本市場的面貌，我堅信今天的滬港通將對香港和內地資本市場產生同樣深遠的影響！

　　接下來，我們會把主要精力轉向完善滬港通的各項制度安排上來，希望各界朋友能夠一如既往地支持我們，多提寶貴意見。

2014 年 11 月 23 日

29 滬港通下一站：探索共同市場

滬港通在順利推出並平穩運行兩個月後，成交量開始穩步增長，南北向交易量也開始呈現更加平衡發展的趨勢。與此同時，我們也在加緊優化完善滬港通機制的各項工作。我們堅信滬港通交易量未來將會迎來更加高速的增長。

隨着深港通準備工作的全面展開，我們已開始思考後滬港通時代的香港證券市場未來發展的方向及路徑。在昨天的亞洲金融論壇上，中國證監會主席肖鋼提出了建立亞洲財富管理中心的宏偉設想，突出強調了中國資本市場開放對國際金融秩序有可能帶來的深遠影響。在今天的香港交易所工作坊中，也有許多朋友問我對中國資本市場開放的看法。我想借此網誌，提出一些初步的思考，拋磚引玉，希望能與朋友們一起探索未來發展的方向與路徑。

一、香港應如何認識中國資本市場雙向開放中遠期的路徑選擇？

二、滬港通模式的本質是什麼？

三、滬港通下「共同市場」開放模式有什麼主要特點？

四、「共同市場」的模式為後滬港通時代香港金融的發展帶來什麼啟示？

一、香港應如何認識中國資本市場雙向開放中遠期的路徑選擇是什麼？

任何一個大國的崛起，必然離不開一個發達的資本市場，而資本市場的發展壯大又離不開「開放」二字。中國內地資本市場逐漸走向雙向開放，是歷史發展之必然。

　　具體來説，雙向開放無外乎「請進來」與「走出去」，即把國際的產品、價格或投資流量「請進來」，或讓自己的產品、價格或投資流量「走出去」，其終極目的是使自己的市場更加國際化，使國民的財富投資配置更多元化、國際化；使中國更有效地參與和影響國際定價與標準的制定。

　　鑒於內地與國際市場在制度法規與交易體制上尚有重大差別，大規模的「請進來」在中短期內可能還不能完全實現；而允許國內資本大舉出境也會帶來重大的政策與監管憂慮，因與國際市場長期隔離而缺乏了解的國內資本也很難有迅速走出去的動力與實力。滬港通正是在這一大背景下，由中央最高層拍板宣佈、由兩地監管機關支持、由滬港交易所共同推出的一個歷史性的創新型突破。

二、滬港通的本質是什麼？

滬港通最重要的意義在於**創設了一個兩地機構共同營運和監管的「共同市場」**，它探索出了一種創新性開放模式，在中國資本市場的「請進來」與「走出去」還未完全到位前，讓世界和中國合適的產品在一個「共同市場」匯聚，從而可以在中國的時區、中國的監管要求下，通過兩地交易所和結算公司的聯接，讓國內的投資者、資本在「共同市場」與國際投資者、資本對接博弈，進而逐步形成中國投資資本走向世界的大潮流。

受此啟發，如果我們未來把滬港通框架延伸至現貨股票之外的其他產品，如股票衍生品、商品、定息及貨幣等各大資產類別，把國際投資者喜歡的中國產品放進「共同市場」（相當於把國際投資者「請進來」），同時裝進中國投資者需要的國際產品（中國投資者「走出去」），這樣我們就可以在不改變本地制度規則與市場交易習慣的原則下，大規模地、迅速擴大中國資本市場的雙向開放，大幅擴大中國標準和定價的國際影響力，為中國資本市場的全面開放贏得時間與空間。同時，由於跨境資金流動全部是人民幣，「共同市場」也可成為人民幣國際化的加速器，可謂一舉多得。這樣的「共同市場」模式起步於與香港的合作，也可延伸至其他市場與地區。

三、「共同市場」開放模式有哪些主要特點？

(1) 兩地監管機構共同監管，執法上密切合作。兩地監管機構的監管半徑得以大幅延伸，在輸出自己核心價值的同時引入外部先進標準，從而可以更大限度地掌控「請進來」和「走出去」的節奏。

(2) 兩地交易所合作共贏，成功走出「零和」競爭循環。利益高度一致的兩地交易所將齊力推進國際化，兩地流量在同一撮合器上互聯互通，實現價格最大發現；價格授權先行則將為後續全面互聯互通贏得時間與空間。

(3) 兩地清算所每日交易總量本地先行結算，淨額跨境結算。在保持兩地現行

清算結算體系完整性和獨立性的基礎上建立了一個既封閉又透明的風險管控架構，並實現儘量小規模的跨境資金流動。

(4) **兩地投資者無需改變交易習慣即可與對方市場投資者在同一平台上共同博弈。** 兩地投資者基本依賴本地市場結構，歸本地監管機構保護，可在「自己家裏」和「自己的時區」按「自己的交易習慣」投資對方市場產品。

(5) **對於兩地股票發行者和衍生品使用者而言，**「共同市場」可使兩地上市公司同時享有雙方投資者創造的流動性，兩地未上市公司有可能將來向對方投資者融資，兩地企業可利用對方市場衍生品和商品進行風險管控，最終兩地衍生品價格或流量的互聯互通將有利於整體提升實體經濟利用金融市場進行融資和風險管控的有效性。

(6) **對於兩地中介機構而言，**「共同市場」有效防範了本地投資者在「走出去」的過程中與本地中介機構「脫媒」，可更好地激勵本地中介機構服務於本地投資者的「國際化」，進而全面促進本地中介機構業務「國際化」進程。

綜上，滬港通開創的創新型模式本質上造就了一個中國市場與香港、國際市場互聯互通的「共同市場」，通過一個雙向的、全方位的、封閉運行的、可擴容的、風險可控的市場開放架構，把「世界帶到中國家門口」，體現了市場各方利益的高度一致性，有助於實現最大市場化程度的國際化進程，以最低的成本為最終的全面開放贏得了時間與空間。

四、「共同市場」的模式為後滬港通時代的發展帶來什麼啟示？

滬港通是「共同市場」的序幕，即股票版 1.0 版本，「共同市場」的潛力肯定遠不止於此。有哪些內涵能夠放入「共同市場」呢？這首先取決於大家能否對「共同市場」的發展模式取得一定的共識。在一定的共識之下，我們還須在政策制定者與監管者的引導下對模式的方方面面進行詳盡的可行性研究與風險評估。在下面有限的篇幅裏，我想不揣冒昧，提出一些初步思考。

滬港通後股票類市場的互聯互通已順理成章

上海與深圳是中國股票市場發展的「兩條腿」。滬港通這條腿邁出去後，深港通這條腿的邁出已無懸念。我們須考慮的問題僅僅是推出時間以及是否會在滬港通的基礎上有所升級與豐富。

如果滬港通與深港通是中國股票市場雙向開放的「兩條腿」的話，股指期貨通就是聯動兩條腿的胯與腰。股指期貨的互聯互通有助於滿足兩地投資者風險對沖的需求，這也是 MSCI 考量是否將 A 股納入其新興市場指數的關鍵因素之一，也是內地市場掌握 A 股衍生品市場國際定價權之必要條件。

簡單而言，股指期貨通可以有快捷版和全通版兩種實現形式：快捷版就是結算價授權先行，中國價格先行「走出去」，國際的價格也可以「請進來」；全通版則是類似滬港通、深港通的產品互掛、流量互通。至於採用哪種模式，可以視乎中國在海外市場確立股指期貨定價權的全盤考慮而定。

中國商品期貨市場國際化刻不容緩

客觀而言，如果不計時間因素以及境內外制度磨合成本，中國期貨市場國際化的最優選擇無疑是流量「請進來」，就地國際化。但是在當前中國資本項下尚未全面開放且法律框架、交易規則與境外尚存巨大差異的情況下，大規模的流量「請進來」可能還需相當長的時間。

中國是世界上最大的大宗商品進口國，是最大的消費者，但卻不是真正意義上的定價者。實際上，中國商品期貨市場國際化已經到了刻不容緩的時刻，我們認為必須儘早考慮「走出去」迎戰，讓「產品走出去」、「價格走出去」、甚至讓「資金走出去」。

到底怎麼走出去？單獨走還是通過合作的方式走？與誰合作？商品領域中互利、雙贏與可持續的合作模式是什麼？這些問題看似容易，實際極具挑戰。商品期貨市場有別於股票，它的雙向開放肯定須考慮不同的因素，可能會有不

同的方式和節奏。內地交易所與香港交易所的同仁都已開始認真思考這一問題，希望能夠儘早找出合適的開放模式與路徑。我相信「共同市場」的雙向開放框架將會大大拓寬我們探索的思路。

定息與貨幣產品的開放與國際化已經擺上日程

當前，中國匯率與利率市場化改革正朝着縱深方向發展，進程明顯加速。另一方面，在現有制度和國內外經濟金融形勢下推進改革並非一蹴而就，作為金融市場基礎價格標竿的匯率和利率面臨着「牽一髮而動全身」的複雜性，對整體配套的要求比較高，改革過程中也需妥善處理各種風險。

在這樣的背景下，香港離岸人民幣市場這塊「試驗田」將更加重要。它既可遮罩風險向內地市場的直接傳導，又可為繼續穩步推進人民幣匯率和利率改革提供參考。我們希望能在兩地監管機構的指導下，繼續深化香港人民幣貨幣和利率產品創新。與商品期貨類似，定息及貨幣類產品的國際化思路也可以是價格授權和流量互聯互通相結合。

內地資本的迅速崛起與註冊制改革為一級市場的互聯互通帶來了新挑戰和新機遇

一級市場功能是資本市場服務實體經濟的最基本功能，「共同市場」是否也能延伸至一級市場呢？由於內地發行者早已成為香港市場的主要構成部份，內地資本市場改革開放的提速（特別是一級市場改革，例如註冊制）將對香港股票市場的發展前景帶來深遠的影響。

具體而言，香港市場需要仔細分析各類企業選擇不同融資市場上市的動因與制約，重新思考如何在新環境下繼續保持自身的競爭力：

- 內地企業選擇內地上市的主要動因與制約：熟悉自己的投資者和較高的歷史估值，但受限於監管者對市場容量與上市節奏的嚴格管控；

- 內地企業選擇香港上市的主要動因與制約：更加開放和國際化的發行上市制度與環境，以及更成熟的國際機構投資者，但苦於歷史上較低的估值，全流通的限制以及缺乏大規模內地資本的有效參與；

- 國際跨國企業考慮來香港與中國內地融資的動因與制約：既希望吸引大量的中國內地投資者成為其股東，但其一級市場融資又無法適應內地目前的法規與市場結構；簡單在香港進行非融資的第二地上市又無法取得足夠規模的、可持續的二級市場流量。

換句話說，影響香港一級市場最重要的兩大因素是內地市場正在崛起的巨大流量與迅速提速的內地發行制度改革。只有正確認識與判斷這兩大因素，我們才能看清自身獨特定位的優勢與劣勢，儘早開始考慮我們是否需要作出必要的調整，以找到互利、共贏而且可持續的共同發展道路。

總而言之，我認為「共同市場」可以成為中國資本市場國際化的一個新思路與新模式。我今天提出了幾個我們關心的問題，但找到正確的答案則需要我們集思廣益，共同探索。毫無疑問的是，擁有「一國兩制」優勢的香港正是內地打造這一共同市場的最佳初始夥伴。

那麼對於香港來說，這樣的共同市場是不是我們的最佳選擇呢？香港能否在不犧牲或不妥協我們自身的標準和獨特核心價值的前提下參與「共同市場」的建設？我相信，充滿智慧與自信的香港人心中自會有清晰的答案。

2015 年 1 月 20 日

30 | 修橋雜思：
進一步完善滬港通

人在忙碌的時候時間總會過得特別快，在剛剛過去的三個月裏，我們一邊忙着收市競價交易時段和市場波動調節機制的諮詢，一邊忙着歸納和總結市場對於不同投票權架構概念文件的各種意見。當然，還有不斷傾聽市場意見和優化滬港通機制，自滬港通去年推出以來，這一直是我們工作的重中之重。同時，我們也會在緊鑼密鼓地準備深港通。彷彿眨眼之間，一個季度就這樣過去了，讓人不禁感歎：時間都去哪兒了？

值得欣慰的是，在監管機構的配套政策支持下，我們之前所做的各種努力近期已經開始初見成效，我們越來越堅信，我們正在朝着正確的方向邁進。在此，我想抽空回顧一下過去數週的工作進展，也跟大家分享一下我們下一步的工作計劃。

你們或許已經留意到了，今天滬港通下的港股通交易量創下近 60 億元的歷史新高，這也是本週港股通交易量三度創出新高，顯示內地投資者對於香港上市公司的投資興趣不斷升溫。除此之外，今天港股總市值達到 27.3 萬億元，同為歷史新高；而在本週二（3 月 31 日），香港股票市場的成交量高達 1,496 億元，也是 2008 年 1 月以來的最高紀錄。

引發港股通交易量近期激增的因素是多方面的，其中很重要的一點就是中國證監會上週末公佈了內地公募基金通過港股通投資港股的監管指引。這一已蘊釀數月的監管指引不僅允許新設的內地公募基金通過港股通投資港股，而且明確指出現存的老基金可以依據基金合同或在修改基金合同後參與港股通，徹底掃除了內地公募基金參與港股的政策障礙。在過去幾個月裏，我們已與監

管當局及主要基金探討一旦該監管指引公佈後，如何使現有基金快捷有效的召開基金持有者大會批准相關基金章程與合同。

此外，自滬港通開通以來，我們已經在內地舉辦了五十多場不同類型的投資者座談會和培訓活動，上個月我們剛剛優化了滬港通的市場數據服務計劃，為內地投資者提供更加及時和多元化的港股行情服務。未來，我們還將繼續組織針對內地的投資者教育活動，幫助內地個人投資者熟悉香港市場，相信港股通的交易量未來會不斷增長。

與此同時，我們也推出了多項滬股通的優化措施，方便香港和海外投資者投資上海的股票。本週一（3月30日），我們的中央結算系統（香港結算）推出了一項新的服務，解決了一些大型機構投資者參與滬股通的後顧之憂。在此之前，機構投資者在賣出A股前須先將A股從託管商轉移至經紀商以滿足前端監控要求，這也令很多大型機構投資者對滬股通望而卻步。香港結算推出的新服務是為有需要的投資者開設特別獨立戶口，擁有特別獨立戶口的投資者可以在賣盤執行後才將賣出的股票轉移到經紀商進行交收，這與目前他們交易港股的交收方式十分近似。

我們做的另外一件事是向投資者說明香港結算作為滬港通A股名義持有人的角色。在滬股通機制安排下，所有投資者通過滬股通買入的A股都登記在香港結算名下，滬港通推出之初，一些海外機構投資者（尤其是美國和歐洲的基金）擔心這一安排將妨礙他們行使實益擁有權。為此，我們一直與國際基金、投資者和監管機構保持溝通，向他們清楚地講解了在內地與香港法律和監管框架下的名義持有人安排。現在大部份投資者已經認識到，透過滬港通名義持有人持有A股的投資者們對這些A股享有實益擁有權。為了進一步打消海外投資者對於滬股通名義持有人安排的顧慮，香港結算最近還修訂了《中央結算系統一般規則》，明確指出在投資者的要求下，香港結算可以為他們出具A股持有證明，在必要時還可以協助投資者在內地依法採取法律行動。

雖然在優化滬港通機制方面我們已經取得了不錯的進展，但是，我們也深知，作為一項全新的試點計劃，滬港通還有很多需要完善的地方，我們決不

能有絲毫的懈怠。目前，我們正在與深圳證券交易所的同仁們細化深港通的方案設計，希望能夠進一步優化現有的互聯互通機制，也為投資者帶來更多投資機遇。

如同我在之前的網誌中所寫，滬港通只是兩地市場互聯互通的第一步，我們的目標是與內地同行和監管當局合作創建一個連接內地投資者與國際投資產品、連接海外投資者與中國投資產品且涵蓋多資產類別的「共同市場」。滬港通和即將到來的深港通將股票產品帶進了這個共同市場，我們希望不久之後也能把股票衍生產品裝進這個共同市場。假以時日，我們希望能把共同市場延伸至股票衍生品、商品、定息及貨幣等各大資產類別，把世界帶到中國家門口，也把國際投資者「請進來」，為參與的各方創造多贏。

中國內地資本市場逐漸走向雙向開放，是歷史發展的必然，也是一個越來越清晰的大趨勢。緊隨這一趨勢，我們修建了滬港通這座連接兩地市場的大橋，目前我們正在加快輔路建設，並着手完善橋上的指示牌、加油站和速食店等配套實施，為即將使用這座大橋的人們提供更多的方便。未來我們還會修建更多的新橋，方便兩邊的人們來往。我堅信，隨着內地資本市場加速開放，隨着橋兩邊的人們逐漸增進相互了解，我們修建的大橋一定會變得熱鬧起來。如果內地 A 股未來能被納入全球主要股指，數以千億的海外資金將會配置到內地市場；中國內地居民的財富搬家 (從銀行搬向資本市場) 的洪流才剛剛開始，未來數以萬億的人民幣資產將會分流到內地和海外的資本市場。

經常有人問我，如果有一天內地市場完全開放，你們修建的這些橋會不會就報廢了？說實話，我不太明白這個問題。如果中國完全開放了，它就一定會允許更多的連接國內外的大橋和隧道開放。是的，也許開放之後有的人會考慮舉家移民，不再需要過橋了，但是絕大部份人還是會留在原地，只是過橋去另外一邊旅遊更頻繁了。只要我們的大橋足夠方便，又怎麼會沒用呢？尤其是我們修橋的地段都是流量最大的地方，更何況我們不收任何過橋費 —— 通過滬港通交易的投資者只需要跟交易本地股票一樣支付經紀佣金和交易所規費，沒有其他附加費用。因此，我不認為我們的橋到那時會被取代。只是到那時，橋

會越來越多，人流也會越來越大。

　　總而言之，我對共同市場的前景充滿信心，好戲才剛剛開始，無限精彩在前方！

2015 年 4 月 2 日

31 給投資者的一點小建議

復活節之後的這幾天，香港證券市場人氣格外活躍，恒生指數連續大漲，市場總市值、交易總量和滬港通成交均屢創新高。在市場一片歡呼雀躍之時，我被投資者問到最多的問題是：

1. 現在買入港股還來得及嗎？
2. 滬港通額度老爆滿買不進去了可怎麼辦？
3. 交易量放大後港股市場會否更波動？

我並沒有水晶球可以預知市場未來走向，但作為滬港通大橋的諸多修橋者之一，我想給大家一點小小的投資建議。

不要急！

滬港通是一座長期開通的大橋，它是為未來十年、二十年而建的，不存在「過了這個村就沒有這個店」的問題。你完全可以也應該根據自己的實際情況與需求安排準備好再從從容容地開始你的投資之旅。不必急於一時，更不必湊熱鬧。有價值的股票永遠可以找到，但急於求成往往會徒增風險。要知道，如同在假日紮堆旅遊一樣，紮堆過橋很容易引起擁堵或者踩踏事件，也非常影響個人體驗。

不用慌！

滬港通南北雙向均設有總額度與每日額度，其主要目的是在啟動初期確保平穩運營。這兩天港股通額度每日下午就提前用完，讓不少投資者因無法進場而困惑，也有不少聲音呼籲能早日擴容。

在這點上，我也請朋友們不要慌，保持耐心，監管當局一直在密切關注市場發展，會在適當時機考慮擴容。我之所以對此有信心，主要基於以下兩點：

(1) 雖然滬港通這幾天火爆，但經滬港通進出的內地資金總量還是非常小的。到目前為止港股通總額度僅使用了 479 億元人民幣（未包括 4 月 9 日的港股通交易），即 19%。滬股通總額度僅使用 1,157 億人民幣，即約 39%。而內地 A 股市場自去年年底以來，日均成交額一直保持在一萬億以上。相對而言，通過滬港通進出的資金規模實在非常有限。

(2) 滬港通的結算交收全程封閉，監管者可以全面監控風險。所有的交易行為都在交易所、結算公司的系統內進行，均有清晰記錄。在這樣的制度安排下，資金並未無序地大進大出，買賣 A 股或港股的資金在股票售出後又沿原路返回主體市場。

如此封閉和透明的制度設計保障了滬港通可以在兩地監管機構密切監測與審慎風控的前提下承載巨大的跨境交易。

機會無窮、風險常在！

滬港通開啟了內地與香港市場的互聯互通，也讓崇尚價值投資理念的國際機構投資者與散戶佔主導的內地投資者羣體開始了「歷史性的交匯」。這一歷史性的交匯一定會催發出大量「化學效應」，開創一個中國資本市場發展的新時代。

在這樣一個新時代裏，機會無窮，但風險常在。對香港投資者來說，新增的內地投資者為市場帶來了活力和前所未有的交易機會。與此同時，雙方在

投資觀念和風險意識上的差異也為香港投資者（尤其是散戶），帶來新的挑戰與風險。如何在亢奮的市況中保持冷靜和謹慎，是每一個投資者都必須思考的問題。

而對於很多內地散戶投資者來說，滬港通是他們投資海外的第一步，自然更需要謹慎。投資就像游泳一樣，如果不下水，就永遠也學不會游泳，但初學者第一次下水時往往不免要嗆上幾口水。因此，投資者一定要多做功課，謹慎決策，切忌跟風。

作為市場運營者和監管者，我們深知，不斷攀升的交易量意味着更大的責任，我們會繼續全力確保系統的穩定可靠，我們也會時刻密切監察市場，並在必要時採取適當的風險管理措施，維護市場的有序運作。

滬港通的設計初衷不是為大家提供一個發快財的機會，而是幫助中國早日實現國民財富的多元化國際配置，為大家提供財富長期保值增值的渠道。

我衷心祝願每一位投資者謹記風險、投資順利！

2015 年 4 月 9 日

32 互聯互通，迎接共同市場新時代——寫在滬港通開通一週年之際

　　轉眼之間，滬港通已經成功通車一週年了。在兩地業界的共同努力和支持下，滬港通在過去一年中運作平穩、交易結算機制不斷完善，在此我想向各位業界同仁表示衷心的感謝。

　　截至上月底，滬股通額度共使用約 1,420 億人民幣，佔總額度的 47%，成交總金額約為 14,750 億人民幣，最高單日成交額為 234 億人民幣（今年 7 月 6 日）；港股通額度共使用約 890 億人民幣，佔總額度的 36%，成交總金額約為 7,210 億港元，最高單日成交額為 261 億港元（今年 4 月 9 日）。

　　一些媒體朋友問我，滬港通的成交量並不火爆，你會不會感到失望？記得在滬港通開通第一週，我曾在網誌裏寫道，滬港通是一座天天開放的大橋，而不是一場音樂會，無法用一週或者一個月的上座率來衡量它的成敗。相反，它的價值可能需要兩三年或者更長的時間才能得到驗證。今天，我還是這句話，滬港通成交量並非是我們的關注重點，尤其是在開通初期，運營的平穩和安全才是我們更為關注的指標。在這兩項指標上，滬港通給市場交出了一份令人滿意的答卷，即使是在今年夏天內地和國際股市暴漲暴跌期間，滬港通這座大橋依然結實穩固，成功經受住了風雨的考驗，這一點讓我們倍感欣慰。

　　作為中國資本市場雙向開放的一大創新，滬港通以最小的制度成本，換取了最大的市場成效。其本地原則為本、結算交收全程封閉的設計，讓兩地投資者可以最大限度地沿用自身市場的法律、法規、交易習慣投資對方市場，在監管透明、風險可控的前提下邁出了中國資本市場雙向開放的第一步。

　　我們還欣喜地看到，滬港通的模式為內地市場與其他市場的互聯互通提供

了有益的借鑒和啟發。近期，上海證券交易所、中國金融期貨交易所與德意志交易所集團共同成立了中歐國際交易所；連接上海和倫敦兩地市場的「滬倫通」也進入研究階段；這些舉措體現了中國推動資本市場雙向開放和人民幣國際化的強大決心和魄力，預示着一個全面互聯互通新時代的到來，同時也意味着，在中國的實體經濟崛起為全球具有影響力的經濟體之後，金融資源進行全球化配置的時代正在到來。

這是未來市場發展的大趨勢，也是歷史賦予香港交易所的新機遇。眾所周知，金融市場最本質的功能之一就是定價，定價能力是一個金融中心的核心競爭力。過去幾十年，香港已經發展成為亞洲首屈一指的股票定價中心，其核心競爭力就是為各類上市公司定價。滬港通如同一座大橋，首次將香港和上海兩個股票定價市場相互聯通，同時提升了兩個市場的股票定價能力，我們正在不斷完善滬港通機制，也在繼續為深港通的開通進行各項準備工作，進一步鞏固香港作為股票定價中心的競爭力。

下一步，我們將致力於將香港發展成為全球大宗商品的定價中心，進一步提升香港作為國際金融中心的競爭力。

隨着中國經濟的崛起，中國已經成為全球最大的商品消費者，但由於資本市場尚不發達和開放，中國並未在全球大宗商品市場上獲得與其經濟實力相匹配的定價權。通過大力培育商品市場和構建商品互聯互通機制，香港完全可以為內地企業提供國際認可的商品定價工具，幫助中國在全球市場上贏得相應的商品定價權。

三年前，正是懷着這樣的遠大夢想，香港收購了全球最大的有色金屬交易所 —— 倫敦金屬交易所（LME）。如今，中國資本市場開放步伐加快，人民幣國際化進程不斷深入，為我們創造了更多實現這一夢想的條件。不久前，在習近平主席和英國首相卡梅倫的見證下，香港交易所與 LME 簽署倫港通合作備忘錄。我們的初步設想是，首先在倫敦和香港這兩個國際金融中心的期貨市場間建一座跨海「大橋」，把 LME 的產品直接「空降」到香港，讓香港市場上的投資者可以如同交易香港交易所的期貨產品一樣方便地交易 LME 產品，兩

邊的產品與流動性相互流通，可以加速香港商品市場的發展，在最短的時間內
將香港培育成一個商品定價中心。在未來條件成熟時，香港可與內地市場修建
「商品通」大橋，聯通兩地商品市場的產品與流量。背靠全球最大的商品消費
市場，憑藉「一國兩制」創造的優勢和香港人同時熟悉國內和國際市場的獨特
能力，香港完全有能力再造一個全球商品定價中心！

　　長期來看，在成為商品定價中心後，我們還有一個更加宏偉的奮鬥目標，
那就是發展成為一個具全球影響力的貨幣定價中心。在過去，這是我們想都不
敢想的事，因為香港採用的是聯繫匯率制度，貨幣政策的獨立性有限，也沒
有主權債券，一直以來都缺乏發展固定收益類產品的土壤。但是，隨着人民幣
國際化進程的不斷加速，人民幣總有一天將在國際舞台上扮演和美元、歐元一
樣重要的角色，成為全球通用的貿易、投資和儲備貨幣。人民幣國際地位的提
升，不僅意味着無數國際投資機構都將增加人民幣資產的配置，也意味着人民
幣作為一種國際貨幣的定價需求將不斷增長，持有人民幣資產的海外機構需要
管理人民幣的匯率和利率風險。作為全球主要離岸人民幣中心，香港如果能順
應這個大趨勢，大力發展人民幣匯率、利率產品，配合在岸人民幣市場發展，
就可以將香港發展成為未來全球最重要的離岸人民幣定價中心。

　　從滬港通的成功起步，無論是離岸商品定價中心的建設，還是離岸貨幣
定價中心的建設，都離不開內地的巨大流量，離不開香港已具備的國際市場地
位，離不開香港與內地市場的互聯互通。滬港通的成功為兩地市場帶來了雙
贏，也為我們探明了一個可複製、可擴容的互聯互通模式，未來，我們將利用
這一模式與內地修建更多大橋，不斷提升香港的金融競爭力，為內地資本市場
的開放與國際化、為香港金融市場的繁榮與發展開創新的明天。

2015 年 11 月 29 日

33 | 解讀深港通

　　深港通終於要來了！8 月 16 日，國務院正式批准了《深港通實施方案》，中國證監會與香港證監會也發出了深港通聯合公告，顯示出中國進一步對外開放資本市場的堅定決心，也為香港交易所的互聯互通戰略迎來了新的里程碑。

　　在週二晚上的深港通新聞發布會上，我向大家介紹了深港通的主要特點，傳媒朋友已經進行了廣泛報道。不過，最近幾天，仍有不少朋友向我詢問深港通的細節，大家關心的問題似乎都差不多，因此，趁週末得空，我把大家常問的一些問題答案整理出來，謹供有興趣了解深港通的朋友們參考。

一、作為互聯互通機制的升級版，深港通有哪些升級之處？深港通對於香港交易所有何意義？

　　首先，深港通為投資者帶來了更多自由和便利，總額度限制取消是一個重大進步（滬港通的總額度亦已即時取消）。雖然目前總額度還有剩餘，但對於機構投資者（尤其是海外機構投資者）來說，總額度限制始終是制約他們投資內地股票市場的一大顧慮，取消總額度限制可以讓他們更加放心地投資，從長期來看，一定會鼓勵更多海外機構投資者參與深港通和滬港通。

　　第二，深港通為投資者帶來了更多投資機會。深港通下的深股通涵蓋了大約 880 隻深圳市場的股票，其中包括約 200 隻來自深圳創業板的高科技、高成長股票，與滬股通投資標的形成良好互補。深港通下的港股通涵蓋約 417 隻港股，比滬港通下的港股通新增了近 100 隻小盤股（包括恒生小盤股指數成份股

及深市 A 股對應的 H 股）。投資標的的擴容，可以滿足不同類型投資者的投資需求。

第三，更豐富的交易品種。除了現有的股票，深港通未來還將納入交易所買賣基金（ETF），為投資者提供更多選擇。

需要指出的是，深港通與滬港通採用同樣的模式，這一模式最大的特色是以最小的制度成本，換取了最大的市場成效。通過這一模式，兩地投資者都可以盡量沿用自身市場的交易習慣投資對方市場，可以最大限度地自由進出對方市場，但跨境資金流動又十分可控，不會引發資金大進大出，實現了最大幅度的中國資本市場雙向開放，為兩地市場創造了共贏。

對於香港交易所而言，兩年前開啟的滬港通是我們互聯互通戰略的第一步，為我們開創了一種全新的資本市場雙向開放模式，今日的深港通則以實踐證明，這種模式是可複製和可擴容的。深港通的推出將是一個質變，意味着我們的「股票通」戰略在二級市場層面已基本完成佈局。

二、 既然已經取消了總額度限制，為什麼深港通仍然保留了每日額度限制？

深港通下的港股通每日額度為 105 億元人民幣，與滬股通下港股通的每日額度相同。由於內地投資者既可以使用滬港通投資香港股票，也可以通過深港通投資香港股票，深港通的推出實際意味着將現有的港股通每日額度擴容一倍。深港通保留每日額度限制，主要是出於審慎風險管理的考慮。雖然從滬港通目前的運行情況來看，很少有每日額度用罄的情況，設置此限制似乎有些過慮，但在設計這個機制的時候必須從全域考慮，每日額度限制有點像減速器，它的作用主要是在資金流動過於猛烈的時候給市場一個緩衝，穩定一下節奏。

深港通《四方協議》簽約儀式。

三、 為什麼深圳創業板的股票僅開放給機構專業投資者，香港和海外散戶投資者如何投資深圳創業板股票？

在深圳創業板上市的股票通常比在深圳主板和中小板上市的股票市值小，波動性往往也更大，可以說是高風險高收益類的股票。所以深圳創業板在內地也不是開放給所有投資者，而是設有一個參與門檻。為了保護中小投資者，中國證監會有一套完整的投資者適當性管理方法，比如內地投資者在開通創業板投資權限時，必須簽署風險揭示書，表示已經充分了解創業板的投資風險。香港目前還沒有這樣的投資者適當性管理體系，因此，目前可通過深股通買賣深圳創業板股票的投資者僅限於香港的機構專業投資者，不包括香港股票市場上的散戶。不過，我相信深港通推出後，香港會有基金公司推出更多投資深圳市場的基金產品，如果有興趣，香港的散戶可以通過購買相關的基金來把握深圳創業板的投資機會，而這也會為香港的業界帶來新的發展機遇。

另外，據我了解，香港方面也在研究如何在香港推出適當的風險提示程序，幫助散戶認知風險等，未來這樣的風險提示程序出台之後，香港的散戶或許也可以通過深港通投資深圳創業板股票。

四、 深港通的推出會否為香港市場帶來增量資金？

這問題可從兩方面考慮：一方面，由於深港通下的港股通投資範圍較滬港通下的港股通更廣，多出近 100 隻港股，相信會吸引一些對這些股票感興趣的內地投資者。另一方面，深交所也會加強對港股通的推廣和投資者教育工作，隨着內地投資者對於港股市場的了解加深，相信港股通將增添活力。

五、 如何看待 A 股和 H 股價差，深港通的啟動是否會縮小甚至消除兩地價差？

AH 股價差的根源，是因為兩地市場的投資者在風險偏好和投資理念上有很大差距。內地市場的投資者主要由散戶構成，比較情緒化，而香港市場則由機構投資者主導，更加理性和注重價值投資。儘管同一個公司的 A 股和 H 股是同股同權，理論上內在價值應該趨同，但由於兩邊的股票並不能自由流動和互相替代，套利機制不存在，所以 AH 股價差才會長期存在，即使在滬港通推出之後。深港通的推出應該無法消除兩地價差，但長期來看，因為兩邊的投資者都有了更多選擇，肯定會有助於縮小兩地價差。

六、 深港通是否會增加 MSCI 指數納入 A 股的概率？

中國已經是世界第二大經濟體，中國內地的 A 股將來一定會被納入國際主要指數，只是早晚的問題。深港通的推出，將為海外投資者開放更多內地股票市場，尤其是總額度的取消將給他們帶來更多投資自由和便利，一定有助於推動國際主要指數編制機構未來將 A 股納入這些主要指數。

七、你怎麼看市場對於深港通公佈的反應？

如同我在滬港通開通時所説，滬港通和深港通都是一座天天開放的大橋，而不是一場音樂會，它的價值可能需要兩三年或者更長的時間來檢驗。作為一項創新的互聯互通模式，滬港通和深港通着眼於長遠和未來，其升級和完善將是循序漸進和持續的。短期市場的波動主要取決於投資者的情緒變化，我們無法也無意預測，但是我堅信，長遠來看，互聯互通機制一定會給兩地市場都將帶來十分深遠的影響。

八、深港通未來是否還會延伸和擴容？互聯互通下一步還有什麼計劃？

可延伸和可擴容是深港通模式的一大特色，比如，深港通未來將加入 ETF 這一新的投資品種，我們預計有望在深港通運行一段時間後加入。之所以需要等待一段時間，主要是因為滬深港三個交易所在 ETF 的清算交收和 ETF 與股票的互換上有不同的機制，所以三個交易所和兩地監管者需要共同研究相關細節，希望能夠盡快推出。

此外，不斷完善滬港通和深港通的交易機制也是我們的一項重要工作，未來我們還將繼續與兩地監管機構和中介機構緊密溝通，盡量優化假期安排和做空機制，減少互聯互通機制休市的時間，為廣大投資者提供更多便利。

深港通和滬港通都屬於股票通，我們相信，在不久的將來，這個模式可以延伸到更多資產類別，比如債券通和貨幣通，為境內外投資者和兩地業界帶來更多機遇。

2016 年 8 月 21 日

34 迎接互聯互通 2.0 時代 ——寫在深港通開通前夕

今年的 12 月 5 日註定將是一個特殊的日子，因為我們翹首以盼的深港通在那天就要開通了！如果說兩年前開啟的滬港通是香港交易所互聯互通機制的 1.0 版本，為我們開創了一種全新的資本市場雙向開放模式，那麼今日的深港通將引領我們進入互聯互通 2.0 時代。香港交易所通過與深圳證券交易所、上海證券交易所的互聯互通，共同搭建起一個總市值 70 萬億人民幣的龐大市場，為內地和海外投資者提供了更多選擇。

我衷心感謝各位同仁和朋友們的大力支持，因為你們的辛勤努力，我們的互聯互通機制才能不斷完善和壯大，邁入 2.0 時代。

與兩年前開啟的互聯互通 1.0 時代相比，互聯互通 2.0 時代有哪些優化和升級之處？

第一，交易機制更加便利：總額度限制取消免除了機構投資者的後顧之憂，將鼓勵更多機構投資者（尤其是海外機構投資者）參與滬港通和深港通。我們去年推出的特別獨立戶口（SPSA）服務，允許投資者在賣盤執行後才將賣出的股票轉移到經紀商進行交收，大大降低了海外機構投資者通過滬港通和深港通投資 A 股交收過程中的對手方風險，更加方便基金公司的參與。

第二，投資者准入不斷擴大：滬港通剛剛推出時，內地基金公司和保險資金還不能使用這一投資渠道。在過去兩年中，我們欣喜地看到，中國堅定不移地推進資本市場雙向開放：自 2015 年開始，內地基金公司獲准使用港股通，

2016 年 12 月 5 日，深港通開通。

如今它們已經成為港股通投資者的重要一員；保險資金也在今年獲准參與港股通，香港交易所一直與中國保險資產管理協會和主要的保險資產管理機構保持密切的聯繫，許多內地保險公司都對港股通表現出了濃厚的興趣；這些進步顯示了內地監管機構對滬港通作為內地資金海外資產配置重要渠道的信心，相信在不久的將來，有更多機構投資者會選用港股通作為海外投資的渠道。

　　第三，投資標的擴容：深港通下的深股通將為海外投資者開放一個全新的市場 —— 深圳股票市場，作為中國的創新之都，深圳聚集了很多高成長的創新企業，深股通涵蓋的大約 880 隻深圳市場的股票，將與滬股通投資標的形成良好互補。深港通下的港股通涵蓋約 417 隻港股，比滬港通下的港股通新增了近 100 隻小盤股（包括恒生小盤股指數成份股及深市 A 股對應的 H 股）。投資標的的擴容，可以滿足不同類型投資者的投資需求。內地投資者資產配置國際化和多樣化的需求巨大，因此，我們對於港股通的未來充滿信心。

　　當然，北上深股通的投資標的擴容較南下港股通更加明顯，南下標的擴容帶來的增量有限。深港通與滬港通下的港股通分別連通內地兩大不同的交易所，是兩條平行的通道，並不能互通，自滬港通下港股通渠道買入的港股必須

經原路賣出，不能經由深港通賣出。因此，我們預計深港通開通初期，南下的港股通不會迎來很大的增量，短期內可能會遠少於北上深股通和來自上海的港股通增量。

第四，未來將會納入 ETF（交易所買賣基金）：除了現有的股票，互聯互通機制還將把 ETF 納入投資標的。中國投資者投資海外股票的需求越來越多樣化，例如他們可能希望投資美國的生物科技股、中東的能源股、德國的製造業股票、南美的礦業股、澳洲的農業股等品種。目前，香港的股票市場還不可能吸引所有的海外公司都來香港上市，但是我們的「貨架」上有來自世界各國的 ETF 產品，南下投資者可以通過這些 ETF 來投資世界各地不同類型的股票。

當然，由於兩地現有市場結構、銀行間資金清算和交易假期安排等方面均存在較大差異，互聯互通要納入 ETF 面臨諸多挑戰，執行起來並不容易。但因為其戰略意義非凡，我們一定會全力以赴，爭取盡早納入 ETF。

互聯互通 2.0 時代意味着什麼？

對內地投資者而言，坐在家裏投世界。互聯互通意味着他們足不出戶就可以通過本地券商、按自己的交易習慣投資海外市場，進行多元化的資產配置。

對於國際投資者而言，坐在香港投中國。互聯互通意味着他們可以坐在家裏按照國際慣例、透過國際券商投資內地市場，分享更多中國經濟成長的機遇。

滬港通和深港通的交易模式以最小的制度成本，換取了最大的市場成效。兩地市場存在巨大差異，而這種差異短期之內不可能消除。滬港通和深港通採用了「本地原則為本，主場規則優先」，這樣投資者可以在自己熟悉的環境下交易對方市場的股票，但同時遵循對方市場的一些特殊規則。在互聯互通的過程中，投資者、業界和監管機構都就對方市場的運作和文化有了更加深入的接觸和認識，雙方市場在互動的過程中，會發現有可以相互學習和借鑒的地方。毫無疑問，這將推動兩地市場的共同變革和進步。同時，滬深港三地市場因為投資者羣體、交易和結算體系等的差異，市場運行的差異會在較長時期記憶體

在，這恰恰賦予了廣大投資者更多的選擇，他們可以充分運用他們的智慧，進行更為多樣化的投資。

對於香港而言，互聯互通 2.0 也有着十分重要的意義。

從整個中國資本市場開放的進程來看，在不同的發展階段，香港始終發揮了十分重要的作用；二十多年前，香港把握住了中國改革開放的機遇，為內地企業籌集資金，並成功轉型為一個國際金融中心。

20 年後的今天，互聯互通的不斷延伸為香港資本市場的發展提供新的動力，讓香港的國際金融中心地位更加鞏固；一方面，海外投資者可以通過香港投資更廣闊的中國市場，另一方面，內地投資者也可以通過香港邁出海外投資第一步，令香港市場的財富管理功能將會更加強化。不難預見，香港市場的籌資能力和發展條件也會因此而提升。

換句話説，過去是中國的「貨」在香港，世界的錢來香港買中國的「貨」。互聯互通時代將會帶來巨大的變化：今後世界的錢到了香港，經過香港這一跳板買國內的「貨」，然後中國老百姓的錢到香港買世界的「貨」。這意味着香港將來的作用和任務，就是把世界的「貨」帶到香港來，讓中國人在家裏投世界。香港將從中國的集資中心，發展成為中國的國民財富國際配置中心，國內外的資金和金融產品都將因此匯聚在香港，這樣香港就不可能邊緣化。

互聯互通是權宜之計，還是具有長久生命力的市場機制？

互聯互通最大的生命力在於，它解決了一對似乎不可能解決的矛盾：既要交易便利，又要風險可控。

由於互聯互通機制採用本地市場原則，投資者交易對方市場的股票很方便，如同交易本地市場的股票一樣自由。但是，結算交收的閉環設計，可以保證買賣對方市場股票的資金最後都原路返回，而不會以其他資產形式留存在對方市場，巧妙地實現了風險的全面監控，贏得了監管機構的信任。

有的朋友也許會問，在中國資本項下完全開放以後，監管機構不再管控資

金流動的風險，互聯互通機制會不會馬上失去生命力？

即便資本項下完全開放，市場仍然是由具有不同風險偏好和投資需求的投資者構成的。無論是北上還是南下的投資者，大致都可以分為三種，我把他們分別比喻作研究生、大學生和中學生：有能力、有資源、有渠道、有意願並早已自行投資海外的好比「研究生」；無論是否開放都對境外市場不感興趣的投資者好比「中學生」；而比「中學生」有較多資金、投資經驗、有需要將一部份資產配置在海外、卻不敢直接投資海外市場的則好比「大學生」。

互聯互通就像是一所大學，一些「大學生」將畢業成為「研究生」，但又有一部份「中學生」會成長為「大學生」，而滬港通與深港通正好為這些「大學生」提供了一個方便的境外投資渠道。儘管中國資本市場會越來越開放，但不同市場的交易制度和文化的巨大差異不可能一下子完全消失，因此針對大學生羣體而設立的互聯互通機制將長期存在，並為更多中國投資者提供資產配置的解決方案。

在互聯互通 2.0 時代之後還有哪些發展目標？

在互聯互通 2.0 時代，我們的「股票通」戰略在二級市場層面已基本完成佈局。接下來，我們會進一步優化「股票通」，並按照既定規劃將這一模式延伸到更多資產類別，例如 ETF、新股、商品、債券等等。我相信，在不遠的將來，我們將會一起書寫互聯互通新篇章、迎來互聯互通 3.0、4.0、5.0 時代！

2016 年 12 月 1 日

第四章

定息及貨幣產品市場
與人民幣國際化

35 「養兒子」與「育兒園」——關於人民幣國際化的再思考

在過去一年半裏，香港離岸人民幣市場得到了飛速發展：自 2010 年 7 月至今，香港人民幣存量從 1,000 多億增長至 6,200 億，人民幣貿易結算總額則增長了近 15 倍。人民幣國際化大大提速，步伐快了、思路廣了、措施多了。

與此同時，國內外關於人民幣國際化的討論也日趨熱烈。特別是在過去半年裏，由於離岸人民幣存量的增速有所放緩，離岸人民幣市場首次出現貶值預期，各種質疑的聲音也開始出現。這些質疑歸納起來主要是圍繞着以下幾個大問題：

(1) 中國在利率、匯率與資本項下管制等頂層設計改革還未完成之前和當前貿易格局之下，是否應該和可能推進人民幣國際化？

(2) 在現階段承受人民幣國際化的風險與成本值得嗎？

(3) 在港人民幣存量已大部份回流國內，香港真的能為人民幣國際化起到作用嗎？人民幣國際化又真的能為香港帶來實際利益嗎？

對這些問題，許多學者與業界人士都做過深入的理論分析與實證研究。立場不同，視角不同，支持與質疑的聲音並存。這些討論與爭論是正常的，而且對推動人民幣國際化發展是非常必要的。在此，我想以一個更直觀與通俗的視角，用「養兒子」和「育兒園」這樣一個比喻來對以上問題談談自己的淺見。

中國人有一個傳統觀念，就是不惜千辛萬苦生兒育女來養老，期盼將來兒女成羣，頤養天年。人民幣國際化之於中國，就好比是「養兒子」的過程：人民幣國際化就是讓人民幣這個「兒子」從體內出生、成長、慢慢走出去，在貿易結算、計價、投資、儲備等領域為國際所用。

之所以要養這個「兒子」，是因為中國已經成為世界第二大經濟體，而其

相對封閉的貨幣已經開始制約經濟的進一步發展，是不可持續的。反觀美國，雖然它開始陷入長期結構性困境並顯現出沒落衰老的跡象，但是，它仍然是歐債危機告急時國際資本的避險地。為什麼呢？就是因為美國有美元這個「兒子」養老。無論美國的債權國有多麼不願意，短期內也無法實現資產的「去美元化」。而美國完全根據自己的宏觀經濟需要所制訂的貨幣政策會挾持整個世界經濟格局。所以說，美國今天的霸主地位與它有美元這一個主導國際貨幣體系的「兒子」直接相關。**中國若要實現獲得更大的政治經濟話語權這一長期目標，就需要讓其貨幣更加積極地參與到國際貨幣體系中**。也就是說，中國已經到了必須考慮養兒育女的時刻。

只有在這個長期目標的大框架下，我們才能認清人民幣國際化的必要性、可行性與長期性，並以平常心來看待「養兒子」過程中可能遇到的挫折與所需承擔的風險與成本。下面我對前文提到的三個問題作簡要的回答。

一、中國在利率、匯率與資本項下管制等頂層設計改革還未完成之前和當前貿易格局之下，是否應該和可能推進人民幣國際化？

我的答案是肯定的。

中國 30 年來的改革開放從來就不是一個自上而下、邏輯完整、計劃周全的進程。無論是八十年代的價格改革、所有制改革，九十年代的資本市場發展，還是過去十年的金融體制改革，都是衝破現有體制制約，突破傳統思維，在缺乏現成完善的經濟學理論支撐下以及充滿矛盾的體制中，由市場力量推動、自下而上地逐步摸索展開的。

儘管中國當前尚未完成利率、匯率與資本項下管制等頂層設計改革，但是，頂層設計改革不應成為人民幣國際化開始推進的先決條件。這就好比父母的經濟條件、工作狀況、住房大小不應是生孩子的必要條件一樣。如果事事求全，那我們可能永遠找不到生孩子的最佳時點。

而且，人民幣國際化這個「兒子」是中國經濟與市場力量發展至今的必然

結果。隨着中國財富的聚集、中國經濟全球化的提速，封閉的資本項目已經不能適應中國經濟發展的需要，這個「兒子」實際上已經到了不生不行的時候。與其爭論「能不能生」，不如給予「准生證」，承認它的存在，推動它的發展。

當然，頂層設計的長期缺位會最終制約人民幣國際化的進程。長期來看，「小房子」養不出「大兒子」，父母還是要改善基本生活與教育環境，使「兒子」能健康地成長。同時，「養兒子」往往也給父母帶來積極的動力與壓力，促使他們努力改善「兒子」的成長環境。也就是說，人民幣國際化也可以「倒逼」必要的頂層設計改革。

至於說中國結構性的全球貿易順差是否會使人民幣國際化根本不可能，我的看法是：需要動態地、有差別地看待中國貿易結構。首先，中國今天的全球貿易格局是中國勞動力、土地等要素特點、市場力量發展與政府宏觀貨幣政策選擇等多種因素綜合所致。隨着這些因素的變化和互動，中國的貿易格局不會一成不變。特別是在金融危機與歐債危機之後，全球性的經濟格局變化與中國國內經濟結構的歷史性轉型必然會對持續了 20 年的國際貿易結構帶來深刻、持久的變化。順差與逆差之間、升值與貶值之間的動態變化會成為常態。而不斷變化的全球貿易結構也為人民幣國際化創造有利條件。

其次，中國貿易在各地區內表現並不一樣 —— 整體呈現順差，但對很多區域（特別是東南亞國家等新興經濟體）保持逆差。人民幣區域性走出去是完全可能的。所以說，雖然人民幣這個「兒子」剛生下來的時候不可能馬上遠走高飛，但可以放在身邊、就近養。不能因為「兒子」暫時無法獨立闖天下而否定「兒子」存在的合理性。

有人可能會問，日本與泰國的貨幣國際化均遭遇失敗，中國是否會步其後塵。其實，中國作為世界第二大經濟體，其貨幣由相對封閉走向開放，本來就無先例可循。單純從經濟學、貨幣學理論出發討論其他國家的貨幣國際化經驗與中國的可比性並不一定恰當。我們不應該因為別人家沒養出好「兒子」就得出不養「兒子」的結論。我們更應該從別人身上虛心學習經驗教訓，把自己的「兒子」養得更好。

二、 在現階段承受人民幣國際化的風險與成本值得嗎？

既然人民幣國際化這個「兒子」既必須養也可以養，我們就要給予充分的營養和呵護，這將會花錢、費時、擔風險。總結近期對人民幣國際化的風險與成本的討論，主要有以下兩點：

（1）中國在當前貿易順差的格局下輸出人民幣會導致外匯儲備短期內增長加快以及造成匯兌損失。

（2）離岸人民幣回流可能會降低國內貨幣政策和匯率政策的有效性。

這兩個問題的確是現實存在。恰恰就是因為貿易順差與境外產品匱乏，人民幣國際化的初期必須允許一定的資本項下進出，才能使人民幣開始境外之「生命」，這是一條繞不開的必由之路。「生孩子」從來不是一個免費、無痛、簡單的過程。

對於風險與成本的討論應放在人民幣國際化的長期目標這個大框架下來進行。相比於這個長期目標，有一些風險與成本是短期的，有一些風險與成本也不是不可以承擔的，有一些風險也不是不可以控制的。最重要的是，我們一旦認清人民幣國際化的長期目標，就要堅定信念，勇於承擔。

2001 年中國加入世界貿易組織（WTO）也是經歷了數年激烈的爭論，我們曾在長期利益與短期成本之間舉棋不定。然而，如今「入世」十年，在貿易領域開放的帶動下，中國經濟從當年的世界排名第六位晉升至第二位。人民幣國際化和資本領域的開放是歷史賦予中國的下一個重大機遇。儘管中間過程可能荊棘滿途，但一旦成功，給中國帶來的將是下一個質的飛躍。所以，我們對於人民幣國際化也需要持有同樣的決心與毅力。

像養孩子一樣，我們對人民幣國際化還要給予耐心與呵護，而不是一味地求全責備。有觀點質疑人民幣國際化收效甚微（例如外匯儲備不減反增，與初衷背道而馳），甚至由此而得出應該暫緩人民幣國際化進程的結論。這就猶如責備一個三歲的兒子一事無成一樣。人民幣國際化是一個長期過程。人民幣成為被廣泛使用的國際投資貨幣可能需要大約十年；而人民幣從投資貨幣發展為

重要儲備貨幣則需要更長的時間。如果我們過早地對人民幣國際化的效用抱有不切合實際的期望，反而會阻礙了它的發展。

　　總而言之，我們要養好人民幣這個「兒子」，就得接受在中短期內成本可能會大於利益這個現實，並做好在未來較長一段時間持續供應營養、提供教育的準備。「生孩子」從懷孕、分娩到養育這一過程是一個充滿艱苦、困難與挑戰的過程。成本與風險是不可避免的。但我們不應該因為這些風險與成本而輕易放棄「養兒子」帶來的幸福、快樂以及最終養老之益。

三、 在港人民幣存量已大部份回流國內，香港真的能為人民幣國際化起到作用嗎？人民幣國際化又真的能為香港帶來實際利益嗎？

　　人民幣國際化必然從香港開始，因為香港擁有「一國兩制」的優勢，既與內地有密切的政治經濟聯繫以及相近的文化，亦有其相對獨立的法律體制、成熟的金融業與伴生的服務業。**香港是內地推行人民幣國際化的不二選擇。在人民幣國際化的開始階段，香港可以成為人民幣國際化的「育兒園」，為其提供安全可靠的試驗場所。**

　　近期，香港人民幣存款在經歷一年多的飛速增長後出現放緩的趨勢。而且，由於缺乏豐富的離岸人民幣產品，這些存款絕大部份存放在香港銀行體系，並隨即被存放在央行或投資於國內銀行間債券市場。也就是說，人民幣走出去後，繞了一圈又回到境內。有朋友可能會質疑香港這個「育兒園」意義何在。

　　儘管大部份的香港人民幣存款以現金形態又回到境內，但這些存款是境外居民換匯而來，其屬性已經發生了變化，因為它們已經成為「香港戶口」，持有「特區身份」。今後境外產品一旦豐富發展起來，它們就可以「多次往返」，自由出入了，也就是國際化了。就像養孩子一樣，上「育兒園」時是每天接送回家，上中小學時住校就週末回家，上大學就可能只是暑假回家，一旦成人工

作了，就常駐海外、節假日再回家 —— 這是發展之必然。今天，香港人民幣直接回流是因為離岸人民幣產品匱乏，而不是「兒子」沒出息。今後，一旦離岸人民幣生態環境建好了、國際對人民幣的信心提高了，境外人民幣自然就可以在海外留得長一些了。這也是為什麼香港交易所努力推動人民幣股票在香港加速發展。日後，香港交易所還將積極推動人民幣利率與匯率產品以及商品衍生品市場的發展，以進一步擴大離岸人民幣的使用範圍。

另一方面，在港人民幣的形式至今主要是替換存量港幣，人民幣國際化的「育兒園」並未為香港的金融帶來實際增量活動、額外收入或直接利益。也有朋友發問：香港「忙」這一圈又有什麼意義呢？我個人認為，這個意義可大了，既有短期的，也有長期的。

短期來看，利益是無形的，並不一定與人民幣國際化直接相關。香港由於為人民幣國際化辦了「育兒園」，在國家發展最重要的時刻貢獻了巨大的與不可替代的價值，香港自然也會在許多方面取得國家有形無形的政策支持與機會，譬如在 CEPA 協議框架下兩地金融合作更緊密、市場相互開放提速、鼓勵境內企業來港上市等。

長期來看，香港一旦成功辦好人民幣國際化的「育兒園」，隨着人民幣國際化的深化，香港就可以進一步辦人民幣國際化的「小學」、「中學」乃至「大學」。人民幣國際化最終從香港「畢業」走向全球，香港就變成真正意義上的人民幣離岸中心。日後，即使人民幣國際化走遍全球，但第一站永遠在香港，海外的「家」也永遠在香港。縱觀今日倫敦，位處大西洋上的英倫孤島，800萬人口，卻能成為傲立全球的金融中心，其核心競爭力就是成為全球美元的離岸中心 —— 其離岸美元交易量已超出在岸美元的交易量一倍有餘。同樣地，香港沒有理由不成為當之無愧的人民幣離岸中心。因此，今天辦這個「育兒園」對香港來説商機無限。

四、結語

人民幣國際化的長期目標是為了重建國際貨幣體系，使中國獲得更大的政治經濟話語權。認清了這個長期目標，對其過程的長期性與所要付出的成本與代價就會有所準備，多份理解。人民幣國際化之於中國有如「養兒子」：不能不養，又不能急於回報，而且要花錢、費力、擔風險。香港在初始階段可以成為人民幣國際化的「育兒園」，讓這個「兒子」快速成長，成為它日後走出國門的第一站。

一句話，**人民幣國際化的規模可以大一些或小一些，但絕不能沒有；速度可以快一些或慢一些，但絕不能停下來；溫度可以熱一些或冷一些，但一定要保證它的健康成長。**

2012 年 1 月 3 日

36 場外結算公司：拓展 新資產業務的重要一環

　　今天是個值得慶祝的日子，香港場外結算公司正式開業，為香港交易所的長遠發展迎來一座重要的里程碑。場外結算公司專門負責場外衍生產品交易結算，可能未必引起普羅大眾的關注，但對於我們來説，這卻是實現香港交易所集團長遠戰略目標的重要一步，也是助力香港成為綜合性國際金融中心的重要舉措。

　　我們發起成立全新的場外結算公司，既是為了迎接監管環境的變化，也是因為業務發展的需要。從監管角度看，2008 年的全球性金融海嘯將場外衍生產品市場的種種問題暴露無遺，例如交易對手風險、透明度不夠、市場實際虧損金額和虧損方不明朗等問題一一浮現。危機過後，20 國集團認識到場外衍生產品市場的規模日漸壯大，必須加強監管。因此，20 國集團要求所有標準化的場外衍生產品合約必須經中央對手方結算，以降低整個金融體系的系統性風險。這為包括香港交易所在內的全球交易所和市場基礎設施建設者們帶來了巨大的機遇。

　　從業務發展的角度看，香港交易所集團的長遠目標是發展成為亞洲時區首屈一指的全資產類別且縱橫向整合的交易所，而場外結算公司的成立對於實現這一目標具有十分重要的戰略意義。我們早已為集團旗下的場內交易證券及股本衍生產品市場設立了自營結算所，可以説，我們為香港市場交易生態圈的所有環節——產品、交易及交收——提供服務。如今，我們正在開拓和發展新的市場及資產類別，如定息產品及貨幣產品，而這些市場則往往涉及許多場外交易。如果要服務好這些市場，為其提供場外結算所可以説是必經之路。

2013 年 11 月 25 日，場外結算公司正式開業。

更重要的是，全球市場使用人民幣作為結算貨幣和投資性貨幣的趨勢正與日俱增，為香港這個主要的離岸人民幣中心和香港交易所創造了獨一無二的機遇。經過近幾年的努力，香港已經成為全球首屈一指的離岸人民幣中心之一，從基礎設施到市場份額都頗具規模。隨着人民幣國際化進程加快，以人民幣計價的衍生產品合約結算業務增長潛力巨大。成立場外結算公司處理人民幣計價合約的結算有助於發揮我們的優勢，進一步提升集團的競爭力，使我們從一眾競爭對手中脫穎而出。

不過，正如上文所述，場外結算業務目前仍屬相對比較新的市場，許多全球性監管問題仍有待解決。因此，我們預計場外結算公司的業務量將有一個循序漸進、穩步發展的過程。我們今天開始提供的場外結算服務將首先涵蓋以下幾種產品：港元、人民幣、美元及歐元四種貨幣的貨幣利率掉期，以及美元/人民幣、美元/台幣、美元/韓元及美元/印度盧比四種貨幣對的不交收遠期外匯合約，未來會逐漸擴展到其他產品。

值得高興的是，我們引入了 12 家創始股東共同參與這項業務，它們大部

份是國內或世界領先的銀行，它們同時將成為場外結算公司的結算會員。雖說場外結算公司並非一個公眾設施（Public Utility）——因為它只涵蓋某些特定產品、特定服務；但是，它也並非創始股東的「專屬俱樂部」，這些創始股東只是場外衍生品交易的主要參與者，我們熱切期盼更多業界參與者陸續加入。

那麼，對於場外結算公司的結算會員，場外結算有何好處？場外結算公司將提供國際風險控制措施、更高的安全性以及更有效的資本配置，而且場外結算公司的費用也相對較低。

對於香港而言，場外結算公司又意味着什麼？在短期而言，場外結算公司將協助香港市場滿足 20 國集團的監管要求；長遠而言，場外結算公司的業務日後有望拓展至更多類別的場外衍生產品結算。此舉不僅能為香港留住一些原本可能流失的業務，也有助於提高整個資本市場及銀行體系的安全性。

我知道，場外結算公司算是冷門話題。然而，交易所很多最重要的工作環節恰恰是依靠有一羣人在背後默默耕耘方能完成，才能確保我們能夠保持長遠的競爭力。猶如一幅美麗的大拼圖由無數個小塊組成，每個小塊缺一不可，場外結算公司正是香港交易所長遠宏偉藍圖中不可或缺的一塊。場外結算公司並不是孤立存在，而是成就我們推行定息產品及貨幣市場垂直整合業務模式的重要一環：它不僅提高了香港市場的透明度，也為重要的場外衍生產品市場提供集中風險管理。雖然場外結算業務的蓬勃發展尚需時日，但我們對其長遠潛力保持樂觀。

2013 年 11 月 25 日

37 | 繫好「安全帶」，迎接人民幣雙向波動時代

　　在過去幾週裏，人民幣匯率的走勢恐怕讓不少的投資者和進出口商都心跳加速。2 月中旬以來，人民幣兌美元匯率連續多天跳水，在短短三週之內曾經下跌 1.4%，創下 2005 年人民幣匯率改革啟動以來的最大跌幅，引發了全球市場的各種揣測和聯想。

　　人民幣匯率為何突然走軟？這究竟意味着什麼？這是很多人都在關心的問題。在我看來，第一個問題其實並不重要，重要的是第二個問題。無論是什麼原因導致了人民幣匯率過去幾週過山車式的運動，有一點是非常確定的：人民幣匯率單邊上揚的時代已經一去不復返了，管理人民幣匯率風險的需求會日趨緊迫。隨着人民幣國際化的進程不斷推進，越來越多國家和地區接受和使用人民幣，參與人民幣匯率市場博弈的供求方將日漸增多，人民幣匯率的波動風險勢必增加。

　　實際上，在過去幾年中，人民幣國際化加速的跡象十分明顯，人民幣早已不是只在中國流通的貨幣。國際清算銀行最新調查報告顯示，2013 年人民幣在全球的交易額已經排名第九。香港的人民幣業務發展尤為迅猛，人民幣存款和結算金額均出現激增。在 2010 年，香港的人民幣存款總額還僅有 600 多億元，今年 1 月已經增加至 8,930 億元。香港銀行自 2009 年開始接受人民幣貿易結算，在短短幾年內，經香港處理的人民幣貿易結算總額已經攀升至 2013 年的 3.8 萬億元。

　　隨着人民幣越來越普及，倫敦、新加坡及台北等城市近期也爭相發展人民幣產品，有意轉型為離岸人民幣中心。香港在幾年前就未雨綢繆，早早做好

基礎設施方面的配套,抓緊人民幣國際化帶來的機遇。以香港交易所為例,我們在 2011 年已開通人證港幣交易通,以方便企業在香港以人民幣發行股票;2012 年我們推出了全球首隻人民幣可交收貨幣期貨合約。目前在香港交易所旗下市場交易的人民幣計價產品已經達到 112 隻。

今天,我們早期所做的準備工作已初見成效,最明顯的例子就是人民幣貨幣期貨。自 2012 年推出以來,香港交易所的人民幣貨幣期貨成交量和持倉量一直穩步增長。上個月,人民幣貨幣期貨的日均成交量創下了 1,461 張合約(名義金額為 1.46 億美元)的新高,較 2012 年的日均成交量增長五倍。2 月 25 日,人民幣貨幣期貨的成交合約張數創下了 5,970 張(名義金額為 5.97 億美元)的最高紀錄。與此同時,人民幣期貨未平倉合約數量也增長了六倍多,在 2 月 14 日創下了 23,887 張(名義金額為 24 億美元)的新高。隨着越來越多投資者意識到人民幣匯率不再只漲不跌,他們管理外匯及利率風險的需求將與日俱增,尤其那些是以人民幣進行結算或投資業務的公司。

去年 11 月,中國領導人在十八大三中全會上明確提出了「讓市場在資源配置中起決定性作用」的改革思路,具體落實到金融改革方面,中國將完善人民幣匯率市場化形成機制,加快實現人民幣資本項目可兌換。我們預計未來數月內人民幣的國際化將進一步提速。對於我們來說,這既是機遇,也是挑戰。在人民幣離岸業務的競爭中,要想持續領先,我們就必須與時俱進,提供滿足市場需求的多種風險管理工具。

因此,我們最近正在籌備多項工作計劃。例如,在收市後期貨交易時段新增人民幣貨幣期貨交易,近期已獲香港證監會批准,預計在 4 月 7 日即可推出。開通這一時段的人民幣貨幣期貨交易,有助位於歐美洲時區的投資者利用這一產品管理人民幣匯率風險。

而且,我們正在研究延長人民幣期貨的合約期,以方便投資者做更長線的風險管理。日後的計劃還包括將合約期延長至超過 16 個月、推出人民幣貨幣期權以及進一步延長收市後期貨交易時段。另外,我們日前宣佈推出內地市場數據樞紐,這是我們首個在中國內地的基礎設施據點,目前會先向內地的客戶

提供指數和證券市場數據，今年稍後將再加入衍生產品市場數據。

最後，我們在人民幣業務方面的準備工作已延伸至場外交易的衍生品領域。去年我們推出了場外結算公司，場外結算公司現已開始為單一貨幣利率掉期合約、單一貨幣基準掉期合約、不交收利率掉期合約及不交收遠期外匯合約進行結算，在另一領域為投資者管理人民幣相關匯率、利率風險提供服務和支持。

多年以來，人們一直認為人民幣被低估了。自 2005 年中國啟動人民幣匯率改革以來，儘管人民幣匯率偶有波動，市場一直預期人民幣將持續升值，幾乎沒有貶值的風險。但是，最近兩週的市場走勢確實給我們上了一堂生動的風險教育課——沒有什麼是恆久不變的，人民幣匯率不可能只漲不跌。

在人民幣匯率雙向波動的時代，所有人民幣用家和投資者都需要繫好「安全帶」，做足匯率風險管理。我們的人民幣貨幣期貨產品，恰好為這類風險管理提供了工具。我們已經朝正確的方向踏出了第一步，不過，這還遠遠不夠，隨着人民幣逐漸走向海外，市場需要更加豐富的人民幣計價產品。未來，我們將繼續努力，銳意進取，密切關注市場需求，準備好迎接人民幣國際化帶來的各種機遇。

2014 年 3 月 9 日

38 | 豐富產品組合，滿足未來市場需求

作為資本市場雙向開放的一大創新，滬港通不僅為滬港兩地市場帶來了新鮮的流動性，也將潛移默化地改變兩地市場的生態和投資需求。

在滬港通平穩運作一週年之後，我們不斷思考如何根據市場需求完善滬港通配套設施。此外，為迎接未來內地 QDII2（合格境內個人投資者境外投資）試點啟動帶來的機遇，我們也在大力豐富資產類別，為內地投資者進行全球資產配置提供更多選擇。

我們總結了港股通下南向投資者的交易情況與經驗，從中選出港股通交易最活躍的股票，新增了 34 隻股票期貨，擬分兩批於 11 月 30 日及 12 月 7 日推出。兩批期貨合約推出後，將覆蓋港股通下佔約 60% 交易量的股票，香港市場股票期貨合約的數量也將由 37 隻擴容至 71 隻，為廣大投資者提供更加豐富的選擇。

相對股票而言，股票期貨能為投資者創造更高的資本效率。如果使用得當，股票期貨也能為投資者提供更靈活的交易策略以及更全面的風險管理手段。當然，任何一種衍生產品都具有高槓桿和高風險的特徵，投資者在參與前應仔細研究合約，並對自己的風險承受能力有一個清楚的認識。我們也將與期貨經紀商一起陸續推出多項投資者教育活動，幫助投資者了解這些新產品。

我們相信，隨着南向投資者的增多，市場對於管理股價波動風險的需求將與日俱增。不過，新產品的發展和市場培育往往需要時間，因此，我們將以長遠的眼光來培育和發展這一市場，而不會在意市場的短期表現。

此外，繼去年推出首批金屬期貨小型合約後，我們還將於 12 月 14 日推出

第二批倫敦金屬期貨小型合約 —— 倫敦鎳期貨小型合約、倫敦錫期貨小型合約及倫敦鉛期貨小型合約。

自 2012 年收購倫敦金屬交易所以來，我們一直在豐富我們的產品線，尤其是加快發展股票以外的其他資產類別產品，希望能夠成為中國客戶走向世界以及國際客戶進入中國的首選全球交易所。

我們深知，無論是此次新增的 34 隻股票期貨合約，還是倫敦金屬期貨小型合約的擴容，都只是我們邁出的一小步，未來我們還將不斷聆聽市場的聲音，大力開發各種衍生產品、固定收益類產品以及商品類產品，及時滿足市場需求，提升香港作為國際金融中心的競爭力。

2015 年 11 月 29 日

39 | 再談繫好人民幣匯率「安全帶」

　　猴年臨近，人民幣匯率的走勢也再次「猴動」起來。尤其是最近一週，人民幣兌美元匯率的急跌讓很多人心驚肉跳，不少做貿易的朋友焦急地問我該如何對沖匯率風險。就連我的助理都開始心神不寧了，因為過去幾年她把自己很大一部份的儲蓄從港幣換成了人民幣，閉着眼指望人民幣升值。

　　他們之所以如此措手不及，皆因過去十多年來人民幣幾乎一路升值，所以很少人會考慮人民幣貶值的可能性，更沒有養成主動管理匯率風險的習慣。記得 2014 年我在網誌〈繫好「安全帶」，迎接人民幣雙向波動時代〉（見本章第 37 節）中曾經說到：人民幣匯率單邊上揚的時代已經一去不復返了，管理人民幣匯率風險的需求會日趨緊迫。今天來看，這一趨勢已經十分確定，隨着人民幣國際地位的提升，人民幣匯率形成機制將更加市場化，雙向波動將成新常態，所有人民幣用家和投資者都需繫好「安全帶」，做足匯率風險管理。

　　為什麼這麼說呢？首先，央行推進人民幣匯率市場化改革的決心十分堅定，近幾年來，央行已經不斷放鬆人民幣匯率管制，放寬人民幣匯率波動區間。既然要由市場在資源配置中起決定性作用，由市場供需來決定人民幣匯率，人民幣匯率勢必雙向波動。縱觀國際，任何一種匯率市場化的貨幣，都是有升有貶，人民幣自然也不能例外。就像我們要適應美元、歐元、日圓等國際性貨幣的大幅波動一樣，對於人民幣，我們也應該習慣這樣的新常態。在金融市場上，匯率風險並不可怕，可怕的是完全不進行風險管理。

　　其次，人民幣不斷國際化，越來越多國家和地區開始使用和持有人民幣，參與人民幣匯率市場博弈的供求方日漸增多，人民幣匯率的波動必然更加靈

活。2015 年 11 月 30 日，國際貨幣基金組織（IMF）宣佈將從 2016 年 10 月 1 日起將人民幣加入 SDR 貨幣籃子。人民幣納入 SDR 貨幣籃子象徵着國際貨幣基金組織對人民幣國際化進程的一種認可，也意味着未來主要中央銀行和投資者需要逐步增加對人民幣資產的配置。

再次，在此前的多次人民幣匯率市場化改革中，央行一再強調人民幣匯率不再緊盯美元匯率，而要保持對一籃子貨幣在合理均衡水平上的基本穩定。去年 12 月 11 日，中國外匯交易中心首次發布根據人民幣兌 13 種外幣匯率加權平均計算得出的 CFETS 人民幣匯率指數，意味着觀察人民幣匯率不能再只看人民幣對任何一種貨幣的匯率了。央行的意思很明顯：作為一種國際化程度不斷提高的新興市場貨幣，人民幣有責任保持幣值基本穩定，但是沒有義務對某一種貨幣保持匯率穩定，人民幣已經長大，今後不會再追隨美元或者其他任何單一貨幣的步伐，準備要秀自己的獨立個性了！這也預示着儘管未來人民幣兌一籃子貨幣的匯率保持基本平穩，但是兌美元的匯率彈性很可能會超出我們的預期，管理人民幣兌美元匯率風險的需求格外迫切。

對於香港而言，在人民幣國際化的大背景下，人民幣匯率雙向波動不僅帶來了挑戰，也帶來了全新的歷史機遇。人民幣升值的時候，香港可以大力提供人民幣計價的各種資產，因為升值有利於持有人民幣資產；當人民幣匯率貶值時，香港可以大力提供各種人民幣債務產品，因為貶值有利於人民幣的借債方，我最近看到不少公司在壓縮美元債務、增加人民幣債務，這是市場在人民幣貶值階段的必然反應；與此同時，作為全球最大的離岸人民幣中心，香港可以大力發展人民幣匯率、利率產品，這樣不僅為人民幣國際化提供了更多樣化的動力，還可以完善香港的人民幣生態圈，成為未來全球最重要的離岸人民幣定價中心。這幾天，離岸人民幣價格波動比在岸市場更劇烈，這首先表明離岸市場更需匯率風險管理，同時也表明離岸市場上人民幣計價產品的深度和多樣化發展得遠遠不夠，以至於一個不大的政策變動往往導致離岸市場過大的波動。

在這方面，香港交易所曾在 2012 年邁出了第一步 —— 推出人民幣貨幣期貨，這一產品目前已成為全球交易所市場交投最活躍的人民幣期貨。近期，由

於人民幣匯率波動加大，人民幣貨幣期貨成交量和持倉量均不斷攀升，其中，未平倉合約於 1 月 11 日創下 29,352 張新高，1 月 7 日成交量亦創下第二新高達 6,425 張（名義價值 6.43 億美元）。

但是，我們深知，這距離滿足市場的需求還很遠。就在這個週五，我們將為香港的期貨公司舉辦人民幣匯率風險管理論壇，詳解如何繫好「安全帶」，類似的活動還會陸續展開。今年，我們還將陸續推出一系列人民幣匯率類產品，包括與多幣種貨幣配對產品和雙幣計價的貴金屬系列產品，滿足更多投資者對人民幣匯率的風險管理需求。

最後，再說一次，人民幣匯率已經進入雙向波動時代。如果你是一位進出口貿易商或者經常借外幣的中國企業家，你必須要加強匯率風險管理，繫好「安全帶」了；如果你是一位 QFII 或 QDII 基金經理，不繫「安全帶」的時代已經結束了；如果你是一家通過滬港通進行雙向投資的資產管理機構，你恐怕也得時常問問自己：繫好「安全帶」了嗎？別嫌我囉嗦，重要的事情就得講三遍！香港交易所的「安全帶」已經為你們都準備好了！

2016 年 1 月 12 日

40 | 定息及貨幣業務「登山」三部曲

　　光陰似箭，轉眼間我們已經是第三次主辦人民幣定息及貨幣論壇。大家都知道，我們的願景是成為中國國民財富與金融資產的離岸配置中心、定價中心和風險管理中心。要想完成這一使命，香港金融市場就必須要征服股票、大宗商品和貨幣這三座「大山」。

　　二十多年前，我們就開始攀登股票業務這座大山，經過多年的歷練，我們已經非常熟悉香港股票市場這座大山的地貌，如今已登至接近山頂的地方，未來我們的使命是與國內外其他大山互聯、互通和互動；2012 年，我們收購了倫敦金屬交易所（LME），開始爬大宗商品業務這座山，如今已經爬到了商品這座大山的山腰，將來的目標是實現國際國內聯通、期貨與現貨市場聯通；而對於貨幣業務這座巍峨的大山，我們 2013 年才初見其貌，更因為沒有經驗，不敢輕舉妄動，這幾年我們在山腳已完成熱身運動，剛開始向上攀登。我們今天的大會就是聚焦貨幣這座大山。

　　可喜的是，三年之間，人民幣國際化進程不斷深入，人民幣匯率、人民幣定息和貨幣產品市場都發生了翻天覆地的變化。一方面，央行不斷完善人民幣匯率形成機制，人民幣匯率從單邊升值進入了雙向波動時代。今年 10 月 1 日起，國際貨幣基金組織（IMF）將把人民幣納入 SDR 貨幣籃子，意味着人民幣將正式成為一種國際儲備貨幣，也意味着未來主要中央銀行和投資者需要逐步增加對人民幣資產的配置。隨着內地不斷提高直接融資比例（股票和債券市場融資）和內地債券市場的逐步對外開放，目前已位居世界第二大的人民幣債券市場將迎來大發展，不難預見，未來十年人民幣債券市場規模或將增長一倍或

兩倍。

另一方面,內地企業和投資者「出海」投資的步伐也在加快。內地投資者需要全球分散投資配置,包括保險公司、養老金等在內的一些內地投資者已經率先開始關注海外債券市場。不少內地企業在進行海外投資或併購時也需要發行離岸人民幣債券,並管理好匯率風險。

這些變化為香港帶來了巨大的機遇,也為我們攀登貨幣業務這座大山帶來了難得的天時和地利!說到具體的登山戰略,我想大致可以分為三步:一是搭平台,二是發產品,三是互聯互通。

搭平台是最基礎的準備工作。2013 年,我們專為人民幣產品成立了場外結算所,作為一個符合國際監管標準的結算平台,它可以為內地、香港以及海外的機構提供安全便捷的場外結算服務,幫助他們更高效地利用資本和降低對手風險。去年,場外結算所已經具備了為美元/離岸人民幣貨幣互換產品提供結算服務的能力,這一服務擁有廣泛的市場需求,在獲得監管批准後,我們將很快推出這一場外結算服務。未來,我們還將不斷升級和完善場外結算所的服務,為更多人民幣定息及貨幣產品提供結算服務。

登山的第二步是發行人民幣產品,我們要由點及面不斷拓展我們的產品線。匯率產品是我們的第一條產品主線,2012 年我們推出了第一隻匯率產品——美元兌人民幣貨幣期貨合約。自去年以來,我們的美元兌人民幣貨幣期貨合約成交量穩步上升,目前已經成為全球市場上最活躍的人民幣期貨品種。

5 月 30 日,我們馬上就要推出歐元兌人民幣、日圓兌人民幣、澳元兌人民幣及人民幣兌美元期貨四種新的人民幣期貨,滿足更多投資者管理人民幣匯率風險的需求。在今天的論壇上,我們還與湯森路透公司簽訂了合作協議,準備在今年聯合推出全球首隻可交易人民幣匯率指數,綜合反映人民幣對一籃子國際貨幣的匯率變化。去年 12 月 11 日,中國外匯交易中心首次發布根據人民幣兌 13 種外幣匯率加權平均計算得出的 CFETS 人民幣匯率指數,標誌着人民幣匯率不再緊盯某一種貨幣的匯率了。我們即將推出的這隻人民幣匯率指數,將方便市場更加全面和及時地觀察人民幣匯率的變化,稍後我們計劃推出與這

隻指數相關的期貨及期權產品，為市場人士提供更多管理人民幣匯率風險的工具。

除了匯率產品，我們將逐步豐富產品形態，包括在獲得監管許可後於年內推出以人民幣及美元計價的雙幣黃金期貨。這一新的雙幣黃金期貨將進一步豐富我們的人民幣貨幣產品線，讓煉金商、珠寶商以至黃金的投資者更好地管理現貨與期貨、以至人民幣與美元兩種貨幣間的差價與匯率風險。

貨幣產品的另一大主線是利率／信用產品，這條「山路」比較陡峭，我們很難赤手攀爬。我們準備先從債券現貨上入手，我們將利用滬港通的經驗探索與內地市場建立「債券通」，豐富香港的定息類現貨產品。在建立現貨市場的同時，逐步以此為基礎向債券指數授權和債券指數期貨延伸。在此之後，我們會向債券信用衍生品進發。債券信用衍生品將是貨幣大山最為艱難的登頂之路，走完這一步，我們才算真正征服貨幣業務這座「大山」。

任何一種貨幣要成為真正的國際貨幣，都必須經歷國際支付貨幣、國際投資貨幣和國際儲備貨幣三個成長階段。幾年前，人民幣已經成為了一種國際支付貨幣，並逐漸變成一種國際投資貨幣。今年，人民幣即將被國際貨幣基金組織納入 SDR 貨幣籃子，正式成為一種國際儲備貨幣。我相信，這將是我們攀登貨幣業務這座大山的大好時機。登山的過程一定會充滿艱辛，但只要堅持下來，爬上山頂，就能看到最美的風景！

2016 年 5 月 24 日

41 | 迎接香港債券市場的春天

　　隨着香港交易所與中國外匯交易中心的合資公司——債券通有限公司昨日在香港成立，「債券通」開啟的步伐越來越近了。「債券通」是中國金融市場的一件大事，因為它開啟了世界第三大債券市場對外開放和人民幣國際化進程的新時代。「債券通」必將為香港債券市場的發展帶來難得的新機遇，鞏固和提升香港國際金融中心地位。

　　與滬深港通類似，「債券通」以本地原則為本，主場規則優先，儘量讓投資者在不改變交易習慣的前提下便捷參與兩地債券市場，兩地實行監管合作。

　　與滬深港通不同的是，滬深港通同時開通南北向交易，而「債券通」初期先開通「北向通」，即境外投資者經由香港投資於內地銀行間債券市場，未來將擴展至「南向通」。滬深港通設有每日額度，「債券通」無額度限制。

　　由於債券市場的投資者主要是機構投資者，很多參與股市投資的朋友們可能還不太了解債券市場。其實，從全球金融市場的結構看，債券市場才常常是國際資本市場上的「帶頭大哥」，因為它有一項非常重要的功能，就是為利率定價，央行制定的政策利率通過債券市場傳導下來，產生不同期限和風險等級的利率價格，這些利率價格成為人們為其他資產類別（包括股票、房產、大宗商品等）定價的基準。

　　在國際市場上，以債券為主要構成的固定收益類產品是規模最大的資產類別，債券市場的規模遠大於股票市場。統計資料顯示，截至 2016 年底，全球債券市場總規模為近 100 萬億美元，全球股票市場規模為 67 萬億美元。以全球第一大債券市場美國為例，截至 2016 年底，美國債券市場餘額為 39.4 萬億

美元，是當年美國 GDP 總量的 210%。

中國債券市場是世界上增長最快的債券市場之一，在過去五年，中國市場以每年 21% 的年均增長率快速發展，目前已經成為僅次於美國和日本的全球第三大債券市場，總市值達 8.4 萬億美元（截至 2016 年底），僅為 GDP 總量的 75%，無論是參照美國等發達市場的情況，還是根據中國自身的債券市場發展規劃，中國債市未來還有巨大的發展空間。

而且，中國債券市場的對外開放才剛剛起步，截至 2016 年底，境外投資者持有的中國債券總值僅為 8526 億元人民幣，佔比約 2%，遠低於新興市場和發達市場的平均水準。2016 年人民幣正式納入國際貨幣基金組織的特別提款權貨幣籃子（SDR），佔比達到 10.92%，這不僅為國際投資者參與中國債券市場了提供新的動力，也提供了一個有趣的參照指標：如果中國債券市場上的境外投資者持有佔比上升到 10.92% 的水準，中國債市的開放無疑就大大提升了一步；同時，境外投資者持有佔比的上升，也會相應擴大中國債券市場價格的國際影響力，因為這些境外投資者的交易和結算都是最終在中國在岸的金融基礎設施完成的，真正能通過吸引外資來提高中國在岸債市基礎設施的國際參與程度。

從美國、日本等國貨幣國際化的國際經驗來看，一個成熟開放的債券市場往往是大國金融市場崛起的一大標誌，也是一國貨幣國際化成功的基礎條件。因為，只有當國家具備一定的經濟和金融實力後，外國投資者才有信心大規模地投資其債券市場；也只有當外國投資者廣泛使用該國貨幣作為金融市場（尤其是債券市場）上的投資計價貨幣和儲備計價貨幣時，一國貨幣才能成為真正的國際貨幣。

可以說，中國債券市場的成熟和開放，是中國經濟和金融實力不斷提升的標誌，也是助推人民幣國際化的「神器」。即將推出的「債券通」，將創造性地方便更多海外投資者通過香港投資內地債券市場，為內地債券市場的開放和發展迎來一大突破，也為人民幣國際化做出重要貢獻。

也許有朋友會問，中國人民銀行不是早已向境外機構投資者開放了國內銀

行間債券市場嗎，「債券通」還有什麼價值？確實，目前海外投資者既可以通過申請 QFII/RQFII 額度的方式投資國內債市，也可以向人民銀行備案後直接在內地開戶參與內地債市，但這種直接進入中國內地債券市場的開放方式主要適用於為數不多但體量巨大的機構，例如央行、主權投資基金和大型金融機構等，它們有足夠的人力、財力和經驗直接參與在岸人民幣債券交易。而「債券通」像滬港通一樣，主要是為全世界數量眾多但體量較小或對中國了解不多的國際機構投資者服務的，方便他們從海外參與中國債券市場。

過去幾十年，香港一直成功扮演着連接內地與海外市場的「轉換器」作用。這一次，也不例外。內地債券市場已經是全球第三大債券市場，但交易機制與海外成熟債券市場存在巨大差異，今時今日，讓內地債券市場完全照海外市場進行改造是不可能的，另一方面，讓廣大海外投資者改變長期形成的交易結算習慣、並完全改用內地現行的債券市場交易機制和慣例也很困難。更加現實的解決方案是為雙方提供一個求同存異的轉換器，「債券通」正是這樣一個轉換器。在此機制安排下，海外機構像投資國際債券一樣通過各大國際債券交易平台進行交易，債券的結算和託管則通過香港進行，香港金管局 CMU 將作為債券的名義持有人提供結算服務，中國外匯交易中心與香港交易所共同負責投資者入市及交易服務。

也就是說，直接參與在岸債券市場的交易模式與「債券通」模式各司其職、互為補充，會長期共存下去。打個簡單的比喻，能夠直接在內地開戶投資債券市場的大型海外投資機構好比是研究生，而通過「債券通」投資內地債券市場的大量海外投資機構則好比是大學生，暫時沒有興趣或能力參與中國債券市場的海外機構是高中生。雖然一些大學生會畢業而成為研究生，但也總有一些高中生會陸續進入大學成為大學生，因此，「債券通」仍將為連接中國內地與國際市場創造獨特的價值！

對於香港而言，「債券通」也將為香港金融市場的發展迎來難得的機遇。眾所周知，香港過去一直是企業上市融資的中心，股票定價能力很強，但要成為真正有競爭力的國際金融中心，香港還必須具備為債券、貨幣和大宗商品等

債券通開通儀式。

多種資產定價的能力，必須為投資者提供豐富的風險管理工具。由於香港特區政府一貫財政穩健，財政盈餘充沛，香港本地主權債市場規模一直比較有限。幸運的是，我們背靠中國這個高速發展的大市場，可以近水樓台先得月。

如前所述，海外資金將通過香港進入內地債券市場，因此，「債券通」必將為香港帶來更多國際「活水」。而且，未來「債券通」開通南向通後，內地資金也會投資香港離岸債券市場，啟動香港債券市場的發展。這些增量資金進入香港之後，不僅有投資債券市場的需求，也有管理匯率風險和利率風險的需求。通過大力發展人民幣貨幣期貨、貨幣期權、人民幣國債期貨等定息及貨幣產品，我們可以為海外資金提供豐富的風險管理工具，讓他們更加安心地投資內地資本市場。

短期來看，「債券通」不會給香港交易所帶來即時的經濟效益，因為債券交易大部份是在交易所以外的 OTC 市場進行，但從長遠發展來看，「債券通」將為我們逐步完善包括匯率產品、利率產品和大宗商品在內的香港離岸人民幣產品生態圈提供重要的契機，持之以恆，我們就能把香港建設成為國際風險管理中心。

債券通開通儀式。

　　值得欣慰的是，我們幾年前佈局的一些風險管理產品已經初見成效，例如，人民幣貨幣期貨成交不斷活躍，目前已經成為國際交易所市場流動性最好的人民幣期貨產品。今年我們推出了人民幣貨幣期權和人民幣國債期貨試點，下半年還計畫推出支持實物交割的人民幣美元雙幣種黃金期貨、LME 貴金屬合約和人民幣指數期貨。未來，我們還將不斷豐富產品系列，滿足市場需求。

　　當然，新產品和新業務的發展往往需要時日，在剛剛推出時成交一般不會很活躍，「債券通」開通初期成交量可能也不大。但是，積沙成塔，集腋成裘，我們今天邁出的一小步，也許將會成就明天的一大步。

　　「債券通」已經為我們播下了一顆寶貴的種子，讓我們共同努力，辛勤耕耘，等待收穫的季節！

2017 年 6 月 8 日

第五章

大宗商品市場

42 | 香港密鑼緊鼓迎辦 LME 亞洲年會

距離倫敦金屬交易所（LME）亞洲年會僅四星期，香港交易所與 LME 上下同仁為這個首度在香港舉辦的盛會而興奮不已，正密鑼緊鼓地作好各項準備。LME 亞洲年會源於 LME Week，後者是 LME 數十年來的傳統盛會，每年 10 月在倫敦舉行，全球金屬業界佼佼者匯聚一堂、指點金屬商品市場的江山。

今年 LME 亞洲年會首度在香港舉行，已經得到業界的熱烈反響和踴躍支持。我們希望這個盛會能夠成為本地以至亞洲區內商品市場參與者認識 LME 的良機，也希望為全球金屬市場參與者提供一個了解亞洲商品市場發展動向的平台。為了迎接 LME 亞洲年會的開幕，我打算在未來數星期內撰寫一系列「商品專題」，一來可讓大家知道我們發展商品業務的最新情況，二來闡明新業務如何配合香港交易所現有本身的業務。

對香港而言，商品衍生品交易或許仍是新鮮事物，要發展成功的商品業務更是一項巨大挑戰。收購 LME 已讓香港在全球商品衍生品市場中佔有重要的一席之地。今天，我想跟大家分享一下我們如何在此基礎上進一步發展 LME 業務，以及如何借力 LME 撬動更大的商機。

簡單而言，我們的商品戰略主要分為三大目標：

(1) 降低參與 LME 業務的門檻與壁壘，特別是對於亞洲參與者而言；使 LME 現有基礎金屬業務發展得更蓬勃更快速；

(2) 推出及拓展 LME 的新能力和新業務，尤其是進駐香港，將商品交易與結算平台由倫敦延伸至香港；

(3) 在亞洲特別是中國發展戰略夥伴關係，將基礎金屬業務擴寬至其他商品。

　　我們該如何達成這三個目標？我想繼續延用上一篇網誌的房子比喻來一一闡述。上一回説到，我們近年來為香港交易所整體業務所採取的投資及戰略方針就像為房子加固地基、建立防火牆、搭建新樓層和新平台、裝修內部微結構，力求吸引各方賓客的到來。收購 LME 猶如為我們的房子添蓋一棟新翼——「商品翼」。接下來我想談談我們如何整合這座新翼，確保 LME 成為集團整體不可分割的一部份。

　　LME 這座新翼已有相當完善的「西翼」，也就是以倫敦為本的 LME 已發展成全球首屈一指的基礎金屬交易中心。雖然 LME 也逐漸發展起亞洲業務，但它尚需要構建一個可與整棟「房子」完全連通的「東翼」，具體需要體現在：

- 亞洲（特別是中國）參與者若要買賣 LME 現有產品尚受到倫敦及亞洲的政策限制及其他制約；

- LME 基礎設施及產品發展能力尚未足以支撐其在亞洲發展成功的、可持續的業務；及

- LME 尚未與中國商品衍生品市場的主要參與者建立起戰略合作關係。

　　因此，我們戰略的第一階段是改善 LME 現有的亞洲業務。這個階段並不需要大量財務投資或重整基礎設施，而是要讓 LME 的交易平台更簡便、更容易進入、更能配合亞洲客戶的需要。我們的首要工作是制定一個亞洲時區的價格發現機制，更好地服務亞洲用戶。這項工作進展良好，希望不日就可以跟大家分享更多詳情。除價格之外，亞洲客戶進駐東翼的第二重門檻是必須在倫敦設有辦事處才能成為 LME 會員。針對這個問題，我們正積極研究接納亞洲參與者的更佳途徑。同時我們亦籌備在倫敦推行人民幣結算及在亞洲拓展倉庫網絡。我們還要加大力度向亞洲用戶推廣 LME——這正是 LME 亞洲年會的目標。

　　第二階段的戰略就是要構建香港交易所商品新翼的基礎設施。這個階段需時較長，亦要花上相當的人力與財力，但將來的回報不僅是財務上的，而且是戰略意義上的，所以我們認為這是值得的。其中一項重要投資，就是成立 LME 結算所（LME Clear）。目前，LME 要通過另一家結算所進行結算及交收，換句話説我們不能完全掌控自己的發展步伐，特別是新產品推出的速度和

靈活性。LME Clear 成立後，我們在結算上有自主權，因而新產品推出市場的時間便可提速，相關收入也歸我們所有。從財務方面看，LME Clear 明年第三季度一經推出亦可產生可觀財務收益，預期很快就可收回建設成本。此外，我們計劃將 LME 的資訊技術由現行的外判制改為「自營制」，提高效率之餘，亦希望將 LME 的資訊技術變成一項戰略資產。最後，我們亦會斥資打通香港本地的交易及結算系統，確保商品交易在香港暢通無阻。

說到這裏，我們就進入商品業務的第三個戰略 —— 吸引更多客戶。這個戰略包括建立跨地域產品相互上市、授權的安排，以及與其他商品市場的主要參與者和機構建立戰略夥伴關係。特別是，基於內地市場龐大以及內地走向國際的需要，內地商品市場的主體機構將是我們的主要合作夥伴之一。在這些夥伴關係的帶動下，我們的發展始於基礎金屬，但還會進一步延伸至其他商品。現時我們在探索發展的產品有焦煤及鐵礦等金屬相關產品；長遠來看，我們希望拓展到軟商品和農產品方面的產品。屆時，我們的「東翼」對新舊客戶的吸引力將大大提高，使用率自然也將進一步提升。同時，借助我們的「東翼」，我們的戰略夥伴可以實現提速國際化、躋身全球商品業前列的願望，所以我深信這個戰略能創造雙贏合作並取得成功。

以上是我們發展 LME 亞洲業務的藍圖。對香港來說，這個藍圖究竟有什麼意義？我們認為意義非常重大，因為在這個藍圖裏，香港是東西交匯的中心點：投資者通過香港由內地進入國際市場，或由國際進入內地市場。雖則細節尚待研究，但在亞洲時區的定價，以至亞洲時區的結算業務，均繞不開香港；而這邊廂我們引入的商品種類日益增加，相信會為香港以至亞洲的投資者及市場參與者帶來巨大商機。

在我們如火如荼地準備下個月的 LME 亞洲年會之際，我還有很多想法與您分享，並希望能夠藉助這個平台多聽取您的意見，助我們與 LME 攜手在香港共建成功的商品市場。

2013 年 5 月 27 日

43 提升 LME 亞洲基準價

我在上星期的網誌中提到我們銳意擴展倫敦金屬交易所（LME）的亞洲業務、擴建金屬亞洲業務「東翼」的藍圖。今天，我們很高興宣佈其中一項「提升工程」—— 提升 LME 亞洲基準價的釐定方法。我們希望這類改善措施可以令中國以至亞洲地區的「客人」更容易使用 LME。我想藉此談談這次提升 LME 亞洲基準價釐定方法的具體內容和背後用意。

我們收購 LME 的一個推動力源自亞洲（特別是中國）在全球商品市場與日俱增的話語權。雖然 LME 位處倫敦，與亞洲存在較大時差，但透過每天運作 18 小時的 LME 電子交易系統，LME 的早市與亞洲下午交易時段重合，亞洲區參與 LME 的交易活動逐年遞增。亞洲的交易量越高，在亞洲時段發現的價格自然越需要反映亞洲的供求狀況。因此，我們有必要好好審視亞洲基準價的釐定機制。不過，在深入探討前，我想先簡單介紹 LME 的商品定價模式 —— 畢竟在香港這還是頗為嶄新的題目。

LME 是全球工業金屬的主要定價中心，全球有色金屬的期貨合約交易逾八成在 LME 的平台進行。為金屬定價，向來是 LME 的核心功能之一。LME 正式牌價及 LME 收盤價等主要基準價，都是在 LME 交易日的固定時段內，透過位於倫敦的公開喊價交易大廳 —— 交易圈（Ring）進行買賣而「發現」出來。交易圈的公開喊價環節歷史悠久，最早可追溯至 1877 年，到今天流通量仍然非常高。

在 LME 平台「發現」的價格，用途廣泛，既是全球實物交易的指標及基準，也用來計算商品指數以至金屬 ETF（交易所買賣基金）等組合的估值。您

手上的 iPad 所用的鋁、您家中電線所用的銅等等，很大可能就是參照在 LME 的鋁價和銅價進行交易的。此外，LME 的授權倉庫網絡遍佈全球，亦有助確保 LME 價格會反映着全球商品現貨的供求趨勢。

這個基準定價機制經歷多個年代的考驗，時至今日亦一直運作暢順。另一方面，近年來中國經濟騰飛，轉眼已成為世界第二大經濟體，產生了龐大的金屬需求 —— 中國現時在全球有色金屬市場中已成為最大生產商兼最大消費者，佔全球總量約 40%；再加上亞洲其他地區的市場增長，已經直接影響到 LME 的交易模式。近幾年，於亞洲時段內（即倫敦的大清早）透過 LME 電子交易系統 LMEselect 進行的交易量不斷上升。2010 年，於倫敦時間上午 7 時前（亞洲午市時段）透過 LMEselect 成交的交易量較前一年倍增，促使 LME 為 LME 鋁、LME 銅及 LME 鋅推出亞洲基準價，向亞洲地區的 LME 用戶打開方便之門。我們相信，亞洲基準價可更準確地反映亞洲時區內的金屬供求量。2012 年，於亞洲時段內透過 LMEselect 平台成交的三個月期貨交易按年增 13%，促成了今天公告的提升措施。

公告中的提升措施有二：一是 LME 亞洲基準價的公佈時間將作出調整，配合亞洲其他主要商品期貨交易市場的收市時間；二是 LME 將定價時段由 15 分鐘縮減至全新的 5 分鐘時段，使流通量集中在更短的交易時間內，完善亞洲基準價的發現過程。

隨着時間的推移，我相信 LME 亞洲基準價的意義將進一步顯現。首先，隨着中國大宗商品用戶對於為進口工業用金屬而對沖國際價格風險的需求愈來愈大，中國參與 LME 的程度將有增無減。這再加上預期亞洲區內 LME 會員數目將不斷上升，LME 亞洲基準價的重要性將與日俱增。另外，在我們為亞洲用戶度身設計地研發新產品時，也能夠以 LME 亞洲基準價作為這些產品的定價基準並用於結算。所有這一切，都是 LME 修整「東翼」業務的有機部份，目的是要加強 LME 對亞洲用戶的吸引力。

提升 LME 亞洲基準價只是 LME 邁進亞洲的重要一步。展望未來，我們將繼續與 LME 的用戶攜手並肩，協助他們進一步拓展在中國內地以至亞洲其

他地區的業務；同時不斷以創新的方式服務於亞洲區內大宗商品用戶，更好地
滿足他們實物交易與風險對沖活動的需要。

2013 年 6 月 3 日

44 | LME 成交量持續上升的由來

倫敦金屬交易所（LME）最近的公告指 LME 的成交量一再刷新紀錄：LME 表示，5 月的合約成交張數打破了一個月前才創下的歷史高位。相比 2012 年 5 月，今年 5 月鋁成交量升 7%，銅升 10%，鋅升 13%，鎳升 19%，鉬的成交量更是大增了 30%。

我們正在實施計劃將 LME 業務拓展至亞洲及全球，這消息對於我們來說的確是個好兆頭。那麼，究竟是什麼在推動着 LME 成交量上升？何以商品價格下滑，成交量卻不跌反升？

首先，影響 LME 成交量的因素其實會隨着時間而改變。一直以來，金屬業界均借助 LME 來管理金屬價格波動；他們進行遠期銷售時所訂立的傳統合約往往是按固定價格交易，因此交割前往往涉及價格變動的風險，也為此需要對沖。事實上，這正正是 135 年前驅使 LME 成立的主要原因。及至現在，許多公司都會商定以交割月份的 LME 現貨平均價作為交割價，惟由於買賣雙方均難以預測將來的價格，結果雙方都要為在交割時可能出現的不利價格進行對沖，這時候亦得用上 LME 合約。隨着現今監管尺度大大收緊，加上銀行和股東紛紛施壓要求減低信貸風險，促使更多公司採用 LME 合約進行對沖，我們相信這方面的業務還有進一步增長的空間。

其次，金屬一直被其他市場人士視為可以對抗通脹的工具。在低息環境中，金屬提供了一個或會帶來額外回報的投資機會。此外，由於發展中國家往往需要耗用金屬來發展基礎設施及製造業，市場亦往往視金屬為相當於投資中國及金磚國家（BRICS）的工具。

此外，電子交易在推動成交量增長方面所發揮的作用亦越趨明顯。透過 LMEselect 系統進行的電子交易持續增長，亦令電子交易量佔 LME 總成交量的百分比與日俱增。

以上種種因素都是 LME 成交量近年來長期保持增長的原動力。近期商品價格波動加大，LME 成交量的升勢還更凌厲。

的確，近期商品價格下跌，LME 成交量不跌反升。這一點對習慣股票市場的投資者來說可能有點費解，但實情是金屬價格下跌對 LME 的影響，跟股票價格下跌對股票市場的影響並不是同一回事。這裏我試着解釋一下。

在股票市場，每日成交高低很受市場氣氛牽動；例如市場轉為熊市，想投資的人可能就減少，成交量也就下跌。不過，LME 的金屬交易卻截然不同。交易員和投資者若認為當前價格太高，預期日後價格會下跌，他們就會出售期貨，若認為價格偏低，就會買入。換言之，無論 LME 金屬的價格是升或跌，對 LME 來說都有商機。另外，市況波動，從事生產、航運和製造業的公司自然更需要運用 LME 期貨合約進行對沖。這是推動成交量上升的其中一個主因。

在我們着手開展 LME 業務計劃之際就看到 LME 成交量一再創新高，對我們來說是莫大的鼓舞。我們的長遠目標，是將 LME 的業務進一步商業化、引入來自亞洲的新會員、推出新產品及建設自家營運的結算平台。隨着我們按部就班落實各項計劃，我深信 LME 的龐大潛力將一一展現。

2013 年 6 月 13 日

45 | 又一亞洲地點成為 LME 核准倉庫

倫敦金屬交易所（LME）亞洲年會即將在下星期開鑼。在此前夕，我們不僅密鑼緊鼓地籌備盛會，對於 LME 的其他工作也毫不鬆懈。例如，我們今天宣佈了 LME 董事會已認可台灣高雄港作為交割地點，也就是說高雄的倉庫今後可以申請成為 LME 的核准倉庫。高雄是亞洲第九個獲 LME 認可的交割地點，其他地點分別位於新加坡、馬來西亞、韓國及日本。

為什麼高雄及其他亞洲地點的倉庫對 LME 如此重要？雖然大部份期貨合約都不會持倉到交割，而是在合約到期之前已經平倉並以現金結算，但是 LME 的倉庫網絡正是 LME 在金屬業界中的價值核心所在，它確保了 LME 期貨合約的價格始終與現貨掛鈎。在進一步闡釋之前，請容我在這裏先說說 LME 的倉庫網絡——這畢竟是 LME 極具競爭力的重要資產之一。

LME 的七百多個核准倉庫遍佈全球 14 個國家 36 個地點，主要位於北美和北歐，現時東亞地區的核准地點亦不斷增加。LME 核准倉庫是全球最龐大的期貨交易所倉庫網絡，但每個核准地點都經過嚴格的申請及審批程序。要獲得 LME 認可為交割地點，申請港口在處理量、交通配套等等多個範疇均須要達到指定要求。港口獲認可後，港口中的倉庫營運商尚要自行申請個別的倉庫牌照，亦要達到一定標準。LME 會每年覆核這些港口是否仍然達標。整個體制設計嚴密審慎，維持和覆檢工作一絲不苟。

LME 建立和維持這個龐大網絡需要動用許多人力物力，為什麼 LME 願意為此投放大量資源呢？那是因為這個倉庫網絡是將金屬期貨價格與現貨價格聯繫在一起的關鍵一環。當期貨合約越接近到期之時，期貨合約價格就越趨向現

貨價格。儘管最終只有小部份的期貨合約以實物交割來結算,但正是這個進行實物交割的可能性,確保了金屬期貨價格不致過分偏離現貨價格。

我的前一篇網誌曾提到期貨價格緊貼現貨價格十分重要,因為這是生產商與消費者有效對沖金屬價格波動的基礎。也就是說,正正是這個倉庫網絡將LME 與金屬業實體經濟緊緊地捆綁在一起。

那麼,LME 為何選擇在現在這些位置布局倉庫地點,而不是在非洲、拉丁美洲這些主要金屬供應國?這是因為作為交割地點,倉庫必須毗鄰金屬消費方,即利用金屬製造下游產品的地點。過去,北美和北歐是全球主要製造中心;今天,愈來愈多的製造業已遷往亞洲。與此相應,LME 在亞洲的核准地點漸多,比如高雄等地。中國經過 20 年的經濟騰飛,如今成為全球第二大經濟體,已經是名副其實的「世界工廠」,亞洲的倉庫要比世界其他地方的倉庫更能為亞洲、特別是中國的金屬消費者提供便捷的服務。當然,最理想的還是最終 LME 能夠直接在中國內地設立核准倉庫。

下星期就是 LME 亞洲年會,相信屆時對 LME 倉庫這個課題會有更充分的討論。長遠來看,我們對於亞洲商品業務有着許多計劃,但倉庫對 LME 定價的根本角色將不會動搖。

2013 年 6 月 17 日

46 從 LME 亞洲年會看倫敦、香港及中國內地商品行業的未來

　　本週，首度在香港舉行的倫敦金屬交易所（LME）亞洲年會隆重開幕。來自世界各地的金屬生產商、用戶、交易員、內地交易所及 LME 會員雲集香江。星期二晚舉行的重頭戲 —— LME 亞洲年會週年晚宴 —— 更是一票難求，售出逾 900 張門票。對於並非傳統商品市場的香港來說，這是極具意義的時刻。

　　我們與業界聚首一堂，互相交流意見，一片熱鬧歡騰，但在交流過程中，我們也認識到前路還遠，要完成的事還有很多。我們當初收購 LME，就是因為我們看到在這個平台上有廣闊的發展天地。如我在 LME 亞洲年會上所說，我們要把倫敦、香港及中國分別定位成三大支柱，即三個 P：產品（Products）、平台（Platform）及參與者（Participants）。

　　第一，LME 有着成功發展商品業務所必要的專才、知識產權、品牌及會員。因此，從定位來看，LME 將是我們開發商品產品的核心基地。今天，LME 已是全球金屬交易業界的佼佼者，也是全球商品市場的定價中心，市場份額高逾 80%，這在自由市場中可謂極之罕見，但 LME 卻做到了。這證明了 LME 的業務模式有着很強的核心競爭力。作為 LME 的新東家，我們想進一步鞏固這個領先地位，包括建設自營結算所 LME Clear、將 LME 的 IT 系統從外包轉自營、建立交易資料儲存庫，以及大幅提升全球投資者、交易員、生產商和用戶進入 LME 平台的電子交易平台。所有這些，都是為了獲得更靈活的空間來發展更多種類的產品。

　　第二，我們的香港業務從定位上則是在亞洲推出新商品產品的交易與結算平台。我們的短期目標是確保香港市場作好發展商品業務的準備。鑒於發展實

物交割基礎設施的複雜性與長期性，我們近期計劃在香港主要着眼於現金結算及每月到期的商品合同。這樣我們就可以充分利用現有的股票衍生品交易與結算平台，稍加升級改造便可延伸到商品期貨。從基礎金屬的選擇上，我們將追求增量，重點考慮可以與其他市場互補的產品，同時在合適時機推出基礎金屬以外的其他商品產品。在合適的商品上構建實物交割能力將是我們努力爭取的長期目標。

第三，與內地市場的互聯互通是我們發展流動性和參與者的動力，而後者是發展商品業務所必需的。換言之，我們希望可以把香港逐步打造成一個國際產品可以接觸內地流量、內地產品可以接觸國際流量、國際流量與內地流量逐步實現互聯互通的商品發展平台。

我清晰地認識到這個計劃非常進取，甚至可能是帶有理想主義，但作為一個樂觀主義者，我不認為這是不可實現的。

也正因為這樣，市場對我們的戰略計劃不無懷疑與憂慮。具體來説，倫敦、香港及中國內地可能分別有三個 C 的疑問：

(1) 倫敦的 LME 會員顧慮我們會從根本上「改變」(Change) LME 的業務模式；

(2) 我們香港的同行和朋友可能對這個「鴻圖」的成功缺乏「信心」(Confidence)；

(3) 內地交易所的同仁可能會審視我們的計劃是會帶來「競爭」(Competition) 還是「合作」(Cooperation)。

我下面想就這些顧慮一一給予誠懇的回答。

對於倫敦的 LME 會員，我想強調一點：我們並無意根本改變 LME 的業務模式，因為這模式已經證實非常成功；我們並無計劃改變交易圈 (The Ring) 的模式；我們亦同意至少在 2015 年 1 月前都不會增加收費或改變會員架構。大家知道，LME 並不是一個傳統意義上的交易所。在收購之前，它是一個由基礎金屬生產、交易、融資、倉儲等方方面面參與者共同擁有的交易平台。LME 的整個生態環境具有高度競爭力與黏力，是世界上最成功的業務模

式之一。大家可能對 LME 生態圈有不同觀點，甚至有人認為它不公平，但無可置疑，它已是最廣泛地被全球基礎金屬界所接受的定價中心，LME 的價格也成為了全球基礎金屬的基準價格。

正因如此，香港交易所選擇對 LME 進行全面收購。這一收購定價的前提是，香港交易所將會把 LME 從一個會員擁有的非盈利機構轉型成一個完全商業運作的交易所。儘管我們不會從根本上改變 LME 生態圈的內在邏輯，但是我們要為已經付出的投資爭取合理公平的商業化回報。因此，我們會在 2014年 9 月推出商業化運作的結算服務，並在 2015 年 1 月後重新審視現行收費。不過，我們亦會兼顧 LME 生態圈裏最重要的 DNA，並投資交易系統提供更好的服務與支援。我們收購 LME 絕不是要破壞價值，也不是要重新分配價值，而是要創造價值。我相信我們可以做出更大的餅，使 LME 整個會員社羣與LME 共同獲益。

除了基本金屬外，我們將會在保留 LME 生態圈 DNA 的基礎上，開發基礎金屬以外的其他商品。特別是在亞洲拓展時，我們會對 LME 原有業務模式作出必要的調整，以更好地符合亞洲模式。相信這些新機遇也是 LME 會員的新增長點，我們期待與他們並肩共同開拓新市場。

對於香港的同行和朋友，我想就他們經常問到的問題談談自己的想法。經常有香港朋友問我，為什麼我有信心內地與國際市場互聯互通的第一站會在香港。其實我的道理非常簡單：內地之所以選擇香港，是因為它的「一國」；國際投資者之所以相信香港，是因為它的「兩制」──「一國兩制」正是香港的核心競爭力所在。

內地逐步開放的過程不無風險，需要在相對可控、放心的環境下進行嘗試。香港作為中國的一部份，正好提供了一個理想的試驗場所。而且，香港已經在內地證券市場開放中擔當過排頭兵的重任，對於內地而言是「一塊熟悉的石頭」，所以可以放心地「摸着熟悉的石頭過河」。

另一方面，在「兩制」之下，香港有着公平兼具透明度的監管環境、引以自豪的公平法制、完善的經濟制度以及發展金融所需的專才和能力 ── 這些

都是香港倍受國際投資者青睞的原因，因為他們認為香港能給他們提供保護和便利。

所以說，「一國兩制」是香港的核心競爭力：若沒有真正的「一國」，香港就不會有承接內地發展東風的機會；若沒有真正的「兩制」，香港就不可能保持其獨特的優勢。所以說，我們既要堅持「一國兩制」，也要對自己的角色和地位抱有信心。

對於內地的交易所同仁，我理解他們對於香港市場發展所帶來的潛在競爭的顧慮。我的看法是：若只談合作而不承認競爭，那是不誠實的；但若只著眼於競爭而忽略了合作，那是缺乏遠見的。從香港交易所來看，我們的戰略非常清晰，即不在存量上競爭——因為這是惡性競爭，只會讓雙方都「不爽」。我們要共同合作、創造增量，因為在這樣的增量上競爭才是雙贏的競爭。打一個通俗一點的比喻，人家碗裏的肉，我們不搶；人家鍋裏的飯，我們不爭；人家田裏的菜，我們不想；甚至人家未來的地，我們不看。我們只希望與內地同仁共同尋找那些只有雙方共同創造才能得到的增量，並且在這種增量中形成雙贏競爭。

談到競爭，不可迴避地大家會問及如何看待內地與香港建設國際定價中心。我認為，真正的定價中心必定是買賣雙方能充分參與的中心，而且所發現的價格是在實體經濟中貿易活動可參照引用的價格。

中國在過去多年來一直是量的製造者、價格的接受者，但苦於未能找到把量的影響力轉換成對價格的影響力的場所。若能夠在香港這個「家門口」建成一個買賣雙方都能參與並發揮各自影響力的生態環境，則有望大幅提速內地國際化的進程，讓內地儘早實現與國際的聯通。而且，一旦有了切實可行的香港模式，就可以把它複製到國際其他主要市場。也就是說，中國市場的國際化既可以在條件允許的情況下直達其他國際市場，也可以在條件尚未成熟的時候選擇中轉香港；香港是中國市場國際化的起點站而不是終點站。如此，中國這趟列車就可以由香港開往紐約、倫敦、芝加哥等等。最終，隨着主體市場的發展和制度制約的逐步消除，定價中心自然會回歸離消費者最近、成本最低的內地

市場，而香港在此當中的歷史作用是不斷成為內地逐步推動開放進程時的前沿
陣地，並最終與內地主體市場相互呼應、互為補充，同時也取得了發展的空間。

　　總之，歷史的巨輪給予香港獨特的機遇。我對香港的錦繡未來充滿信心。

　　最後，我想再一次感謝抽空出席香港首度舉辦的 LME 亞洲年會的各位來
賓，我很榮幸能與大家會面交流，從中獲益良多。我希望大家都渡過了美好的
一週，也喜歡香港這個城市。這是香港日後舉行 LME 亞洲年會的序幕，我相
信在各位的支持下，這個一年一度的盛會只會越辦越好！

2013 年 6 月 28 日

47 | 淺談 LME 倉庫排隊提貨現象——表象與現實

在過去一星期的倫敦金屬交易所（LME）亞洲年會中，LME 全球倉庫網絡仍然是話題焦點之一。說到倉庫，LME 剛剛認可了台灣高雄港為亞洲第九個交割地點，市場對此反應非常熱烈。加上 LME 近期提升了亞洲基準價，我相信 LME 在致力鞏固其環球金屬市場定位的同時，將可以為亞洲用戶提供更好的服務。

另一方面，也許大家多少都聽說到近來市場對 LME 倉庫網絡有一些批評的聲音，特別是有關從倉庫提取金屬排隊時間太長的問題，大家已經討論了好一段時間。我很高興地告訴讀者，LME 董事會已經公佈 LME 將正式進行市場諮詢，廣泛徵求業界的意見，共同尋求合適的解決方案。

我們在倫敦的會員及業界對這個話題已經比較熟悉了。雖然市場至今還沒有共識，甚至對是否需要諮詢、應該諮詢什麼也可能有不同的意見，但是大家對這個問題以及建議方案是比較了解的，我誠心希望諮詢反饋能最終幫助 LME 作出正確的決定。

對於香港交易所的非金屬期貨業界的朋友們，關於倉庫網絡的辯論可能還是一個嶄新的話題，其中的技術細節也有點晦澀難懂。雖然對亞洲的影響可能比其他地方小，但事關 LME 乃至香港交易所這個新東家的聲譽，所以我想在這裏稍微闡述。下文將按照六大問題來一一作答，讀者可以按照自己的了解程度與興趣來選擇閱讀。

一、什麼是 LME 倉庫排隊提貨現象？

由於 LME 在全球金屬行業中扮演着重要的角色，所以許多金屬生產者及持有者願意把金屬放在 LME 核准的交割倉庫中，換取交割倉庫向客戶開具的倉單證明。如果金屬持有人想提貨，就得註銷倉單輪候出貨。LME 對核准倉庫的金屬出貨量有一定的最低要求，大型倉庫地點每日須至少運出 3,000 噸金屬。不過，有些倉庫由於歷史原因積存了大量金屬（下文會進一步解釋大量金屬積存的原因），當很多金屬持有人同時要註銷倉單提貨時，他們就得排隊了。在一些地點，註銷倉單等待提貨的要求越來越多，輪候時間越來越長，甚至長達數百日。

值得注意的是，在等候提貨的長隊當中，大部份都並非工業用家，而只是金融投資者為了減低倉租而把金屬從一個倉庫轉到另一個倉庫來。事實上，在 LME 的核准倉庫網絡中，只有 5 個大型的重要倉庫地點因為上述情況而出現倉庫提貨輪候時間達到百日以上，其他絕大多數的倉庫地點是完全不需要排隊的。亦正因如此，才引起了金屬用家和新聞媒體的注意。

二、導致倉庫排隊提貨的宏觀因素有哪些？

在經濟衰退的環境中，例如 2008 年的全球金融海嘯，市場對金屬需求量自然地會下跌。為應對需求放緩，生產商可以選擇減少產量，但是對於鋁冶煉廠等難以馬上停產的公司來說，更經濟可行的做法是繼續生產、把賣不去的金屬寄倉，然後等需求量逐步回升之後再消化掉多餘存貨。

近年，全球經濟放緩，極端低息的環境又持續多年，令生產商可承擔更長的存貨期。再者，鋁的期貨升水（即期貨價格高於現貨價格，「Contango」）吸引了金融投資者利用金屬投資來套利。這種種因素加起來，使得各金屬——特別是鋁金屬——不尋常地積存了大量存貨。而 LME 這個遍佈全球的核准倉庫網絡正正是持有金屬存貨的最佳地點。

三、倉庫排隊提貨現象是否是一件壞事？如果是的話，對誰而言？

就我所見，市場對於倉庫提貨輪候問題的憂慮主要集中在以下三個方面：

第一，輪候時間過長，令工業用家無法及時提貨，進而影響實體經濟生產；

第二，輪候時間推高金屬整體成本，特別是鋁價；

第三，在個別倉庫地點，要求即時提貨所要付出的溢價佔金屬總價格的比例過高。

這三點憂慮是否成立，我下文將逐一回應。但從對實體經濟的影響程度來看，這三個問題的嚴重程度從重到輕，所採取的應對措施也應該有所區別。我曾戲言，視乎以上三點是否成立，我們需要分別用上「火箭炮」、「外科手術」或者「中藥」來解決。這裏讓我嘗試逐一分析，看看需要動用哪一種解決方案。

關於第一個問題，這裏有一個很重要的概念，就是金屬輪候出倉的問題從來都沒有妨礙實際金屬用家取得金屬。事實上，真正需要使用金屬的廠商，一般不會在 LME 倉庫提貨，而是直接從生產商買貨，再利用 LME 對沖價格（詳情見我之前關於 LME 成交量的網誌）。換句話説，誰想要金屬都可以拿到金屬。堆在 LME 倉庫裏的金屬，往往只是後備供應，以供不時之需，也確保 LME 價格能夠真實反映實體經濟的供求情況。其實，從來沒有人跟我們説無貨可用，只是消費者需要支付溢價，這個我回頭再談。所以説，第一點其實並不成立，我們並不需要「火箭炮」。

至於第二點，所謂輪候時間推高金屬價格，這其實亦不是問題。鋁價現時大約低於歷史高位 47%，反映鋁生產商供大於求；其他金屬價格亦同樣低於各自的歷史高位。沒錯，你或會説若生產商沒有囤貨，金屬投資者又沒有插手參與投機，金屬價格或者會更低。不過，長遠而言，待市況好轉，金屬價格可能再度攀升；忽冷忽熱的金屬價格未必對市場有利。總而言之，金屬價格與輪候時間其實沒多大關係，金屬市價始終是取決金屬在公開市場上的供應量。所以，「外科手術」也是不需要的。

最後來看第三點，即剛才提到要付予生產商的溢價。這溢價一般只是

LME 價格的一個很低的百分比，只有那些需要在個別地點即時提取金屬的參與者才需要支付；溢價高低視乎個別地點的供求而定。溢價向來是市場的特徵之一，波動幅度可以很大；但近年輪候提貨者有增無減，溢價也就一漲再漲。舉例來說，現時美國中西部的鋁合同溢價已經高達 12%，比三年前高出一倍。

　　金屬溢價與倉庫輪候時間關係密切。過去，LME 倉庫網絡對金屬溢價起到了平衡器的作用：若生產商索取的溢價過高，套利者就會在 LME（按 LME 的價格）購買金屬，然後從 LME 倉庫提貨，從而避免給生產商支付過高溢價。這樣，LME 價格就會上升，溢價的幅度就會收窄。不過，當前由於若干主要倉庫都排起長隊，所以這樣的套利行為較難實現，導致了溢價持續上漲。溢價的走勢去向，市場上有廣泛的追蹤報導，金屬的「總成本」（LME 價格加溢價）是比較透明的。但我們完全理解用戶希望 LME 價格與「總成本」之間的價差愈少愈好 —— 畢竟，溢價是不易對沖的。

　　我一直認為，過高溢價對市場健康發展很不利，就像紅腫一樣，雖無生命之虞，但讓人非常不舒服甚至可能產生長期後遺症。但這個既不是「火箭炮」、也不是「外科手術」能解決的。相反，在這個情況下，「中藥」可能更為合適。這也就是本次市場諮詢的思路。

四、LME 迄今做了哪些措施來緩解市場顧慮呢？

　　之前，LME 的共識是倉庫排隊提貨現象是宏觀因素所導致，LME 不宜過多干涉。雖然市場有一些聲音認為 LME 如果能夠更早地採取強硬措施，就不至於形成今天的局面，但我相信 LME 一直是依照其歷史沿襲下來的準繩來處理，也無可厚非。當前最重要的是香港交易所、LME 與市場一起向前看、積極尋求解決問題的方案。

　　不過，我們也不妨在這裏溫故而知新，回顧一下近幾年 LME 針對緩解倉庫排隊提貨現象所曾採取的措施。

　　在 2010 年，LME 委託 Europe Economics 對 LME 的倉庫規則進行獨立評

估。評估報告中提出了幾點針對排隊提貨現象的緩解方案，並建議 LME 每六個月進行一次對出貨率的常規評估。LME 聽取了 Europe Economics 的意見，並修改了 LME 的倉庫規則，例如在 2012 年 4 月，把某幾個最大的交割地點的最低出貨率提高了一倍；今年 4 月，進一步增加了針對鎳和錫的最低出貨率；我們又引入了新的要求，讓倉庫運營公司重新審視某些金屬的提貨隊伍對其他金屬出貨的影響。

五、 LME 在本次市場諮詢中有哪些建議？

雖然我們對於「LME 倉庫排隊提貨現象是由宏觀因素造成」的觀點不無認同，但隨着排隊問題，我們想重新看待這個問題。因此，LME 董事會提出了一個縮短倉庫出貨時間的建議，並儘量平衡 LME 會員、倉庫營運商及金屬業界各方的不同需要。

我們的目標是給出一個不同於目前制度安排的備選方案，這個方案可以：

(1) 通過引入新的基於「入貨率」而調整的「出貨率」，防止當前的排隊現象繼續惡化；

(2) 針對「註銷倉單」重新排隊的行為，引入新的「出貨率」，從而減少當前的隊伍長度；

(3) 通過「祖父條款」(即豁免新例生效前的行為的條款)，避免懲罰了那些依據之前 LME 規則來作出投資和運營決定的倉庫公司。

最後，我們希望確保新的制度規則將有足夠彈性，可以消除任何意料之外的後果。如果大家有興趣了解更多的詳情，不妨閱讀諮詢建議全文。

在制定建議的時候，LME 董事會考慮到了實際操作上的局限。例如，金屬都是重物，運送不易，要求倉庫大增出貨率是不可行的；無論運送卡車還是鐵道吞吐量等等，都難以在短時間內配合出貨率的大幅提高；而現有 LME 規則對倉庫的結構布局以至地點分布都有清晰的要求，因此調整倉庫也非一朝一夕就可以完成。在充分考慮到這些現實的局限後，我們現在建議，只要求提貨

輪候時間逾 100 日的倉庫必須將進貨率與出貨率掛鈎，以確保提貨輪候問題不會惡化。假以時日，我們希望提貨輪候問題可以得到改善。

我想重申，董事會現時對任何調整方案並無定見；只有市場願意，我們才會進一步落實這個備選方案。

六、我們期待市場做什麼？

我們現階段雖則未有具體立場，但仍決定開展市場諮詢，是因為我們認為聆聽市場意見、及時回應市場人士的關注乃十分重要。作為 LME 的新東家，我們希望可以積極回應用戶訴求，處理市場關注的問題。

只有在我們誠心誠意地聆聽市場聲音的前提下，市場諮詢才有意義。事實上，我們正是持着這樣的態度開展諮詢的。我們特別希望能聽到深思熟慮的、詳盡的反饋意見。如果您對建議方案持反對意見，我們希望能夠聽到為什麼反對、如果推行了可能有哪些後果等等。如果您是投贊成票，那麼您的詳盡理據也會是非常有益的。

所有的反饋意見將會被嚴格保密（除非您希望公開）。在香港，市場諮詢這一操作由來已久，並且對我們不斷改善市場結構起到了重要的作用。我們希望在倫敦能延續這一傳統。

現在來判斷諮詢結果會否帶來任何改變還言之尚早。但我深信，無論結論如何，LME 的規則總是本着一個原則，就是向金屬業界提供最佳的實物供應網絡，並且以此來保障真正全球的、真實的 LME 價格。

2013 年 7 月 1 日

48 新產品，新起點——
LME 亞洲年會來了

時間過得真快，去年香港交易所首次舉辦倫敦金屬交易所（LME）亞洲年會的盛況仍然歷歷在目，轉眼我們就迎來了 LME 亞洲年會 2014。今年的 LME 亞洲年會規模比往年更大，活動更多，相信一定也會更加精彩。

通過 LME 亞洲年會，來自世界各地的生產商、消費者、金屬交易商、內地交易所以及 LME 會員濟濟一堂，共同探討全球、特別是中國的市場發展趨勢，尋求新的機遇。隨着商品市場的重心逐漸從西方轉向東方，香港擁有足夠的優勢，成為一座連接買賣雙方、連接亞洲與世界、連接中國內地與海外市場的橋樑。

今年 LME 亞洲年會的主題正是「環球聯繫」，本週將有一系列豐富多彩的活動圍繞這一主題展開深入探討。去年我曾在這裏談及三個 P：產品（Products）、平台（Platform）及參與者（Participants）。LME 能為我們帶來專才、會員和「產品」，香港則是為亞洲市場提供產品的「平台」，而「參與者」則來自中國內地。香港將成為促成這種互聯互通的樞紐。

今天，我們將向實現這一願景邁出關鍵一步。我們的亞洲商品團隊下午剛剛宣佈了即將推出四隻新的商品合約的好消息，這四隻產品分別是：倫敦鋁期貨小型合約、倫敦銅期貨小型合約、倫敦鋅期貨小型合約以及 API 8 動力煤期貨。在全球金屬市場，倫敦的鋁期貨、銅期貨及鋅期貨一向是成交最為活躍的品種，正因如此，我們才選擇了它們作為首批在香港交易結算及交收的商品產品。

第四隻新合約是 API 8 動力煤期貨合約。對於香港交易所而言，這一期貨

合約的推出極具戰略意義，因為中國內地現時已是國際市場上其中一個最大的動力煤用家，與亞洲其他用家合計佔全球消耗量逾五成＊。我們希望更多動力煤用家可以使用這一新產品作為管理風險的對沖工具。

如果得到監管部門批准並且市場準備就緒，這四隻合約預計將在今年下半年推出。其中，三隻金屬期貨將以人民幣交易，動力煤期貨則以美元交易，全部實行現金結算，在香港進行結算及交收。如上所述，這些產品長遠來看具有重大的戰略意義和市場需求，儘管短期可能仍然需要一個培育市場的過程。我們有耐心、也有信心能夠在香港做好這些產品。

值得高興的是，目前已有跡象顯示商品業務開始在香港扎根，例如渣打銀行去年曾將其全球金屬交易主管由倫敦調來香港；在我們收購 LME 之後，一些銀行也開始擴大在香港的金屬交易部門，他們認為中國在商品市場的定價權未來將與日俱增，而且中國的金融市場將逐漸加快開放步伐，滬港通計劃的宣佈再次佐證了這一點。

不久前，中國證監會與香港證監會發佈聯合公告，原則上批准了滬港通計劃，允許內地投資者和香港投資者分別投資香港和上海市場的合資格股票。滬港通計劃的推出不僅標誌着中國金融市場對外開放邁出了一大步，也是香港金融發展史上的一個重要里程碑。

滬港通如果得以成功推出，它必將也會對中國內地商品市場的國際化和香港未來商品發展的方向提供重要啟示。隨着中國經濟不斷崛起，中國已經成為全球很多大宗商品的主要消費國和進口國，如何讓中國內地商品市場走向國際化，如何讓中國在國際商品市場擁有與其經濟實力相匹配的話語權和定價權，已經成為所有中國期貨界人士追求的共同夢想。

＊註：　根據 2013 年 BP 能源調查報告及一家主要證券商的研究報告，全球五大用煤國（中國、美國、印度、俄羅斯及日本）的用煤量已佔全球煤炭消耗總量的 76%。中國也是動力煤的主要用家，佔全球消耗總量逾 45%（中國國家統計局數字），與亞洲其他主要經濟體（例如日本及南韓）合計佔全球消耗量逾五成（美國能源信息署資料）。

中國內地商品市場的國際化路徑有多種選擇，可以選擇讓自己的產品和價格「走出去」、將國際市場的產品和價格「請進來」（產品互掛），也可以選擇將國際用家和投資者「請進來」（例如原油期貨），或者讓內地用家和投資者「走出去」（例如滬港通模式）。香港交易所將一如既往，繼續與我們的合作夥伴一起通過優勢互補尋求共贏方案，以期為內地商品市場國際化提供多樣的可能性，推進兩地商品市場互聯互通。

重新說回今年的 LME 亞洲年會，此次 LME 亞洲年會的規模盛況空前。本週四即將開始的金屬研討會是年會中非常重要的一項活動，今年將有六百多位來自世界各地的貴賓出席這一研討會，而參加 LME 年會晚宴的貴賓更超過1,400 人，較去年的參會人數多出近五成。

在研討會上，香港特區政府財政司司長曾俊華先生將以香港作為商品交易中心的角色為主題發表演講。而專題論壇也格外熱鬧：除了各位業界專家會探討金屬交易在今後十年的大趨勢以外，我們還有幸請到了上海期貨交易所理事長楊邁軍、大連商品交易所理事長劉興強、鄭州商品交易所副總經理巫克力和中國期貨業協會副會長彭剛參加當天的「行政總裁專題論壇」。他們將與大家分享關於中國內地商品市場如何進一步國際化的真知灼見。如同滬港通一樣，我們將基於共同的夢想、共同的理念、共同的利益與在座商品期貨界的領軍人物探索中國商品市場國際化的藍圖，讓這一偉大夢想在幾年內成為現實。

在本週五年會活動的最後一天，我們將首次舉辦商品界女士午餐會（Women in Commodities Luncheon）。這一活動旨在表彰商品界女性精英的貢獻，並為她們提供與業界其他優秀女性聯誼的機會。

以上只是今年 LME 亞洲年會活動的精彩一瞥。去年 LME 亞洲年會首次在香港成功舉辦，為我們開了一個好頭。我相信今年的年會必將更上一層樓，為香港成為國際性商品中心畫上濃墨重彩的一筆。

2014 年 4 月 22 日

49 | 開啟 LME 新時代

　　最近幾週，倫敦金屬交易所（LME）備受媒體關注：先是美國聯邦地區法院駁回了所有針對 LME 鋁倉庫的訴訟；上週 LME 自建的結算所 LME Clear 成功推出了自己的結算平台 LMEmercury；今天，LME 公佈了明年的收費調整計劃，朝着真正的商業化運作邁出了重要一步。在此，我要向所有 LME 的同事們表示衷心的感謝和祝賀：謝謝你們的辛勤付出，讓我們在商品領域的戰略規劃正一步一步如期實現！

　　也許很多朋友還不太了解大宗商品業務，更不明白 LME 為什麼要大費周折自建結算所，這大概得從我們的戰略規劃說起。在《戰略規劃 2013-2015》中，香港交易所集團明確提出要發展成為一家提供全方位產品及服務、且縱向全面整合的全球領先交易所，並作好準備以把握中國資本項下審慎、加速開放的種種機遇。產品方面，香港交易所集團將覆蓋現貨股票、股票衍生產品、定息產品及貨幣以及商品等資產類別，同時就每個資產類別建立從產品至交易、及至結算的垂直整合業務模式。LME 正是我們進軍商品業務的先鋒。

　　眾所周知，LME 是全球最大的基礎金屬交易所，但在香港交易所集團收購之前，它並不是一個傳統意義上的商業化運作的交易所，而是一個由會員擁有的盈利有限的機構。在不根本改變 LME 業務模式的前提下，如何讓 LME 在全球激烈的交易所競爭中立於不敗之地，如何讓 LME 在集團的發展戰略中發揮應有的作用，如何讓 LME 和香港交易所的優勢互補尋找新的增長點？這些是我們在收購 LME 之前曾經反復思考的問題，我們的答案是大力投資 LME 的平台建設，將其轉型為一間完全商業化運作的交易所，並充分利用它和香港

金融中心的優勢實現內地與國際商品市場的互聯互通。

延伸到整個集團的商品戰略規劃，大致可以分為三步：第一步，實現 LME 的商業化運作，進一步鞏固其在全球金屬業界的領先地位；第二步，將 LME 的商品業務向東拓展，在香港建立商品業務平台；第三步，利用香港平台的優勢實現內地與國際商品市場的互聯互通。

作為 LME 的新東家，我們大力投資 LME 的基礎建設，付出了不少真金白銀，包括建設自營結算所 LME Clear、將 LME 的 IT 系統從外包轉自營、建立交易資料儲存庫，以及不斷完善全球投資者、交易員、生產商和用戶進入 LME 的電子交易平台等等。因為，所有這些都是關係 LME 未來的重要投資，它們會為我們將來豐富產品線打下堅實的基礎，創造更靈活的發展空間。例如，LME 的 IT 系統從外包轉向自營，可以大大提高交易系統的穩定性和可靠性，並為未來系統升級創造主動性和便利。

上週 LME Clear 順利推出自建的結算平台 LMEmercury，則是 LME 商業化運作的重要前提，它標誌着 LME 成為一家提供全方位服務的金屬交易所——不僅能夠提供先進的交易服務，而且提供一流的清算交收和倉位風險管理服務。從此以後，LME 可以自己掌握結算的「命運」了，產品結算再也不用依賴其他的結算所。

更重要的是，LME Clear 將為 LME 和香港交易所集團的商品發展戰略帶來前所未有的靈活性和空間，尤其是我們在推出新產品或者把產品拓展到新的貨幣品種或時區的時候，這些靈活性將顯得彌足珍貴。

除了 LME Clear 的順利開張，LME 公佈的最新收費計劃也是 LME 商業化運作的重要一步。在調整收費標準時，我們充分考慮了用戶的意見，儘量確保新的收費標準公平合理，並在業界具有足夠競爭力。我們無意改變 LME 生態圈的重要 DNA，我們也十分重視用戶的感受，但作為一家上市公司，我們也需要對股東負責，需要為已經付出的投資爭取合理公平的商業化回報。這些回報也是 LME 未來為廣大用戶提升服務的經濟保證，利用這些增加的交易費收入，我們可以繼續加大投入，推出更加多元化的產品，不斷拓展業務和完善

我們的服務。

值得高興的是，經過一年多的努力，我們的付出初見成效。目前，我們已經基本實現了集團商品戰略的第一步，並且正在積極向第二步和第三步邁進，今年 12 月，我們的亞洲商品平台將會如期推出。

我們堅信，LME 的商業化運作將在我們實現內地與國際商品市場互聯互通的進程中發揮巨大作用，也會使這間擁有一百多年悠久歷史的交易所煥發出新的活力。作為中國資本市場雙向開放的首班車，滬港通的各項準備工作進展順利，預計將在下月啟動。我相信，隨着滬港通的成功運行，它的模式有望在未來拓展至商品和其他資產領域。

當然，LME 的商業化如今才剛剛邁出了第一步，實現集團商品戰略的路還有很長，更多的精彩在前方。讓我們一起努力，共同開啟 LME 的新時代！

2014 年 9 月 29 日

50 | LME 進入收穫季

還有幾天，我們就要在香港迎來第三屆倫敦金屬交易所（LME）亞洲年會了，我想在此跟大家聊聊我們的商品業務。

昨天我們剛剛發佈今年的一季報，非常可喜的是，我們的業績在今年一季度增長明顯，這不僅源於成交量的增長，更加源於我們前幾年播下的一些「種子」逐漸開花結果，進入收穫季節。其中最大的一顆「種子」就是我們 2012 年對倫敦金屬交易所（LME）的收購。我還依稀記得當時人們對這項收購頗多懷疑，有人懷疑這一收購是否值得，也有人擔心香港是否適合辦商品交易所，還有人質疑收購價格是否太高。

現在回頭來看，我完全可以理解當時人們的質疑。不過，自我們醞釀收購 LME 的第一天開始，我們的團隊其實一直對 LME 的定位和轉型有着清晰的規劃。LME 是香港交易所集團長期國際戰略規劃的重要組成部份，在中短期內，它也肩負着豐富我們的產品綫和盈利來源的重要使命。在收購完成三年後，LME 正在朝我們之前規劃的方向順利轉型，如今的 LME 已經成為香港交易所大家庭的主要盈利貢獻者之一。

首先，從盈利數據來看，LME 各項業務營收明顯增長，今年首季營收達到 6.47 億港元，佔集團一季度總營收的 23%，較去年同期增長超過一倍 2014 年首季營收為 3.15 億）。其中 LME 營收 4.47 億港元、LME Clear 則貢獻了 2 億港元的營收。LME 營業收入大幅增長主要是因為今年一月份調整了交易費收費標準和去年 9 月份新開業的 LME Clear 運作良好，此外，隨着 LME 的相關訴訟陸續被駁回，LME 的訴訟費用驟降。

不僅倫敦的商品業務蒸蒸日上，我們在香港的商品業務也在不斷進步。早在 2012 年，香港還是全球商品地圖上的無名之輩，如今，香港即將迎來規模最大的一屆 LME 亞洲年會，來自全球 300 多家公司的代表和近 100 名記者將參加今年的 LME 金屬研討會，130 桌 LME 晚宴的席位已被預訂。這是 LME 亞洲年會舉辦以來 LME 晚宴席位預訂最多的一次，屆時近 1,600 位商品界的朋友將歡聚一堂。在今年的 LME 亞洲年會期間，商品業界也將同時在香港舉辦更多交流活動，並圍繞亞洲金屬交易的未來展開深入探討，可見香港的大宗商品交易生態系統正在萌芽。

雖然香港的商品生態系統才剛剛萌芽，但是發展潛力卻無限巨大。去年12 月我們推出了首批在香港交易和結算的金屬期貨合約 —— 倫敦鋁期貨小型合約、倫敦銅期貨小型合約、倫敦鋅期貨小型合約，不久後我們還將推出更多新產品。儘管目前這三隻小型金屬合約的成交量很小，但它們的推出標誌着香港交易所亞洲商品業務的一大突破，如同春天種下的一棵棵樹苗，只要精心呵護和培育，它們終會茁壯成長，變成一片茂密的森林。

現在，已經有越來越多的亞洲公司（特別是中國公司）成為 LME 會員：去年廣發金融交易成為首家中資背景的 LME 圈內交易會員，11 月份招商證券（英國）獲批成為 LME 第二類交易會員，中國銀行和中國工商銀行旗下子公司也都已經是 LME 第二類交易會員，不久前韓國的 Sorin Corporation 和香港的利記集團也成為了 LME 第五類交易會員。利記集團行政總裁陳婉珊女士還是 LME 鋅鉛委員會和 LME Clear 風險管理委員會的成員。

與此同時，中國對外開放的步伐正在加快。就在上週，中國宣佈取消對國企參與海外商品衍生品交易的審批，百多家國企將獲准參與海外商品衍生品交易，參與家數較以往超出三倍。這是中國加速對外開放的又一信號，必將為大家帶來新的機遇。

尤為重要的是，LME 賦予了我們與內地交易所夥伴一起開拓共同市場的能力。滬港通已經成功邁出了連接兩地證券市場的第一步，深港通目前正在緊鑼密鼓的籌備之中。如同我之前所說，共同市場讓兩邊的投資者都可以「在自

己家裏」和「按自己的交易習慣」投資對方市場產品，這一市場可以從股票延伸到股票衍生品、商品、定息和貨幣產品等多種資產類別，未來發展空間巨大。

具體到我們的商品戰略，長遠來看我們在互聯互通上主要有三大需要努力的方向：一是連通中國的產品與海外的交易量，也把國際的產品與國內的交易量連通起來；二是使海外交易量與中國交易量充分連通；三是使國內現貨市場與國際期貨市場充分連通。展望未來，通過 LME 和互聯互通機制的歷史性突破，我們擁有能夠幫助中國加速商品市場國際化進程的絕佳優勢地位。一個國際化的商品市場不僅能為投資者和商品用家提供方便，也能更好地服務中國的實體經濟，更重要的是，它能夠幫助中國取得與其經濟實力相匹配的國際大宗商品定價權。在中國商品市場國際化的進程中，我們希望能夠同樣扮演一個低調但重要的連接器角色。

最後，我熱烈歡迎每位來香港參加 LME 亞洲年會的嘉賓。下週，LME 亞洲年會將舉辦一系列豐富多彩的活動，為各位提供深入探討全球金屬市場熱門話題的機會。我十分期待下週的 LME 亞洲年會，也期待着香港商品生態系統的茁壯成長。

2015 年 5 月 14 日

51 | 再談中國大宗商品市場「實體化」

英國脫歐的消息震動全球資本市場，也引發了不少朋友對於倫敦金屬交易所業務和我們未來商品戰略的關切：英國脫歐事件對於倫敦金屬交易所（LME）的經營會產生什麼影響？會否影響香港交易所商品戰略的實施？LME 是否還會繼續爭取在內地認證倉庫？這對服務內地實體經濟有何意義？……在此，我想一併回答這些朋友們關心的問題，也跟大家分享一下我近期的一些思考。

關於英國脫歐

英國脫歐事件對於倫敦金屬交易所（LME）的經營會產生什麼影響？

英國公投的結果給全球金融市場都帶來了不小的震動，未來英國如何脫歐、何時脫歐目前還存在很大的不確定性。英國離開歐盟，對 LME 的影響比較有限。首先，LME 的金屬交易是全球性的，並非局限在英國或歐洲，會員及客戶遍佈全球，而且，LME 來自歐洲的交易相對比較穩定，未來的成長將主要來自亞洲地區。其次，LME 的交易基本以美元計價，其經營收入也以美元計算，而非英鎊，只有營運成本是以英鎊計價的。LME 有上百年的成功運營經驗，早在英國加入歐盟之前，它已經是全球首屈一指的基本金屬市場，我們有信心維持 LME 系統和市場的正常運作，確保市場公平公正。

英國脫歐會否影響「倫港通」項目的推進？會否影響香港交易所其他商品戰略的實施？

由於「倫港通」涉及倫敦和香港兩地結算所的未平倉合約結算，實施「倫港通」需要與香港、倫敦與歐盟三地市場的監管機構進行相關討論。英國脫歐後我們如何繼續與歐盟監管機構跟進出現了新的不確定性。因此，我們目前正在等待政策進一步明朗，再做下一步推進。

但是，英國脫歐不會影響香港交易所實施其他的商品戰略，比如，我們目前正積極爭取在內地設立 LME 認證的交割庫，以降低內地實體企業參與國際金屬交易的成本，服務實體經濟；也正在積極推進前海大宗商品現貨交易平台的建設，促進內地大宗商品市場的「實體化」。

關於實物交割

LME 為什麼要在中國「設」倉庫？這對於服務中國實體經濟有什麼意義？

首先需要說明的是，LME 不是要在內地建倉庫，而是認證內地倉庫供我們的客戶在實物交割時使用。我們在全球不擁有任何一家倉庫，但是認證了600 多家倉庫。

因為 LME 是全球最大的基本金屬市場，它每天產生的價格也是全球金屬買家和賣家通用的基準參考價格，比如，一個中國的銅冶煉廠要從海外採購銅精礦，它跟海外賣家簽訂的採購合同上通常會寫明某年某月某日的交易價格參照 LME 某天的銅價。這個價格是一個期貨價格，而實物交割機制正是期貨價格變動與現貨價格變動同步的重要前提。因為實物交割機制的存在，同時參與實物交割的各方不受人為限制，期貨價格在合約到期日才會跟現貨價格趨同。從很大程度上說，正是因為 LME 遍佈全球 35 個地區的 600 多家認證倉庫能為

客戶提供方便快捷的實物交割，LME 交易產生的期貨價格才能真實反映全球實體經濟對基本金屬的供求狀況。

　　眾所周知，中國已是全世界最大的原材料消費國，很多中國的金屬用家是從國際市場按 LME 價格購買金屬原材料，因此需要通過 LME 的期貨合約管理大宗商品價格波動帶來的風險。目前來自中國客戶的交易已經佔到了 LME 交易總量的三成以上，但是，由於 LME 目前在內地還沒有認證倉庫，內地的客戶在參與 LME 交易後如果需要進行實物交割，就只能把貨從海外 LME 認證倉庫運回國，或把貨運到海外的 LME 認證倉庫，並為此支付高額的海運費用，交割的時間成本和經濟成本都大大增加。有時，由於交割的困難，內地的客戶甚至會被國際交易對手「擠倉」，不得不讓利平倉。簡言之，因為這些額外的成本，內地客戶在參與國際大宗商品的交易和定價時，已經輸在了起跑線上。因此，近幾年來，國內客戶要求我們在內地認證倉庫的呼聲不斷。

　　允許 LME 認證國內保稅區倉庫，不僅可以降低內地企業國際貿易的交割成本，方便內地企業更加充分地參與 LME 交易和定價，長遠來看，也有助於中國在國際大宗商品市場上贏得與其經濟實力相匹配的定價權。因此，它值得監管機構認真考慮，目前，我們正在積極爭取。

關於定價權

定價權是一個熱議話題，中國是全世界最大的原材料消費國，但是它對定價方式的影響力仍比較弱，你怎麼看定價權這個問題？

　　關於定價權，大家的討論不少，但也存在不小的誤區，一種典型的誤解是認為定價平台在哪裏，哪裏就有定價權。其實，所謂的定價權，一定是買賣雙方都能有效參與和都能接受的一種定價，任何一方不能接受的定價權都是沒有意義的。

　　隨着經濟全球化的深入，中國企業與海外企業進行大宗商品交易的機會越

來越多。我們現在討論的定價權主要是中國買家（或賣家）與海外賣家（或買家）之間交易時如何定價的問題，那麼問題來了：一方面，由於外匯管制的限制，中國企業還不能完全自由地參與國際期貨市場的交易去套保，不能在海外市場的定價過程中充分發揮自己的影響力，同時，內地限制境外交易所設保稅區交割倉庫又人為地增加了中國實體企業保稅交割的成本，造成中國企業按國際價格買進或賣出大宗商品時卻不能用實物交割來對沖價格風險、有效地保護自己，猶如自己綁住一隻手與人博弈。另一方面，國內交易所的交易雖然很活躍，但是海外賣家並未參與其中，他們自然不願意接受國內交易所產生的價格作為定價基準，國內產生的定價在全球市場並沒有影響力，最後雙方交易還得使用海外市場的定價。

因此，在我看來，定價權的核心是買賣雙方都能參與和認可定價過程。無論定價平台在哪裏，只要中國人在這個平台有足夠的影響力，海外交易對手也認可這個定價，中國就可以掌握定價權。

通過哪些途徑可以提高中國在大宗商品領域的定價權？

通過雙向對外開放，中國可以逐步提高在國際大宗商品市場上的話語權。首先，中國可以加快開放內地商品期貨市場，引入境外投資者參與。如果由於資本管制等因素的限制，暫時不能在這方面大踏步邁進，那麼不妨創造條件鼓勵內地企業和投資者走出去，參與國際大宗商品市場。

關於大宗商品交易平台

內地有上千家大宗商品交易平台，香港交易所在前海建立平台的優勢是什麼？

目前內地有超過 1,000 家大宗商品交易平台，但信譽良莠不齊，缺乏嚴格

監管，一些平台利用監管規則的漏洞，以現貨交易的名義推出了一些承諾高收益低風險的產品，誘導許多不明風險的散戶投資者參與，但募集來的資金實際流向高風險領域，最終資金鏈斷裂，引發兌付危機。我們希望在深圳前海建設一個規範、透明、可信賴的、有實物交割體系和倉儲體系的大宗商品交易平台，有效服務實體經濟。

在這方面，我們有三大優勢：一是香港交易所在收購 LME 後具有了在大宗商品市場的專業經驗和品牌；二是香港交易所有動力、決心和資源提供可靠的倉儲、物流等配套設施及系統，有效服務實體經濟；三是香港交易所具有公信力。設立在深圳前海的大宗商品交易平台將通過現貨交割服務實體經濟，並形成新的價格基準，日後香港交易所可進一步使用價格基準，在香港發展指數、期貨及其他衍生產品。

你們可能遇到的挑戰是什麼？

說到挑戰，經常有朋友認為內地監管政策是我們即將面臨的最大挑戰，這其實是一種認識誤區。所有監管政策的目標都是保護投資者、維護市場公平和透明、減少系統性風險。現在出現不少問題的交易場所均是不法商人打着大宗商品交易的名號操縱市場，誤導散戶，非法集資，引發金融風險。而我們在前海建立大宗商品交易平台，要做好人、做好事，建設安全可靠的交易與清算體系，規範市場交易，降低中小企業融資成本，服務實體經濟，推動有效投資，完全符合國家利益和監管目標，因此我相信一定會得到監管者的支持。

在我看來，我們最大的挑戰其實來自我們自身，就是我們是否有決心和遠見，是否足夠了解中國市場。LME 雖然在全球商品市場運作方面有非常豐富的經驗，但是中國市場有其獨特性，不可能完全照搬海外經驗，如何成功將 LME 經驗移植到中國並使之適應中國的這方「水土」，可能是我們未來面臨的一大挑戰。

2016 年 7 月 20 日

52 | 走近前海聯合交易中心

一年一度的倫敦金屬交易所（LME）亞洲年會今年再次高朋滿座，和往年有些不同的是，本屆 LME 亞洲年會除了在香港舉辦一系列精彩活動外，還在深圳前海設立了分論壇。約 300 位海內外嘉賓雲集我們正在籌備中的前海聯合交易中心，探討大宗商品界的熱門話題。

很多朋友都對前海聯合交易中心充滿興趣，向我提出不少問題。由於場地有限，無法邀請所有的朋友來參加我們的前海分論壇，在此我將一些大家常問的問題整理了一下，集中作答，試着帶大家走近前海聯合交易中心。

一、香港交易所為什麼要設立前海聯合交易中心？

大宗商品是香港交易所發展戰略的重要組成部份，設立前海聯合交易中心則是我們大宗商品發展戰略的重要一步。

眾所周知，香港資本市場傳統上是一個股票市場，在過去 20 年裏一直是中國企業首選的海外融資中心，隨着中國經濟的快速崛起和內地資本市場的開放與發展，香港資本市場的功能也開始逐步轉型。在今後的 20 年裏，香港將會發展成為集股票、大宗商品與外匯為一體的全方位國際金融中心。

但是，香港人多地少，缺乏發展腹地，也沒有大宗商品發展的傳統，我們需要尋求符合香港實際、發揮香港所長且滿足國家所需的獨特的大宗商品發展之路。

因此，我們開始探索一個「兩條腿並行」的大宗商品發展戰略：期貨戰

略 —— 與內地交易所互聯互通，互掛互惠；現貨戰略 —— 走出去買、走進去建。

繼滬港通、深港通成功後，我們與內地期貨交易所深入探討互聯互通、互掛互惠模式，考慮過各種合作方式。我深信在不遠的將來，我們一定能在期貨領域找到互惠共贏的互聯互通機制，實現「商品通」。

在現貨戰略上，囿於香港的地域狹小，我們的戰略選擇是「走出去買」和「走進去建」。「走出去買」即海外兼併收購。2012 年，我們成功收購了具有140 年歷史的 LME，走出了併購國際金融基礎設施的第一步。「走進去建」就是回到祖國腹地，與內地監管當局和機構深度合作，利用香港的獨特優勢新建一個扎根內地、服務實體、合規守法的大宗商品現貨市場，打通金融進入實體經濟的渠道，彌補市場空缺，助推供給側改革。

「走出去」和「走進去」戰略的最終目標是利用香港獨特優勢，將「走出去」收購的平台與「走進去」建設的平台有效的互聯互通，利用自己的國際金融基礎設施，加快中國大宗商品市場國際化的進程，真正實現國際大宗商品定價的東移。這對於中國發起的「一帶一路」戰略有着重要意義，「一帶一路」沿線涉及 60 多個國家，預計未來十年中國與沿線國家的貿易額將達 2.5 萬億美元，其中相當比例將是大宗商品資源，發展潛力巨大。對內地而言，如何有效管理商品價格波動所帶來的風險也將直接影響到「一帶一路」大戰略下的許多具體項目的順利推進。「一帶一路」沿線有香港和倫敦兩個國際金融中心，如果我們的「兩條腿並行」的戰略順利實施，那麼，我們就可以把倫敦和香港的國際經驗與中國的實體經濟需求以前海為節點進行對接。

二、香港交易所內地商品交易中心為什麼選址前海？

深港兩地一衣帶水，兩座城市在國家的改革開放進程中均肩負特殊使命和獨特優勢。中央政府在深圳設立前海深港現代服務業合作區，為深港兩地更緊密地融合發展提供了平台。深圳前海也是習近平總書記十八大之後到地方視察

的第一站，他特別指出前海應該「依託香港、服務內地、面向世界」。李克強總理在今年兩會上提出了要研究制定粵港澳大灣區城市羣發展規劃，進一步推動粵港澳深化合作。

然而，作為國家經濟發展主要引擎之一的華南地區，尚無一家全國性的商品交易市場，必須加快補足短板。

因此，在深圳市政府的支持下，香港交易所聯手前海的合作夥伴，共同發揮前海的特殊政策優勢，合資籌建前海聯合交易中心。今年是香港回歸 20 週年的重要時刻，前海聯合交易中心若能在今年正式開始營運，將是「一國兩制」優勢下深港合作的一大成果，這也將是香港的主要金融基礎設施首次在內地設立金融平台。

三、前海聯合交易中心會否與內地期貨交易所形成競爭？

前海聯合交易中心是一個大宗商品交易平台，我們的定位是立足現貨，服務實體，與內地期貨交易所形成互補。

自從 2012 年收購 LME 以來，我們對大宗商品市場有了更深刻地認識。全世界的期貨市場，對於企業客戶來說，其實都是一個小眾市場，永遠只有少數的企業能夠通過期貨直接去做套期保值。期貨的標準化和高槓桿，對使用企業的知識、觀念、管理、資金、人才配備等都提出了較高的要求，也註定了不適合絕大多數的企業（尤其是中小企業）參與。因此也就不難理解為什麼國內參與期貨交易的企業，相對於大量有對沖風險需要的企業數量，少之又少。絕大部份的產業企業沒有得到有效的套保避險服務，這些企業多是中小企業，得到有效的套保避險服務，對於他們的穩定可持續發展是至關重要的。可以說，滿足這些企業的個性化的需求，是服務實體經濟的關鍵，對於推動供給側改革和調整經濟結構、疏通金融進入實體經濟的渠道具有重要的作用。

LME 恰好在滿足企業個性化需求、服務實體經濟方面有着上百年的成功經驗，我們希望借鑒 LME 的成功模式和歷史經驗，在內地打造一個能夠有效

服務實體經濟的大宗商品交易平台，有效彌補這一市場空缺，為這絕大多數的產業企業服務。如果前海聯合交易中心能夠有效地服務這些企業，將能促使更多實體企業在前海聯合交易中心平台上進行現貨交易的同時參與期貨市場的套期保值，與期貨市場形成互補和良性循環，促進期貨市場的機構化和實體化。前海聯合交易中心的初心和理想就是希望從現貨入手，幫助實體企業茁壯成長。也就是說，既然在定位上有明確的差異，前海聯合交易中心與內地交易所定能精誠合作。

四、 前海聯合交易中心和 LME 是什麼關係？前海聯合交易中心準備如何服務實體經濟？

前海聯合交易中心既要借鑒 LME 的經驗服務於實體經濟和產業客戶，又絕對不能簡單照抄照搬，必須切合中國國情。

誕生於 1877 年的 LME，是一家非常接地氣的百年老店，積累了豐富的服務實體經濟的經驗，包括天天交易日日交割的個性化合約、遍佈世界的交割倉庫網絡、分級結算的風險管理制度和會員自律管理機制等。舉個例子，與一般期貨交易所僅提供標準化的月合約不同，LME 提供包括日合約、週合約、月合約在內的 195 個合約，緊貼現貨交易的習慣，全天候滿足企業套期保值交易的個性化需求。因此，LME 不是期貨交易所，在國際市場上它通常被同行們公認為是現貨交易所。

我們在前海設立前海聯合交易中心，就是要創造性地借鑒 LME 服務實體經濟的經驗，服務內地實體經濟。經過二十多年的發展，中國的大宗商品期貨市場取得了舉世矚目的成就，交易量已經佔全球市場的一半以上。海外成熟大宗商品市場大多經歷了上百年的自然沉澱，由現貨市場、中遠期市場逐步發展到期貨市場，而中國的大宗商品市場是個「跳級生」，在現貨市場還不發達的時候直接跳級到了期貨市場階段。現在期貨市場很發達，監管充分，風險管控有效，但現貨市場卻極為分散，沒有統一標準，缺乏有效服務，存在市場發展

嚴重失衡、在國際市場上缺乏足夠的定價權等問題。

　　針對這一現狀，前海聯合交易中心將會借鑒 LME 經驗，以服務實體經濟為己任，培育以機構客戶為主的現貨市場。前海聯合交易中心不做小散戶的生意，只服務機構客戶，尤其是中小產業企業。具體而言，首先，我們要打造可靠的倉儲和便利的物流，建立 LME 式的交割倉庫網絡和行業信用；第二，我們將圍繞企業需求，為大宗商品使用者、貿易商、物流商和金融中介等各方提供安全、高效的大宗商品現貨交易、融資、倉儲物流及供應鏈管理等一系列綜合服務；第三，我們將創新服務模式，最大限度降低企業的資金成本和交易成本，尤其要降低中小產業企業套期保值的成本，為他們提供更加個性化的服務。

五、內地有上千家大宗商品交易平台，前海聯合交易中心與它們有什麼不同？

　　目前內地有超過 1,000 家大宗商品交易平台，但信譽和規範程度良莠不齊。這種現象本身反映了國內大宗商品交易服務無法滿足實體經濟巨大需求的現狀，同時，大規模的發展亂象也為有效監管帶來了巨大挑戰。

　　在我看來，一個成功的大宗商品交易平台必須至少具備以下三大核心要素，缺一不可：

　　一是強大公信力 —— 只有強大的公信力才能吸引市場各方在一個可靠的平台上放心地交易；

　　二是強大的實力及對大宗商品交易的深度認識 —— 大宗商品交易的複雜性、規模化與國際化註定了交易平台必須擁有強大的財力、物力和領導力；

　　三是良好的風險管控能力 —— 作為世界上最大的交易所集團之一，香港交易所同時是三家交易所、五家結算所的運營機構和一線監管機構，業務橫跨歐亞市場，運營着市值超過 30 萬億港元的港股市場和每年成交額高達 10 萬億美元的全球金屬市場。在風險管理領域，我們符合所有國際標準的要求，是全球所有主要市場認可的中央結算對手。

今天一千多家的交易平台中能滿足以上條件的平台寥寥無幾，一個交易平台如果沒有強大公信力，則很難整合現貨交易行業生態，無法實現有效服務實體經濟的理想；如果沒有強大的實力和精耕細作的定力，則很容易走上變相期貨交易、誘使散戶參與的不歸路，最終給老百姓造成損失，也為政府和監管者帶來監管挑戰。

香港交易所與深圳前海共同籌建的前海聯合交易中心具有上述三大優勢。我們希望自下而上地建設一個創新型大宗商品交易平台，踏踏實實地立足現貨，樹立新的行業標竿，在充分管控風險的同時，滿足實體經濟中最迫切的需求。

六、如何看待內地清理整頓各類不規範交易平台？

我們堅決支持監管機構清理整頓各類不規範交易場所，只有清除害羣之馬，才能為行業發展創造良好的市場環境。

監管機構工作的核心是清理與整頓，不是簡單的封殺關閉，目的是為了市場能更好更安全地發展，更有效地服務實體。從這個意義上看，前海聯合交易中心生正逢時。

從表面上看，這次清理整頓聚焦在集合競價、保證金交易、中遠期交易或類期貨產品上，反映了這些產品和交易模式在一個散戶主導的不成熟市場很容易被濫用，特別是在欺詐個人投資者方面。因此，現階段大力整治不僅必要，而且及時，有利於防範風險升級。

與此同時，有識之士也都明白，在一個監管到位、以機構為主的成熟市場，中遠期交易是全球大宗商品交易的最基本模式，本質上是中性的交易工具，是服務實體經濟離不開的必要交易模式。

因此，我們相信清理整頓是一個正本清源的過程，我們也相信監管機構的智慧和遠見。在清理整頓工作結束後，相關的監管制度和規則應該會更加完善和有效。

七、目前的清理整頓會不會影響前海聯合交易中心的正常開業？

前海聯合交易中心今天尚未開業，因為我們還有大量的基礎準備工作沒有完成，目前我們正在加緊準備中。

在今天清理整頓的大背景下，前海聯合交易中心一定不忘初心，合規經營，深耕細作，在推出時間、模式、產品等方面做到合法合規，助力內地大宗商品市場的長期健康發展。

媒體朋友經常問我，前海聯合交易中心到底何時開業？這個答案其實並不重要。由於我們着眼現貨，大量投入和準備都有利於未來長期的發展，因此關注第一個產品究竟何時上線其實並沒有太大意義。

有朋友曾給我講過一個關於竹子生長的故事，最後在這裏跟大家分享一下。據說在成長的前幾年，竹子只能長幾厘米，可能還出不了土。進入第五年，竹子就開始以驚人的神速成長，一天就能長幾十厘米，這是因為竹子利用前幾年的時間已經深深地在土壤裏扎下了根，積蓄了豐富的營養。竹子的成長經歷和信念也適用於今天的前海聯合交易中心。我們對未來充滿信心，我們會踏踏實實、一步一個腳印地迎接「第五年」的到來。

2017 年 5 月 11 日

53 ｜黃金雙城記

今天是個大喜的日子，香港交易所集團的三隻「黃金寶貝」分別在香港和倫敦出生了！這三隻「黃金寶貝」就是今天在香港交易所上市的人民幣黃金期貨、美元黃金期貨和在倫敦金屬交易所（LME）上市的美元黃金期貨。

這是我們第一次在旗下的兩大交易所同時推出兩地雙金，也是我們第一次推出人民幣黃金，也是香港第一次推出可以實物交收的黃金期貨。這三個「黃金寶貝」的三個「第一次」反映了我們商品與貨幣發展的重要組合戰略。推出兩地雙金，可以橫跨時區，影響全球商品定價；推出雙幣黃金，可以橫跨貨幣，引領人民幣離岸匯率與利率發展；推出實物交割，可以橫跨期貨與現貨，實現黃金美元人民幣三價完美合一。

在此，我想跟朋友們分享一下這三隻「黃金寶貝」的出生故事：

一、為什麼推黃金？定價黃金，發展商品。

這是我們提升香港交易所集團大宗商品定價能力的重要一步。定價能力是一個金融中心的核心競爭力，要成為一個真正的國際金融中心，就必須具備為公司、商品、貨幣等多種資產定價的能力。作為重要的避險資產，黃金在全球各大金融市場都是一大重要資產類別，扮演着重要角色。香港交易所在《戰略規劃 2016-2018》中明確提出，要發展成為中國客戶以及國際客戶尋求中國投資機遇的全球首選跨資產類別交易所，在大宗商品業務方面，要根據市場需求增加貴金屬產品供應。

統計資料顯示，中國目前已是世界上最大的黃金生產國和消費國，中國內地每年進口的黃金佔全球實物黃金供應量的四分之一左右，而毗鄰內地的香港近年來實物黃金貿易一直十分活躍，已成為中國內地的主要黃金進口地。如此大的進出口貿易量決定了香港市場存在着管理黃金價格波動風險的巨大需求，完全有條件發展成為亞洲的黃金定價中心。

遺憾的是，由於目前亞洲尚無具有國際影響力的黃金定價中心，黃金的國際定價權主要集中在倫敦和紐約，即使從香港出口往內地的黃金貿易，也要參考倫敦或紐約的基準價格來定價。

我們今天在香港推出的人民幣黃金期貨和美元黃金期貨正是為了滿足業界和投資者長期以來的投資和風險管理需求，我們在倫敦推出的黃金期貨則將豐富倫敦金屬交易所的產品體系和提升倫敦黃金市場的價格發現功能與透明度。

假以時日，我們希望能夠把香港建設成為世界最重要的黃金定價中心之一，形成具有國際影響力的黃金基準價格。

二、為什麼推兩地雙金？橫跨時區，影響全球。

此次推出的兩地黃金期貨合約各有特點，旨在滿足兩地市場不同的需求。例如，香港黃金期貨從香港時間上午 8 時 30 分一直交易到翌日凌晨 1 時，而倫敦黃金期貨則可以 24 小時交易；香港的黃金期貨分別以人民幣和美元計價和交易，而倫敦黃金期貨僅以美元計價和交易；香港的黃金期貨是月合約，而倫敦黃金期貨則涵蓋現貨、日合約和月合約。

儘管兩地黃金期貨產品的設計不同、合約不能相互替換，但都是集團發展大宗商品貴金屬業務的重要戰略部署，籌備時間相近，同時推出可以發揮協同推廣效應，有效利用集團資源。而且，兩者同時推出，有望相互促進，相互呼應，提升跨時區場內黃金產品的交易需求和流動性。

三、為什麼推雙幣金？黃金美元人民幣，匯率利率一起抓。

目前市場上的黃金期貨多以美元計價和交易，我們今天在香港推出的黃金期貨特別增加了人民幣計價和交易合約，主要是為了滿足離岸人民幣的投資需求，為投資者提供更多選擇。

眾所周知，香港目前已經是全球最大的離岸人民幣中心，提供包括股票、債券、人民幣期貨、人民幣期權在內的一系列離岸人民幣投資產品。以人民幣計價的投資產品越豐富，香港作為離岸人民幣中心的優勢就越穩固。隨着人民幣國際化進程的推進和債券通的開通，對於離岸人民幣產品的投資需求也會越來越大，這是香港難得的機遇。我們推出的人民幣（香港）黃金期貨將為離岸人民幣的持有者提供新的投資選擇，也將有助於推進人民幣國際化進程。

隨着人民幣國際化的穩步推進，有效管理離岸人民幣匯率和利率對廣大投資者來說都變得異常重要。將具有特殊貨幣屬性的黃金交易放在美元與人民幣一對合約中同時交易，有利於促進離岸人民幣匯率與利率市場化的健康發展與完善。

四、最大看點是什麼？黃金美元人民幣，實物交割三合一。

如前所述，黃金產品在全球金融市場都是交易活躍的品種，很多交易所都推出了黃金期貨。不過，香港交易所是世界上唯一一個同時提供人民幣、美元雙幣黃金期貨和人民幣兌美元期貨的交易所。這應該是此次推出的香港黃金期貨的最大看點。由於人民幣、美元雙幣黃金期貨到期後均以實物交收，而交收的黃金規格是一模一樣的，因此在交收日兩張黃金期貨合約的內在價值是相等的，這意味着將交收日人民幣（香港）黃金期貨的結算價和美元黃金期貨的結算價相除，可以產生一個人民幣兌美元匯率。

理論上，這個匯率應該與當時的人民幣兌美元期貨結算價十分相近。一旦兩個匯率價格偏離，就會出現套利機會，市場上就會有聰明人和聰明錢通過兩邊交收的形式獲利。對於更加專業的投資者，還可以關注雙幣黃金形成的匯率和人民幣匯率期權產品的價格差，豐富自己的投資策略。因此，通過實物交割形成黃金、美元、人民幣的完美三合一應該是此次推出的香港黃金期貨的最大看點。

最後，我衷心祝願這三隻「黃金寶貝」在倫港兩地茁壯成長，希望它們能為倫港兩地市場帶來新的生機和活力！

我年輕的時候選了文科，最怕人家說：學好數理化，走遍天下都不怕。今天工作在交易所，希望有個新的說法：黃金美元人民幣，走遍天下都不怕。

2017 年 7 月 10 日

第六章　其他

54 給聖誕老人的一封信

一年將近，又到了該總結過去展望未來的時候了，可是，寫工作小結這種報告實在不是我的強項。望着窗外滿街的聖誕燈飾和孩子們歡笑的臉龐，我忽然突發奇想，何不像孩子們小時候一樣給聖誕老人寫封信，報報平安，討些祝福？於是，就有了下面這封我這輩子第一封寫給聖誕老人的信。

親愛的聖誕老人：

您好！一年又快要過去了，今年是我們香港交易所大家庭非常不平凡的一年，家裏的事情實在太多了，忙得我都差點忘了給您老人家寄我們的心願清單。現在剛好有了一點空閒，我想跟您分享一下我們家今年的好消息，當然，還有我們希望在今年的聖誕樹下收到的您的禮物。

首先，我終於成功減肥了！由於堅持不懈地參加足球比賽，我今年終於瘦了將近 2 磅，這點重量對於大部份人來說可能不算什麼，但您是知道的，對於我這樣管不住嘴的吃貨來說，減肥是一件多麼困難的事！您知道嗎，這個月我們才跟深圳證券交易所的足球隊踢過一場球，我們 1:1 踢平了，我特別感謝隊長讓我踢了整個半場，儘管全場下來我基本沒碰到球，但我還是感到很自豪。不知道明年您可不可以多送我幾個進球？要不然我在小夥伴面前還挺不好意思的。

事實上，不單只是我，我們交易所大家庭的每個成員今年都應該得到您的獎勵。為了讓我們的家更加寬敞更加漂亮，他們每天都像您手下的小精靈一樣辛勤工作，不辭勞苦，如果您的預算允許的話，請給他們每個人都準備一份他

們最想要的聖誕禮物吧！

以我們的 IT 精靈為例，他們每天都兢兢業業地守護着我們的家園，防患於未然，不讓任何停機紕漏的事故發生，這一點相當不容易，我要為他們的敬業點讚。明年，我們還希望在香港大屋裏做一些重新佈置和裝修，包括搭建收市競價和市場波動調節機制，讓我們的家可以更加堅實牢固。順便說一句，他們已經連煙囪都打掃得乾乾淨淨了，您下來給我們送禮物的時候完全不用擔心弄髒衣服。

今年，我們房子新添的東、西兩個「商品翼」也搭蓋得很不錯；我們花了很大力氣了裝修我們兩年前買下的 LME 餐廳，現在已經完全整合成為我們房子的商品西翼，成為我們大院不可分割的一部份。我們又在房子內新建了一套全新的水暖系統 LME 結算所（LME Clear），讓我們在運作上擁有完全自主權，鄰居們都稱讚不已。

我們還成功搭建起 LME 餐廳的「東翼」廚房 —— 香港交易所亞洲商品平台，這個廚房已於上月推出首批的新「菜品」三個小型金屬合約；新「菜品」往往需要時間才可成為「家常菜」，但我們有信心、不焦急。

對了，為了給顧客提供更優質的食物和服務，今年我們也宣佈將稍微提高一下 LME 西翼餐廳功能表上的價格，但我相信提價之後我們的餐廳仍會門庭若市。將來，我們會利用增加的收益開發更多新的「菜品」，同時也會把我們的「東翼」廚房裝飾得更漂亮，連通西翼餐廳，為新舊客戶提供更有吸引力的餐單。

今年我們最大的好消息就是辛苦籌備了兩年的滬港通大橋終於通車了！這座大橋能夠建成並順利通車，多虧了兩個交易所全體同仁齊心協力，也要衷心感謝香港證監會和中國證監會兩位員警的日夜巡邏，他們的辛勤工作為橋上的交通安全提供了堅實保障。拜託您不要忘了給他們也準備聖誕禮物！

雖然目前這座新橋上的車流量還不多，但我們已經有了一個好的開始，下一步，我們會加快輔路建設，並着手完善橋上的指示牌、加油站和速食店等配套實施，為即將使用這座大橋的人們提供更多的方便。我相信，明年這座大橋

一定會變得更熱鬧，屆時有可能還需要建新的橋，畢竟，沒橋的日子太久了。

　　當然，今年也不是完全順風順水。一羣遠道從阿拉伯來的朋友原本打算來我們家做客，但最後未能如願，主要是因為要接待他們，我們就必須對我們的房子做大的裝修和改造，可是這麼大動作的改造不徵得左鄰右里的同意是沒法開工的。好在我們現在已經開始討論是否應該為了將來考慮必要的改造。儘管有此波折，我們的家今年仍然賓客盈門，熱鬧非凡。

　　除了工作上的好消息之外，我們也有很多生活上的好消息與您分享。今年我們這個大家庭中共有 21 位成員步入了婚姻的殿堂，還有 28 個寶寶誕生，記得給他們準備一點特別的禮物哦！

　　最後，我希望在新的一年裏我們每個人都能健康快樂、工作順利，這也是您能送給我們的最好的禮物。我熱切期盼您的到來，實現我們所有的願望！

　　祝節日愉快！

李小加
2014 年 12 月 22 日

55 | 再給聖誕老人寫信

不少讀者或許還記得，去年我給聖誕老人寫了一封信，果然他給我們送來了不俗的一年。我在想：今年是不是應該再給聖誕老爺爺寫封信祈求些祝福，還是先跟我女兒討論一下此事？畢竟，我們香港交易所已經有很多超級英雄了，上週四我們員工週年晚會的主題就是超級英雄，既然有這麼多超級英雄，是不是就不用祈求祝福了？可是，我女兒肯定會說：「做超級英雄？別逗了，爸爸，你還是先練好足球射門的技巧吧！」是的，香港交易所的同事們雖然非常能幹也很努力，但我們畢竟沒有超能力，而且超級英雄也不送禮物。不說那麼多了，我還是趕快動手吧，不然就趕不上聖誕老爺爺送禮物的雪橇啦！

親愛的聖誕老人：

您好！謝謝您去年送給我們那麼多精美的聖誕禮物。托您的福，今年咱們香港交易所平平安安，我十分感恩。

在即將過去的這一年裏，香港交易所的同事們工作格外辛苦。在他們的共同努力下，滬港通這座大橋已經安全平穩運作一年了（是的，滬港通已經滿一歲了，時間過得真快吧？），各項指標都很正常。目前，他們正在為大橋下一步的擴容和完善而忙碌，同時也已經為新的大橋開通做好了準備。

今年也是倫敦金屬交易所商業化運作的第一年，也是倫敦金屬結算所建成後正式投入運營的第一年，值得高興的是，這兩項大發展都已經初見成效，它們今年都為我們集團的經營業績做出了不少的貢獻。

當然，和往年的天氣一樣，市場總是風雲多變。今年北邊的市場天氣尤其

多變，先是春天來了一場大牛市，燥熱無比，緊接着夏天來了一場大暴雨，氣溫驟降。幸好久經風雨的香港市場能夠處驚不變、沉着應對，順利經受住了這場暴風雨的考驗。經過這場風雨，香港再次向人們展示了一個世界金融中心的成熟和穩健，作為香港金融界的一員，我內心深懷感激，也為香港感到自豪。

今年，我們還推出了好幾種新產品。不久前，我們剛剛推出了34隻新股票期貨和第二批倫敦金屬期貨小型合約，明年我們還要推出更多新的衍生產品來豐富我們的產品線。這些新產品剛推出時交易量都不會很大，但是，就像籌辦嘉年華會一樣，要是我們的嘉年華會只有一、兩個遊樂項目可玩，肯定無法吸引很多顧客光臨。只有提供一定規模豐富多彩的娛樂項目、表演和遊戲，客人才會絡繹不絕。我們近期推出的新產品只是我們的開場白，未來我們推出的產品會吸引更多新的市場參與者，尤其是內地龐大的投資者羣體，到那時，希望我們的嘉年華會將是最大、最熱鬧、最受歡迎的場所！

跟您說說我的心願吧！首先，我希望交易所的所有小夥伴們能夠齊心協力，不斷提升整個市場的質素。就像上週我們員工年度晚會的主題是超級英雄，小夥伴們紛紛扮上各種英雄造型粉墨登場，格外威武。可是，我們都知道，超人只存在於電影裏，我們不能指望超人來除惡揚善，更不能指望他們來維護市場秩序，要維持市場的活力和秩序，還得靠監管機構、市場各界和交易所的小夥伴們全力以赴。今年，我們已經完成了市場波動調節機制及收市競價交易時段的市場諮詢，明年就要正式推出這兩項機制了，希望能夠在您的聖誕樹下收穫市場對這兩項市場機制的大力支持。有了這兩項機制，我們的市場就可以與國際市場更加接軌，也更加堅實穩固。對了，差點忘了告訴您，我們馬上就要宣佈我們未來三年的戰略發展規劃了，大家都對此十分期待。您要是方便的話，明年1月也可以聽一聽，好希望您能幫助我們實現這些宏圖大計！

還有一件事也希望您能幫我們個「大忙」。上週我們和深圳交易所足球友誼賽0:1輸了，隊長拍着我的肩膀讓我下半場時上場多進球，可惜我只有一次機會碰到了球，雖然友誼第一比賽第二，或許您能明年順手讓我們也進一兩個球？

此外，我還希望您能給我們交易所的每位同事都帶來驚喜。一直以來，他們兢兢業業、辛勤工作，正因為他們的汗水和付出，香港交易所才能成為全球一大領先的交易所集團。他們今年工作特別辛苦，您得好好鼓勵他們一下哦，聖誕夜時給他們多送些禮物吧！

最後，我想感謝每一位辛勤工作的同事和支持過香港交易所的朋友們。新年的鐘聲即將敲響，我衷心希望大家在新的一年裏健康快樂、心想事成！

祝節日愉快！

李小加

2015 年 12 月 20 日

56 「妮妲」閒敍

　　托颱風「妮妲」小姐的福，我今早難得不用上班，不僅悠閒地在家吃了早餐，還美美地睡了個回籠覺，10 點鐘才被手機上微信的提示音吵醒。不看不知道，一看嚇一跳，原來小夥伴們一大早就開始在微信羣裏熱烈地討論「妮妲」啦，討論熱度絲毫不亞於我們的董事會的頭腦風暴。

　　香港朋友 A：8 號風球居然還掛着，今天上午有颱風假啦！看來這次「力場效應」這次沒有起作用，哈哈哈……

　　香港朋友 B：還真是，以前香港天文台總是半夜掛 8 號早上又換 3 號，就趕在你要上班前那幾小時臨時降風球，不給人放假的機會。這次居然一直 8 號風球掛到現在，就看能不能堅持到 12 點了，要是 12 點還掛着，我今天都不用去銀行了。好幸福！

　　深圳朋友 A：8 號風球很嚴重嗎，而且不是預報説下午 1 點前就改掛 3 號風球了嗎，為什麼香港下午還要停市？説好的國際金融中心呢，怎麼刮個颱風就停市了？我們深圳不也颱風了，股市還不是正常運作？一下午不交易，這市場得損失多少啊？

　　香港朋友 A：一上來就説錢，俗！人命關天，安全最重要。況且週二就可以享受週末待遇，求之不得啊！

　　香港朋友 B：這是香港特有的一種天氣警告。因為香港是颱風多發地區，所以很早就形成了一套成熟的颱風預警機制來保護市民的人身安全，好像 1884 年就有了，後來經過多次修改，目前香港一共有 1-3-8-9-10 五種風球。8 號已經很嚴重了，如果掛 8 號風球，依照政府規定，學校要停課，電車、纜車

及絕大部份巴士路線停駛，銀行暫停營業、交易所休市、貨櫃停止交收。

深圳朋友 A：其實颱風的風向對於登陸後的影響很大，如果風向不同，即使同樣的風速帶來的影響也是差別很大的，香港目前的颱風預警制度好像主要考慮風速，有時颱風影響並不大，但還是掛着 8 號風球，是不是太死板了，不夠與時俱進啊？應該根據不同的颱風風向進行更科學的預警，現在有些停市也許是可以避免的。

香港朋友 B：香港的颱風預警制度雖然未必最科學，但簡單清晰、可預見性強，上至高官、下至百姓都知道颱風天應該做什麼、不應該做什麼，秩序井然。

深圳朋友 B：刮颱風就可以不開市了？一大早冒雨去公司的深圳人民表示很羨慕，我們證券公司今天完全照常運營，打風放假這種好事我們怎麼從來沒趕上？

香港朋友 A：@ 深圳朋友 B，你們跟我們這邊的市場不一樣：你們內地電子交易化程度高，股民都是手機下單，就算銀行和證券公司不開門，只要交易所和證券公司的 IT 系統正常運營，大家可以照常交易，畢竟停市會影響全國啊；香港市場還是很依賴券商和銀行的，如果銀行券商都停業，@ 小加想開市也開不了啊？小加你說是不是？

深圳朋友 B：有道理。深交所之所以有底氣颱風天開市，最主要還是因為滬深兩地市場是可以統一管理運作的，滬深兩市交易所相當於互相備份，只要深交所數據中心正常運作，即使深圳市因颱風停工，全國股民照樣可以交易深市股票！

上海朋友：小加這麼久沒反應，肯定還在睡懶覺！上交所和深交所的老總們只能表示羨慕嫉妒恨啦，等他來了我強烈要求他給每個早上上班的哥們派發紅包一個。

紐約朋友：同意。

芝加哥朋友：同意 +1。

深圳朋友 C：我這裏怎麼好像要出太陽了？「妮姐」就這樣過去了嗎？感覺這一屆颱風不行啊，哈哈。

2016 年 8 月 2 日

57 | 常懷希望與夢想

時間過得真快，轉眼又快到聖誕節了。2016 年即將成為過去。

回想這一年，我不禁感歎我們老祖宗的智慧與遠見 —— 猴年當下，萬事難測。

今年的「意外」真是太多了：英國選民好好的就突然決定脫歐了；美國人民選着選着就選出了特朗普；而香港今年的路也有些始料不及的變化！借用一句歌詞來形容，不是我不明白，只是這世界變化太快。

值得安慰的是，儘管世界變得撕裂和難測，心懷希望和夢想的人們，只要不懈努力，總會一點一點進步。即便是面對挫折身處逆境，不要絕望，夢還是要做的！

曾幾何時，我也在做着一些遙不可及的夢。比如，當我還是石油工人的時候，有次培訓後和同伴路過廣州中山大學門外，我看着戴着校徽進進出出的大學生們，無比羨慕，多想自己有一天也能像他們一樣背着書包瀟灑地走在校園中；剛剛在北京工作的時候，我很嚮往外面的世界，夢想可以去美國留學，可是簽證三次被拒，我是多麼的絕望與無助；後來到了美國，口袋裏沒有幾分錢，我的夢想是可以隨便走進麥當勞，想點幾個雞翅就點幾個，不用先算計口袋裏的錢；再後來，我就夢想着哪天能擁有一輛不會常常拋錨、不用天天修理的小車，因為在美國汽車就是我的雙腳，而我的二手車實在太破了……

今天看來，這些也許都只是小心願，根本稱不上「夢想」，但在當時，它們卻都是「癡心妄想」。

幾年前，當我們剛剛開始構想和內地市場互聯互通長遠要把更多資金帶入

香港市場時，不少人也認為這是「癡心妄想」：「兩地市場存在巨大差異，怎麼可能直通呢？」幸運的是，香港交易所的同事們不僅是夢想家，更是實幹家，在大家的共同努力下，曾經看上去根本不可能實現的滬港通和深港通機制如今都已經成為了現實。

無獨有偶，以前不少人認為香港發展大宗商品市場無異於是癡人說夢，但我們前幾年收購了倫敦金屬交易所，並逐漸佈局「商品通」戰略，我們也在積極籌備在深圳前海設立大宗商品交易平台。我們現在夢想着要做的事情，今天看來可能遙不可及，挑戰重重，但五年後卻可能是現實。

所以，做人一定要心懷希望和夢想，敢想才能敢做，敢做才有可能實現夢想！若處處自我設限不敢想、不嘗試，夢想未圓又能怪誰？

值此聖誕佳節之際，我想衷心感謝香港交易所全體同仁的辛勤工作，也感謝我們的合作夥伴、兩地監管機構和仲介機構的大力支持。沒有你們的付出，就沒有深港通的順利開啟；沒有你們的付出，我們的很多夢想都不能成真。

祝願所有的朋友新年進步，常懷希望，夢想成真！

至於我的新年願望，容我先賣個關子，明年 1 月再跟你們分享！

2016 年 12 月 19 日

58 | **播種與收穫**

今天是香港交易所一年一度的媒體見面會。每年大約這個時候，我們會跟媒體朋友們一起回顧過去一年的發展，分析這一年發展背後的原因，並展望未來努力的方向。下面請允許我就這幾方面做一些簡要的分享。

2017 年的回顧

在股票市場方面，我們啟動了香港市場近 25 年來最重大的一次上市制度改革，通過市場諮詢集思廣益，我們在迎接新經濟公司來港上市方面實現了歷史性的突破。我們希望今年 6 月底後新經濟公司就可以按新規則申請上市了。

在債券市場方面，我們在香港回歸祖國 20 週年之際迎來了意義深遠的北上債券通。債券通對於中國資本市場有序可控的開放和人民幣國際化都會產生重大的影響。

在定息及貨幣產品方面，我們的人民幣兌美元期貨成交穩步增長，在此基礎上，我們推出了人民幣貨幣期權，為投資者管理人民幣匯率風險提供了新的管理工具。

在大宗商品市場方面，我們推出了香港首對可以實物交收的人民幣（香港）黃金期貨及美元黃金期貨，還推出了首隻黑色金屬產品 —— 鐵礦石期貨。與此同時，我們基本完成了倫敦金屬交易所的商業化改革，確定了中長期的倉儲與交易費用結構。我們也在繼續努力，為前海大宗商品現貨交易平台開業做準備。

我們整個市場這麼多年來的努力也越來越充份地體現在 2017 年的市場表現中：香港交易所的股票現貨平均每日成交金額達到 880 億元，較 2016 年增長超過 30%，今年以來，日均成交額已經超過 1,400 億元；2017 年衍生品市場的每日平均成交量創出了 873,000 張合約的新高，今年以來的 16 個交易日中，已經有 11 個交易日，我們的衍生品成交量保持在一百萬張合約以上；我們整個證券市場的總市值在去年底達到了 34 萬億元，創歷史新高，同比增長 37%。在上週，港股總市值更首次超越了 36 萬億元。

此外，滬港通及深港通的南北向成交量均顯著增加，2017 年北向成交總額達 22,660 億元人民幣，同比增長 194%；南向成交總額達 22,590 億元，同比增長 170%。更可喜的是，滬深港通雙向均出現了大幅資金淨流入，為兩地市場都帶來了新的活力，實現了互利、互惠、共贏的良好局面。

2017 年初獲豐收的原因

之所以能在去年取得這樣的豐收，離不開同事們過去一年的辛勤工作、業界的多方努力與特區政府和監管部門的大力支持，也離不開我們前幾年的辛勤耕耘。當然，更離不開前輩們當年的勇敢開拓，他們打下了香港股票市場堅實的根基，為我們今天在互聯互通、上市改革以及新資產類別方面開拓創新提供了勇氣和底氣。

2016 年，我們制定了《2016-2018 戰略規劃》，首次提出了連接中國與世界、重塑全球市場格局的願景，致力於發展成為中國客戶以及國際客戶尋求中國投資機遇的全球首選跨資產類別交易所。這兩年，我們一直在朝着這一願景努力。一方面，我們傾力打造和不斷完善連接兩地市場的互聯互通平台，為內地投資者和海外投資者提供更多便利和選擇；另一方面，我們推動了上市制度改革的進程，以提升香港作為上市地的競爭力。與此同時，我們圍繞股票、大宗商品及定息及貨幣產品三大資產類別，不斷豐富我們的產品，滿足市場不斷變化的需求。

如今，兩年過去了，戰略規劃中所列的很多目標要麼已經實現，要麼將在 2018 年全面實施，但也有一些還沒有實現或者看上去難以實現，特別是在大宗商品領域中的努力。慶祝成功、歷數成就很容易；梳理困難、直面挫折則很難。也許有朋友會問，既然當初明知這些目標很難實現，為什麼還要朝這些方向努力？為什麼還要把它們放在戰略規劃中？是不是制定規劃的時候太不切實際了？怕不怕市場會失望？

這個問題讓我想到了播種與收穫的關係：播了種、精心耕耘，未必會有收穫；但是如果今日不播種，明天一定不會有收穫。香港交易所的運營頗有點像農夫過日子：農夫最早是種水稻（香港本地股票）的，只要付出辛勤的努力，每年肯定會有收成的，但收成的好壞基本靠天（股票成交量），後來農夫為了增加收成，決定引入北方的麥子（H 股）。而新經濟的到來，又讓農夫意識到不能光種糧食，必須學會種植經濟價值更高的蔬菜（新經濟公司），農夫必須進一步改良土壤（上市制度改革），學會使用農藥化肥和管理使用中的風險。與此同時，為了擺脫靠天吃飯的窘境，農夫開始開溝築渠、改良灌溉（南水北上、北水南下），爭取旱澇保收。

展望未來

農夫並不想止步於此，還想嘗試更多經營，比如開荒種樹，發展更有前景和競爭力的果樹種植和木材業務（大宗商品、債券貨幣）。因為農夫以前從來沒種過蔬菜和果樹，更沒有想過開拓一片森林，難免經歷不少失敗和挫折，但是農夫始終堅持不肯放棄，為了讓子孫後代過上更好的生活，他在種好糧食的同時，必須堅持摸索和學習種好蔬菜和果樹的方法，必須從傳統種植向現代農業進發。

我們要學習這個農夫，在糧食豐收的時候能夠居安思危、未雨綢繆，不僅為當下努力耕耘，更要為將來播下希望的種子，為未來發展及早佈局，營造市場發展的良好生態環境。只有這樣，我們的市場才能不斷壯大和發展。

　　這幾年我們的戰略計劃已經成效初顯。未來，內地將提升金融市場對實體經濟的服務，進一步加大對外開放；與此同時，世界對投資中國的興趣也將穩步提高。新的形勢將帶來新的機會，我們更加堅定了在香港連接內地與世界的信心，今天的耕耘和播種，將為我們明天實現夢想打下堅實基礎。

　　不是每一個農夫都能做這麼大的夢，但擁有「一國兩制」優勢、背靠強大內地市場、面向全球的香港資本市場有能力、有希望、也有責任去追尋這個夢想，實現這個夢想。只有這樣，香港才能成為真正的國際金融中心；只有這樣，我們才能不辜負這個充滿機遇的時代！

　　春節馬上就要到了，讓我們一起繼續辛勤播種，用心耕耘，用信心與希望追求夢想，用淡定與冷靜面對挫折，靜待花開！

2018 年 1 月 24 日

HKEX's Strategic Development

It is a Chinese custom to hand out "red packets" on the first trading day of the Chinese New Year.

01 | HKEx's prospects in the Year of the Snake: Are we too focused on China?

Kung hei fat choi! It's great to be back at work following celebrations to ring in the Year of the Snake.

I hope you had a wonderful time with your families over the Chinese New Year holiday.

Like you, we at HKEx have begun planning ahead for the Year of the Snake, which is filled with opportunities and promises. We can't wait to get down to work!

In the weeks and months ahead, I want to use this space to connect with you. To provide more detail about what we're doing for the market, and give you some insight into what we're thinking and why we're taking the steps we are.

We are at a key transformational moment in Hong Kong's development as an international financial centre. We have grown dramatically over the past 20 years, primarily by meeting China's needs as it's developed into the second largest economy in the world. In the 1990s, we began a long process of helping Chinese companies go international, which was an important step in China's development at that time.

But China's needs are changing, and we must change, too. In many ways, our 2013-2015 Strategic Plan outlines precisely what we believe are the key changes in China that are likely to shape Hong Kong's development in the coming years. At this juncture, I want to address concerns I often hear from some in our markets who believe we are focusing too much on China. People ask me, "Why are we putting so much emphasis on China? Are we putting too many eggs in one basket? Why can't we become more international first?"

Are we too focused on China?

These are all fair and important questions, and the answers lie in part in the questions themselves: Hong Kong is both China and international; it is "one country and two systems". This unique "dual personality", when applied properly, is the secret sauce of Hong Kong past and future success. But China is only one component of our overall plan. Let me explain.

Imagine you are throwing a party, and you want to make it the biggest party in the neighborhood. The key is to make your party attractive to as many guests as possible; people want to go to parties where they can meet people whom they would otherwise be unable to meet elsewhere. In order to ensure you have the most people show up, it is often important to figure out who you would like to invite first so that those who have confirmed their attendance become strong attractions to those who are yet to be invited.

In Hong Kong, we want our party to be the go-to party for everyone, whether they are from China or from international markets. We want our party to be the biggest and the best, and having China attend our party is one way to do that. This happens to serve China, too. China is in a long process of opening its financial and capital markets and wants to do it safely by "crossing the river touching familiar stones" in Hong Kong. With China as a VIP guest, we're more likely to attract other high profile guests. Crowds attract more crowds, and in our case, liquidity attracts more liquidity.

Hong Kong's success of the last 20 years largely reflects such a model; from our beginnings as a regional player 20 years ago, we secured the arrivals of Chinese issuers whose IPOs and listings over the last 20 years brought unprecedented participation by international investors, which transformed HKEx into a leading international exchange in cash equity and equity derivatives. Our growing global appeal has also begun to attract some key foreign issuers as well. For many of our stakeholders, the HKEx success story has been the best party for the last 20 years. Most importantly, the experience we've gained has given us unique advantages that will enable us to throw even bigger parties for our Mainland and international customers. We should ensure that we continue to leverage these advantages.

Why are we chasing fixed income, currency and commodities rather than focusing on our core traditional equity business?

Our first party was a success, and now we're planning an even bigger one, with a potentially much larger and diverse group of guests. The biggest opportunity Hong

Kong faces today is the accelerated opening of China's capital, financial and commodity markets. As China begins the shift from an importer of capital to an exporter of capital, Hong Kong needs to accelerate its own shift from its traditional role as China's offshore IPO and fundraising centre to a comprehensive global financial centre across equity, fixed income, currencies and commodities. Not only can Hong Kong not afford to miss out on such an historic opportunity, we have an obligation and the required capability to ensure that it starts in Hong Kong.

For our next big party on equity, our focus is shifting to inviting Chinese domestic investors to come to Hong Kong, which will attract more international issuers over time; this is the reverse of our equity story over the last 20 years. The structural growth of a strong domestic investor base in Hong Kong will attract even greater participation by our international investor base. For international investors already in Hong Kong, we will try to facilitate their easier access into the domestic Mainland equity markets where they are still largely restricted. Our ability to achieve a broader level of mutual market access will allow Hong Kong to be nexus of this seismic shift of the East finally meeting the West in equities. So we are not ignoring our core equity business; on the contrary, we are making it stronger. But we are also shifting our market structures to accommodate important changes in capital flows.

On the commodities front, we fully recognise how difficult it is to start a commodities party from scratch. Hong Kong is not known for throwing good commodities parties, but history is presenting us with a great opportunity. We know that, as the world's largest producer and consumer of many bulk commodities, China will benefit by accelerating the opening of its domestic commodities markets to international participation and by permitting domestic capital to play a bigger role in the global commodities market. This seismic shift of capital flows both ways will not only extend China's influence in currency and commodities to a level commensurate with its global importance, it will also allow the global commodities markets to participate in the Mainland commodities market via Hong Kong in a fair, transparent and sustainable manner based on international norms and standards.

Our acquisition of the London Metal Exchange (LME) is HKEx's most important step towards building a commodities platform. Although it cannot resolve all our challenges, it is our first shot. It is a sign to the world that Hong Kong can be a world class commodities centre, one that encourages our Chinese guests to venture out and begin their international march.

In terms of Fixed Income and Currency, we know that there is a strong desire for China to internationalise its currency. The broader connectivity in the equity and commodities markets will almost certainly lead to a sizable, incremental outflow of Renminbi (RMB), making the RMB one of the key investment and trading currencies in the offshore market. This will create significant possibilities in offshore fixed income and currency products in both trading and clearing.

Hong Kong is now faced with a monumental opportunity to host this party, but it won't be easy. We need to offer a venue that all of our guests, China or international, feel comfortable in and is consistent with their values, goals and needs. Again, we will start to prepare our guest list by focusing on China.

So coming back to the question at hand: Why China, and why not more international? Our goal is to be both, as one leads to the other. We are leveraging our China connectivity to become a truly global marketplace. We do not aspire to become a domestic China exchange, a strong regional exchange or generic international exchange. As we said in our three-year strategic plan, our vision is to become a global exchange for our Chinese clients and our international clients with a China aspiration.

As we carry out our plan, I look forward to discussing many issues on this blog, including LME integration, the development of our broker community, more on mutual market access, and our progress as we continue our work to build HKEx and Hong Kong into one of the leading financial centres in the world. I'm also always keen to hear your thoughts and opinions. I can be reached by email anytime at ceo@hkex.com.hk.

I wish you, your friends, and your family a healthy, happy, and prosperous Year of the Snake!

14 February 2013

02 | Connecting the dots for a full picture of HKEx's plans

Over the past four years, we have launched a lot of initiatives, made investments, completed a major acquisition, and launched reforms. It's been a very busy time, and people have asked me why we're doing so much, so fast. Others openly wonder whether and when these initiatives will bear fruit. These are good questions. Because we've done so much of late, I feel it's time to take stock of where we are, what these initiatives mean, why we're doing them, and how these dots can be connected to show the larger picture.

Imagine you are undertaking large-scale upgrades to your home. The first task is to strengthen the foundation. We've done this with our AMS/3.8 order matching system and the major investment in our Orion Technology Initiatives, which include the new Data Centre we opened last year, Hosting Services, and the new Market Data Feed. This infrastructure required substantial investment, but it is crucial; we want to build our house on solid ground.

The second thing we've done is to build new floors and rooms in our home. When adding on to a home, you not only need to provide space for today, but also extra rooms for guests or children you might have in the future. That's what we're doing with our house, too: we are preparing for new business. The London Metal Exchange (LME) is obviously the largest new floor, but our joint-venture with the Shanghai and Shenzhen exchanges is another, and OTC Clear — the newest — will be launched soon. In various ways, these initiatives propel us into new asset classes, make us more international, and prepare us better to grasp the emerging opportunities in the accelerated opening of the Mainland capital markets.

Now that we've completed our structural renovations, we need to protect the house. This is where risk management reform comes in. Risk management, like a security system, is not an exciting part of the building process. But it's extremely important, and we've undertaken a number of measures to strengthen risk capital in the market and ensure that the house remains safe.

The next step is to begin furnishing the home. These are the micro structure changes that make our market more efficient and more competitive, such as the recent revamp of our stock options market. Other micro structure changes may require more discussion with the market: one is closing auctions while another is trading anonymity. When we consider these and other reforms, we will of course listen carefully to different voices and various points of view from the market. Over time, we'll improve the decorations to make our house more aligned with international standards and more attractive to visitors wherever they come from.

Last but not least, we can begin inviting guests from both Mainland China and international markets to visit our new home. To do this, we have to make ourselves compatible with the Mainland and international markets. New trading hours and After Hours Futures Trading are intended to align our market with those in the Mainland and the rest of world and to support mutual market access. In addition, the Renminbi (RMB) Equity Trading Support Facility, launch of RMB futures, and the Dual Tranche, Dual Counter IPO model are all designed to ensure we are the most prepared among international markets once the internationalisation of the RMB intensifies and more funds start flowing out of the Mainland.

What are the financial implications of these initiatives? Consider our Data Centre for example. It required heavy investment and before revenue reaches its full potential, the depreciation costs of the completed centre will put pressure on our profit margin. By contrast, the BRICS Exchanges Alliance and our CESC joint venture involved little investment but the brand and strategic effect is substantial. Last but not least, the LME is the largest of our strategic investments. In the medium to long term, it will be one of our key growth engines. However, the LME fee structure has yet to become fully commercialised and the LME is in its most critical stage of infrastructure investment. Its lower profit margin and cost of the acquisition means the LME's financial contribution in the next two years will be limited. Its most tangible contribution will be to help HKEx substantially speed up connectivity with the Mainland market.

But each of these initiatives fits into a larger plan and they need to be viewed as being a part of this whole. When the market opportunities come, they will be powerful drivers of growth for HKEx and Hong Kong.

Take the example of trading hours. When we changed our market opening time to align with the Mainland, the market did not immediately see the value. However, the emergence of the RMB A-share ETF is to a large extent a result of the alignment of trading hours. Once the A-share related stock options and futures come and when we enable mutual market access on an even larger scale, the idea of aligned market opening times with the Mainland will seem natural.

There's no question that what we've been doing is ambitious. Conducting large-scale renovations to a house is not easy. But I believe that the efforts our market has made over the past few years will bear fruit sooner than most people expect.

I sincerely thank the market for the patience it has shown during this transitional period. A house is not built overnight; it must be built in stages over time until it finally takes shape and is ready to move in. As I said, our plan is comprised of many individual initiatives that all contribute to a better, bigger, solid and more efficient house. Once we've executed the plan, we'll have a highly efficient and even more globally competitive market.

8 May 2013

03 | What does the Shanghai FTZ mean for Hong Kong?

The Shanghai Free Trade Zone (FTZ) has been receiving a lot of media attention recently. In Mainland China, the introduction of the special economic zone has raised expectations and ignited people's imaginations about what's possible, especially in terms of economic liberalisation and further opening up. The zone has also received a lot of attention in Hong Kong where it has generated substantial discussion.

I hear a lot of comments in the market, with the most common being: "Can Hong Kong maintain its leading position with the emergence of the Shanghai FTZ? The FTZ will reinforce the status of Shanghai as an international financial centre, which makes it a direct threat to Hong Kong."

I understand why people in Hong Kong might have mixed feelings. On one hand, many people are excited; but on the other, they feel uncomfortable and threatened. They say China already has one international financial centre in Hong Kong, so why does it need Shanghai, too? While I see why people would ask this, I also think it's approaching the issue from the wrong point of view. It's clear that Shanghai will benefit from the FTZ much more than Hong Kong will, but that misses the point. What we *should* be asking is this: What will bring more benefit to Hong Kong, a China with a Shanghai FTZ or a China without?

The answer is undoubtedly a China with a Shanghai FTZ. The Shanghai FTZ will make China more open and internationalised, and Hong Kong is in the best position among all global financial centres to convert the liberalisation of the Mainland market into its own opportunities. The faster China's economy and financial markets open up, the greater the opportunities for Hong Kong.

We have been a huge beneficiary of China's reform and opening up over the past 30 years. When the Chinese economy and financial markets were closed, Hong Kong, with its advantages under the "One Country, Two Systems" framework, became the main bridge between Chinese and world economies. Hong Kong helped Mainland China take its first step in opening up its capital account by facilitating Mainland companies to raise funds through overseas listings. By doing this, Hong Kong became not only the main overseas hub for Chinese capital, but also the most important market for overseas investors to invest in China.

Now China is taking another step forward. The National Pilot FTZ in Shanghai is being designed to innovate with financial policy. It will be a testing ground for China's deeper reform and opening up. If reforms tested in the zone succeed, they may be rolled out to other cities in China, which could accelerate China's overall reform and opening up.

In the future, the development of Shanghai FTZ will influence Hong Kong in two ways. First, Hong Kong has developed into a well-respected international financial centre over the past 20 years. Our success has attracted global institutions, funds and talent from around the world, making Hong Kong the first stop for investment in China. The establishment of the Shanghai FTZ may result in some international institutions, funds and talent flowing out of Hong Kong to the north, but a more open and globalised China will absolutely make Hong Kong an even more attractive transfer station to the world and bring more traffic flow to Hong Kong. In this sense, the establishment of the Shanghai FTZ will benefit Hong Kong.

More importantly, the quick development of the FTZ will make China speed up the opening of its capital market, which will gradually make the traffic between Hong Kong and the Mainland move from one direction to bi-directionally. We expect overseas funds to continue to enter mainland China through Hong Kong, the convenient and well-equipped transfer station. But with the further opening of its capital market, we also expect China to give domestic investors access to international markets. Our well-established legal and credit systems as well as our business friendly operating environment mean Hong Kong has key advantages to becoming the main destination for Chinese funds going overseas.

In my opinion, the faster China reforms and opens, the bigger the opportunity Hong Kong will have as both a transit point and a destination. China now has total financial assets of over RMB 100 trillion. In the last 20 years, H-share companies have raised only HK$1.5 trillion, less than 2 per cent of China's total financial assets. But even this small number has brought huge opportunities for the Hong Kong capital market. So if the

Chinese financial market continues to open up, it will speed up the flow of capital, which will create a larger pie for Hong Kong's financial industry.

Now to the question of whether the Shanghai FTZ will intensify competition with Hong Kong. Our city has been in competition for as long as we've existed as a financial centre and that will not change now. We are competing with our peers, but more importantly, we are competing for growth. A larger pie will bring benefits for all competitors, but it also means a bigger slice for Hong Kong. This is why Hong Kong should welcome this development and root for its success.

As a financial professional who has worked in Hong Kong for many years, I sincerely welcome the establishment of the Shanghai FTZ and I wish it great success because I firmly believe it will ultimately bring many more exciting opportunities to Hong Kong.

14 October 2013

04 | After 15 great years, a new beginning

We are celebrating our 15th birthday today as a listed company with our market friends in Hong Kong and abroad, many of whom have made substantial contributions to HKEx's growth story over the last decade and a half. In that time we've grown from a local exchange into a truly global exchange group across multiple asset classes. While we've prospered tremendously over the last 15 years, milestones like this are good opportunities to take stock and look forward. We are getting set for a new journey that starts now.

More than 20 years ago the Stock Exchange of Hong Kong, the predecessor to HKEx, worked with its Mainland counterparts to launch the H-share regime. It provided a valuable mechanism for Mainland companies to internationalise, and for the Mainland itself to reform and open. Today, Hong Kong is China's number one offshore fundraising centre for Chinese companies, which has helped propel us into being one of the world's top exchanges and the largest exchange by market capitalisation of our own shares.

While it's nice to bask in the glow of past success, we mustn't linger there or we'll miss the opportunities in the present. We are in a time of great change; the world is evolving faster than ever before, China is opening quickly, and Hong Kong must keep pace. So what can we do? Our vision at HKEx is to **reshape the global market landscape and connect China with the world**. We want to build a robust and sustainable **offshore Renminbi (RMB) ecosystem**, eventually developing Hong Kong into **China's premier offshore wealth management centre**. We want to bring the world to Hong Kong, and we want to give Chinese investors access to the world *in* Hong Kong.

This era will be marked by China shifting from a large net importer of capital to one of the world's largest exporters of capital. For years, foreign direct investment has poured into China's manufacturing sectors, making China into the"factory of the world". International capital has also flown into Mainland companies that have raised funds, transforming Chinese listed companies into some of the largest banks, insurance, telecom, energy and consumer companies in the world by market capitalisation. The next era, however, will be marked by fund outflows of historic proportions, driven by China's needs to deploy and diversify its national wealth to the global markets, and this is Hong Kong's next big opportunity.

There is some US$20 trillion locked up in bank accounts in the Mainland earning little return, and even more in real estate and other physical assets which have become increasingly risky. As China implements its interest rate, currency exchange and credit market reforms, that money will begin to be re-allocated to other assets, migrating from the property market and bank accounts to a variety of financial assets such as stocks, bonds, and so forth, many of them in international markets. As the Mainland economy continues to grow, this trend will accelerate further in the coming 10 to 20 years.

The RMB is also internationalising faster than many people expected, with the International Monetary Fund expected to add the RMB to its Special Drawing Rights currency basket soon. The A-share market opening via Shanghai-Hong Kong Stock Connect and the upcoming Shenzhen Connect has made it only a matter of time before A-shares are included in many global emerging markets indices, which will mean foreign investors taking much greater exposure to Mainland equities.

For us to prosper in this new landscape, we need to provide China investment opportunities for international investors and international investment opportunities for Chinese investors. We want international liquidity and Chinese liquidity to come together in Hong Kong with a variety of Chinese and international investment and hedging products on offer. In the past, we were serving two customers: the financing needs of Chinese enterprises and foreign investors looking for exposure for Chinese equity. In the future, we must serve four customers, and serve them all well: Chinese issuers, foreign investors, Chinese investors, and overseas issuers. And the definition of issuers and investors is expanding to include all kinds of financial products issuers, users, risk managers and investors.

What makes us think we can achieve it? The key is our political system: the "One Country, Two Systems" framework that is buttressed by internationally-respected legal and regulatory standards. This system gives Hong Kong a unique edge to experiment with innovative structures such as Shanghai-Hong Kong Stock Connect, where international and Chinese investors are able to trade the other's product electronically with a unified price discovery mechanism, but anchored on local clearing and settlement, and overseen by joint regulation and enforcement regimes. Under such a regime, investors on both sides are able to trade in the other market, but without having to move funds across the border and without having to change their market structures and investment behaviours. The strength of this model depends on robust regulatory, legal, and enforcement cooperation on both sides of the border, something that has been established with Shanghai-Hong Kong Stock Connect.

This model will develop Hong Kong into an offshore RMB ecosystem that provides a variety of investment products for international and Mainland investors. If the ecosystem is liquid enough and convenient enough, more foreign investors will use RMB to invest. As the currency's internationalisation continues its march forward, the Hong Kong RMB ecosystem will continue to grow in a virtuous cycle. If we are able to have Chinese and international investors to "meet" in this offshore ecosystem and if we are able to have each other's products to be available to each other in this offshore ecosystem, we will hopefully transform Hong Kong's traditional role as a pure gateway to a truly global leader of wealth management anchored on China's liquidity, the last closed capital base that is still yet to be globally available and deployed.

Shanghai-Hong Kong Stock Connect was a good first step, and now we're about to take the second step with Shenzhen-Hong Kong Stock Connect later this year. These are, however, just the start, and we still have a long way to go. We need to work together with regulators, market participants and Mainland exchanges to expand the ecosystem as we build Hong Kong into China's premier wealth management centre. It is this path that will ensure Hong Kong's prosperity for the next 15 years and beyond.

22 June 2015

05 | The paths to China's full liberalisation

I wrote last week about Hong Kong's potential opportunities as the Mainland continues to evolve, and this week I want to elaborate with a few details of what we can do to benefit from China's growth and facilitate its further opening. I spoke at the 2015 Lujiazui Forum in Shanghai yesterday and the pace of China's internationalisation and opening was a hot topic. Shanghai-Hong Kong Stock Connect has created unprecedented new opportunities for investors, and the march of Renminbi (RMB) towards internationalisation is gaining speed. But we're also at a point where China will be making some important decisions about how to proceed into the next phase of its development.

The liberalisation of China's financial markets is inevitable

A robust financial market can only truly be realised once it is open and transparent. Open financial markets have more breadth and depth, with more accurate and representative price signals. China is at a point in its development where opening up is the logical next step in order to promote innovation, help Chinese enterprises acquire assets with greater efficiency on a global scale, and assist retail and institutional investors to diversify their asset allocation globally. In our view, China's integration with global financial markets is inevitable, but how it happens is not.

A truly international market must also have a truly international currency, and the RMB's internationalisation is gaining momentum in that direction. But the RMB will not reach its full potential unless there is a diverse RMB ecosystem, both onshore and offshore. A new suite of RMB investment products will help drive the currency's internationalisation, including making it a candidate as a global reserve currency.

What are the different ways China could open?

There are two key ways China could open up: the first involves allowing investors and funds to come in and go out from China in a direct path. The second option is an assisted path that removes some of the stress and confusion of a new environment but still provides the benefits of access.

The first path, "coming in and going out", involves welcoming foreign liquidity and foreign investors directly into China, thus bringing them into China's legal and regulatory scope, while also opening the gates to allow Chinese investors, liquidity and products to go abroad and participate in foreign markets. China is already experimenting with this path via the QFII, RQFII, QDII schemes and even the upcoming QDII2, as well as initiatives like the new trading platforms in the Shanghai Free Trade Zone. While this could be considered the ultimate model of opening without compromising China's regulatory control, the downside is it will take some time for China to reach scale because foreign investors are largely unfamiliar with China's legal and regulatory regime.

The "going out" leg of this path allows Mainland investors and liquidity to go abroad, making them completely subject to the rules of the game in overseas markets. This not only means China would lose regulatory influence over Chinese liquidity and products, but also means institutions and individuals will need to have sufficient knowledge and risk tolerance to venture into foreign markets. It's not for everyone.

Whether coming in or going out, the philosophical differences between the Chinese market and international markets are so vast, and the legal environments so different, that it will take a long time to bridge the gap.

The second path is an assisted path in and out. This is epitomised by the Stock Connect model, which allows Mainland investors and products to tap foreign markets without leaving the comfort of their home market's legal and regulatory regime. Conversely, it allows international investors and issuers to tap opportunities in China without being subject to a completely unfamiliar system.

The Connect programme allows mutual access to each other's secondary markets, or companies already listed in Hong Kong or Shanghai. The next step is to expand the link to include Shenzhen. After that, the programme will, over time, be extended to include equity derivatives, possible primary offerings, commodity futures and fixed income and currency products.

Using this model, China's opening could achieve considerable scale within a short period of time, as the structural and rule changes required would be minimal. For Chinese intermediaries, the model would also allow the expansion of their businesses and client bases internationally.

This path started with Shanghai-Hong Kong Stock Connect and will evolve into other asset classes under our Mutual Market model, supported by a strong offshore RMB ecosystem. The RMB ecosystem in Hong Kong makes it easier for Chinese issuers and investors to "go out" from Mainland China, and also puts Hong Kong and Shanghai in complementary roles as joint promoters of RMB internationalisation. Furthermore, with more openness and efficiency, this path would be able to support the next phase of China's economic growth and transformation with little change to the Mainland's current structure.

China's capital market has been opening up at various levels and in multiple dimensions. The two different paths to full liberalisation can be coordinated and implemented in parallel to suit the needs of different types of investors while also jointly promoting the internationalisation of the RMB. We believe that with appropriate risk control, a higher degree of market openness will lead to more market players and more efficient capital markets. This will help drive a new round of economic growth.

How can Hong Kong contribute?

So what can Hong Kong do to contribute? We can be a buffer between the vibrant global marketplace and China's insulated domestic market. This is already working well under Stock Connect, as international investors can buy and sell A shares in Shanghai via their Hong Kong broker, and have the trades cleared here in Hong Kong. It provides the best of both worlds.

We can also provide value by acting as a testing ground for the Mainland's interest rate reform, which continues to progress. As exchange and interest rate reform can have repercussions on the rest of China's financial system, it's important that they are done cautiously. In this context, the offshore RMB market in Hong Kong is even more important. Experimentation in Hong Kong would support the continued steady progress in the reform of interest rates and the RMB exchange rate, assist the Mainland as it takes its next important steps towards full liberalisation, and develop Hong Kong into a premier RMB wealth management centre.

28 June 2015

06 | Why our strategic plan is important to Hong Kong

We recently unveiled our third three-year Strategic Plan since I've been at HKEX, and it's our most ambitious yet. Our first plan aspired to go into commodities, and we bought the London Metal Exchange (LME). Our second plan aspired to develop mutual market access and connect with Mainland China's equity market, and we launched Shanghai-Hong Kong Stock Connect. Now we have a new plan for 2016 to 2018, and given our bold aspirations, people are asking me why we have such an ambitious agenda and how we can achieve it.

Our goal, as I explained previously, is to transform Hong Kong into a global financial centre that prices companies, prices goods, and prices money. Our new motto is "connecting China with the world, reshaping the global market landscape". We can fulfill this pledge by connecting China with the world in three main asset classes: equities, commodities, and fixed income and currencies (FIC).

Everybody is familiar with our equities business, which is one of HKEX's core strengths. We finished first in IPO funds raised again in 2015, the fourth time in seven years we've accomplished that, but there's more to be done. The launch of Shanghai-Hong Kong Stock Connect has been a good first step, and we now intend to expand it and launch Shenzhen Connect. One of our key aims is also to launch a Primary Equity Connect, which would give investors in Mainland China and overseas the option of subscribing to primary offerings in Hong Kong and the Mainland, respectively, for the first time ever. This is feasible from a technical perspective, but important work remains for regulators to iron out the right regulatory framework.

The Primary Equity Connect is a key tenet of our strategy, because it would enhance Hong Kong's position as a listing destination for large overseas companies. With the Primary Connect in place, US or European household names could list in Hong Kong and tap the hundreds of millions of investors in Mainland China in a familiar, international listing and regulatory regime that follows well-respected global standards.

But are we too focused on China? We are an international and open marketplace that is influenced by global economic trends as well as the Mainland's economy. Even if we don't connect with the Mainland, we will still be impacted by volatility in the A-share market, just as New York and London have been. But if we do connect, we can reap the benefits of China's continued economic growth and development.

The Primary Equity Connect would also not compete with Shanghai's long-term plans for an International Board. It will likely take years for the Mainland to substantially restructure its legal and regulatory regime in order for large numbers of international companies to list on the A-share market. The Primary Connect is the best way for the Mainland to leverage Hong Kong's unique advantages to accelerate the pace for Mainland institutions and individual investors to internationalise their asset allocation while overseas investors would have access to A-share issuances and placements, leading to a more diverse investor composition in the Mainland and a better pricing mechanism. The Primary Connect can be executed in ways to allow Mainland exchanges and intermediaries to partner with their Hong Kong counterparts in effecting primary offerings to achieve a win-win for both sides.

We also plan to expand our equity connect to the derivatives side. There are three ways to do this: through mutual market access (like Shanghai-Hong Kong Stock Connect), by cross-listing futures products, or by developing A-share derivatives products in Hong Kong, a trusted market. The first and second options are difficult given the current conditions on the Mainland, so we are exploring how to realise the third option. By developing A-share index derivatives, we will be meeting the needs of overseas and domestic investors to manage volatility risk and pave the way for international investors to go into the A-share market. Investors are generally hesitant to increase their participation in a market if there is a lack of risk management tools.

The second major asset class is commodities. Our current commodities businesses are largely concentrated in London with the LME, which we acquired in 2012. More than three quarters of global base metals trading, LME is supported by a global ecosystem deeply rooted in the physical markets and heavily supported by the world's largest trading companies, banks and logistics providers. Futures contracts trade on LME-provided global price benchmarks in base metals. Compared with other commodities futures exchanges, participation by pure financial investors on LME is still relatively small. So

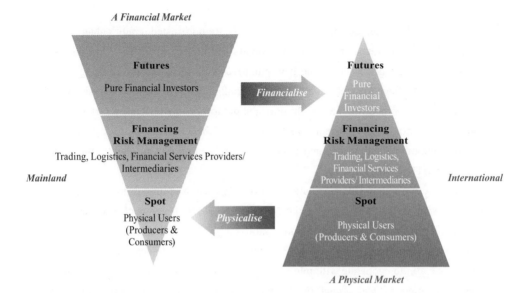

a key part of our strategic plan is to build a London-Hong Kong Connect so that a much higher level of participation by Asia financial investors, particularly from China, can be secured over the coming years. This is what we mean by saying we want to "financialise" LME.

The most significant component of our commodities strategy is our plan to build an onshore commodities spot market in China starting with base metals. This is a bold and unconventional approach largely reflecting our views that (i) Hong Kong proper is not capable of supporting a substantial commodities trading hub due to its size, location and border controls; (ii) standard futures contract trading is strictly licensed and controlled in China; and (iii) substantial service gaps exist in China's spot market infrastructure that HKEX is uniquely capable of filling, which could lay the foundation for HKEX to build into a leading commodities exchange that is deeply rooted in the physical markets just like LME.

Unlike the LME model, China's domestic commodities futures markets are top-heavy, with commodities futures traded primarily by pure financial retail investors with limited physical settlement. As a result, daily trading volumes are multiples of open interest. Although physical users reference their prices, they rarely participate in trading as they cannot easily make or take physical delivery. At the same time, the physical spot markets are highly fragmented, warehouse systems are not trusted, credit enhancements are unavailable, and financing is difficult to obtain and highly expensive. Our plan is to leverage our experience in launching new initiatives with the Mainland to bring LME's successful model there to create an effective spot trading platform, thereby "physicalising" the Mainland's commodities market. In time, the platform could generate a series of truly globally-influential and representative "China price" benchmarks, and would provide a solid foundation for the sustained development of commodity futures trading either onshore or in Hong Kong.

Finally, our Strategic Plan talks about FIC. We will become an interest rate and exchange rate pricing hub built on the foundations of Bond Connect, another of our planned initiatives. Right now Mainland bond trading is mostly concentrated in the OTC market, so we want to help overseas investors participate in the domestic inter-bank bond market. We aim to cooperate with international platforms to allow Mainland investors to invest in a variety of international bonds, with trading and clearing done in Hong Kong. We believe that Bond Connect will attract investors and liquidity to support the future development of more interest rate and exchange rate derivatives.

We know these goals are ambitious, and they won't be realised overnight. It's going to take a long time (probably more than the three years) to accomplish everything. But we won't get to the finish line unless we start the race. Once these initiatives are in place, Hong Kong will become more relevant to both the Chinese and international market users, and cement our position as a comprehensive financial centre across multiple asset classes.

I am pleased to say we have the plan in place, and we're off to a good start.

Also, with the Lunar New Year just a few days away, I want to take this opportunity to wish all of you health, happiness, and prosperity in the coming Year of the Monkey!

3 February 2016

07 | The new journey of Hong Kong's financial markets

There has been some pessimism lately about China's economy that has seeped into concerns about Hong Kong's future. This was underscored recently when major international ratings agencies lowered the outlook for Hong Kong, and just a few days later we fell outside of the top three major global financial centres index for the first time. Naturally, people are getting antsy about our future. My Hong Kong friends wonder if we should tie our future so closely to the Mainland, and some in the Mainland ask if Hong Kong is even needed now that the country has become so strong.

So what does our future hold? Do Hong Kong's advantages still exist? Can we still be a major financial centre in the world? To answer these questions, we need to take a quick trip down memory lane…

Three big things Hong Kong has done for China

Hong Kong's success today is largely attributed to how it has harnessed the opportunities presented by China while also assisting the country as it develops, every step of the way. Hong Kong has made three key contributions to the Mainland's reform and opening over the last 30 years: the entrepot trade ended up bringing China its first bucket of gold; foreign direct investment (FDI) enabled it to become the factory of the world; and fundraising in Hong Kong's capital market gave birth to Chinese companies that have become world leaders in telecommunications, energy, banking and insurance. We were able to do this because we had a unique system in place, one that was different from the Mainland, and one that helped facilitate the Mainland's development. We added value to China, and we reaped the benefits.

One central theme of this development over the last 30 years was capital inflows.

China was poor and needed capital for development, so Hong Kong served as a reliable fundraising centre. China, however, as we all know, doesn't lack for funds anymore. It is a capital-abundant country seeking new investment channels, not more funding ones. So does it still need Hong Kong?

The answer, of course, is a resounding yes. Hong Kong's open markets, legal environment, transparent and internationally-recognised regulatory system, robust markets, and pool of highly-trained and bilingual professionals are unmatched in the Mainland. This unique environment in Hong Kong has been built up over generations. So while cities like Shanghai and Shenzhen may have surpassed Hong Kong in terms of the hardware, it will take them much longer, if ever, to surpass Hong Kong in terms of software.

That doesn't mean we can be complacent, though. China is evolving quickly, and Hong Kong has to leverage these advantages as it adapts to remain competitive. The next 30 years will be marked by the two-way flow of Chinese capital, so it is imperative that Hong Kong adjust, as we've always done, in order to continue to provide value to the Mainland while cementing Hong Kong's position as a global financial centre.

The next three big things Hong Kong can do for China

There are three big things Hong Kong can do to add value to China going forward. First, Hong Kong can become China's global wealth management centre. A few years ago, domestic wealth was mainly held in real estate and bank deposits, and then in stocks and bonds. But China lacks more diverse investment instruments, as the products available on the Mainland don't meet the needs of sophisticated investors who are seeking international asset diversification. Some, with the willingness and know-how, have begun to invest directly abroad, but the rest are still nervous and unsure of their options. As China's capital controls are set to remain for the foreseeable future, we aim to strengthen Shanghai-Hong Kong Stock Connect and launch Shenzhen Connect so Mainland investors can use these secure, reliable channels to invest abroad in a familiar market environment. It's a key step towards developing Hong Kong into an offshore wealth management centre for China.

Secondly, Hong Kong can become the top offshore risk management centre for managing onshore investment risks. Interest rate and exchange rate controls in China mean the debt and currency derivatives markets are not mature enough to meet investors' sophisticated risk management needs. Furthermore, while the Mainland's stock markets are relatively open, the derivatives market is not, leaving international and domestic investors unable to properly hedge and manage risk.

This creates a natural opportunity for Hong Kong, which already has both domestic and foreign investors in an environment that is largely trusted and familiar to both sides. We are seeing the potential now with rising volumes of our Renminbi (RMB) currency futures contract as a way to hedge risks from the two-way fluctuation in the RMB exchange rate. We are now set to launch more RMB currency futures pairs against the US dollar, Euro, Australian dollar, Japanese yen, and many other currencies to meet rising demand.

Thirdly, Hong Kong can become China's global asset pricing centre. As Chinese capital goes global over the next 30 years, Chinese investors will no longer just play a creditors' role in overseas markets. They will buy goods, commodities and acquire interests in overseas companies. As more international equities and commodities are priced in RMB, China will be able to gradually master the RMB exchange rate and interest rate pricing worldwide.

To become an international pricing centre, however, a market must be recognised and accepted by all parties as transparent with a robust legal regime in place. Hong Kong, with its "two systems" formula, is "home" for Mainland investors but also trusted by international ones, making us a great first choice as an overseas pricing centre.

If we can develop these three "centres", we will be able to continue in our unique and irreplaceable role of facilitating China's development while securing Hong Kong's prosperity for another generation and beyond.

"One Country, Two Systems"

Whether Hong Kong can accomplish these goals depends entirely on the sustainability of the "One Country, Two Systems" principle. "One Country" means the Mainland can trust Hong Kong to help the country in the city's pursuit of further growing its financial markets, while "Two Systems" is the lynchpin that gives international markets the confidence to integrate with China through Hong Kong.

"One Country, Two Systems" is not an easy topic to discuss, particularly in the current social environment in Hong Kong. As Chief Executive of HKEX, I am touching on this because it is essential to our markets' future. Realising Hong Kong's potential means the Mainland must feel confident that "One Country" is not challenged and Hong Kong can be relied upon, while international and Hong Kong communities believe "Two Systems" must be fundamentally preserved.

I have been in Hong Kong for more than 20 years, and there has never been serious opposition to "One Country", though there have been recent murmurs in Hong Kong out of anxiety over the future of "Two Systems". My associates on the Mainland sometimes ask me why, if Hong Kongers don't object to "One Country", they don't show more enthusiasm for the nation. "Why don't they speak up against those who have doubts about China?" they ask. Well, my answer is because we have "Two Systems". Hong Kong deals with these issues differently than the Mainland does. Hong Kong should be allowed to handle uncomfortable opinions in the city's own way under "Two Systems".

Conversely, Hong Kong people need to also recognise that "One Country, Two Systems" for Hong Kong is in-line with the country's development. China is already rich and powerful, so it doesn't need another Shanghai, Shenzhen, or Guangzhou — it needs a unique Hong Kong. It is in Hong Kong and the Mainland's interest to preserve the "Two Systems". We should feel confident in our abilities and the soundness of "Two Systems" and not allow our concerns to create the perception of opposition to "One Country", because it is not in the fundamental interests of Hong Kong or the Mainland.

The right questions

Hong Kong is facing many challenges today, and overcoming them requires the collective wisdom of all of us. We can't let negative emotions get in the way. We need to ask the right questions to get the right results.

Therefore, "Could Hong Kong be prosperous without China?" is the wrong question in my view. The correct question should be: "Why would we want to separate a prosperous Hong Kong from China?"

Conversely, on the Mainland, a question may be: "Why does China still need Hong Kong?" Instead, the correct question is: "Will a prosperous, stable, confident Hong Kong make China even better?"

If we ask the right questions, we can get the right answers. Hong Kong and the Mainland are both unique — and different — in many ways. This means they can complement each other to the benefit of both sides. Hong Kong's success in the past is a reminder of how we can achieve prosperity in the future.

20 April 2016

08 | Hong Kong's next 20 years

It's been 20 years since Hong Kong was returned to Chinese sovereignty, and many people have recently been reflecting on the past 20 years and how our city has changed. From a financial standpoint, we have proven to be nimble and competitive, which has allowed us to reap the benefits of Mainland China's economic expansion.

In my view, three key events have contributed substantially to China's success over the last three decades: (i) through-trade, giving China its first "bucket of gold"; (ii) foreign direct investment, making China the "manufacturer of the world"; and (iii) capital market fundraising, making China's big industrial and financial companies among the largest in the world in terms of market capitalisation. In all of these events, Hong Kong has been the instrumental facilitator. In essence, Hong Kong has been China's primary offshore capital formation centre.

By focusing on these three things — each of which are distinct and helped China at different stages of its evolution — Hong Kong was able to leverage its advantages, provide value to the Mainland, and capture the benefits of a vastly expanded capital market. In fact, largely thanks to China's economic boom, Hong Kong finished first in IPO funds raised in five out of the last eight years, a truly remarkable achievement. One common characteristic that defines all three events is that money flows into China. In other words, China has been a capital importer throughout the last two to three decades.

Looking into the next 20 years, however, big changes are on the way. China has plenty of capital now and its growing challenge is to find a fast, yet safe, way to deploy that capital globally. Therefore we need to diversify and grow our financial market to grasp the opportunities that will come from the trillions of dollars China is expected to deploy globally over the next two decades, and we will do this in three ways: by becoming China's global wealth management centre, Renminbi (RMB) pricing centre, and risk management centre.

To become a wealth management centre, we need to offer Mainland investors a full variety of products to meet their international investment and diversification needs. With the unprecedented access provided by both Shanghai Connect and Shenzhen Connect, Hong Kong now has a unique opportunity that we must capitalise on.

On the risk management side, we aim to become the top offshore risk management centre for managing cross-border investment risks. We have rolled out a number of products, such as our RMB Currency Futures contracts, to provide investors with hedging tools to manage currency risks.

And third, we aim to become China's global asset pricing centre. Chinese capital is already being deployed in overseas markets, and this trend is expected to accelerate over the next decade. As that happens, the RMB will be used more and more as the main settlement currency, which opens up an opportunity for Hong Kong to leverage its transparent regulatory and legal regime and trust of Mainland China and international market players to become a global RMB asset pricing centre.

There's also one more thing that presents an opportunity for Hong Kong, and could be a catalyst as we develop our three major "centres": the Belt and Road initiative.

I know this concept is still a bit fuzzy to most people outside China, and even to many people inside the country! Chinese leaders often emphasise "Belt and Road", but it still seems remote and not directly related to our daily lives. But the scheme is real, and too important to ignore.

In a nutshell, the Belt and Road initiative is a co-development scheme for about 60 countries along two key transport routes stretching from China to Europe. It is a concerted effort by China's Central Government to aid in the development of emerging economies, many of which are in desperate need of infrastructure investment. These investments are designed to result in economic growth and prosperity across the region.

China is willing to be the first mover and help fund some of these projects, even taking on the initial risk of loss. But for this initiative to succeed in the longer term, China cannot fund everything alone. Foreign investment and know-how will need to be leveraged, too.

This is where Hong Kong fits in.

At the moment, many of the Belt and Road countries are resource rich but cash poor. They require big investments into ports, roads, bridges and railroads before they can realise their full economic potential. There are a number of ways Hong Kong can play an important role in coordinating the investments and diversifying risk: (i) by helping to list the resource assets in Belt and Road countries to raise capital to fund development of the resources; (ii) allowing the listed entities to help their country to raise debt financing for infrastructure projects by providing the listed securities as collateral; (iii) providing tools to manage commodity price risks, given the enormous demands of these infrastructure projects; and (iv) by leveraging Hong Kong as an RMB pricing centre.

Initial funding for Belt and Road infrastructure projects could take the form of equity, debt and even asset securitisations. Hong Kong, having served as China's primary offshore capital formation centre for the past several decades, is ideally suited to performing these roles.

Widespread infrastructure development in emerging economies could also have the potential of triggering a commodities boom, which leads to commodity pricing risk. Six metals contracts denominated in RMB are already trading on our Hong Kong platform, and we could provide other ways for investors and end-users to manage risk and control pricing power. As the leading metals exchange in the world and with a whole host of hedging and risk management options, the London Metal Exchange would also have a role to play.

Finally, with China being the primary driving force behind Belt and Road, it is expected that many of these investments will be priced in RMB. However, to ensure the RMB is widely accepted by Belt and Road countries and by investors, there must be a range of interest rate and FX hedging tools available — which is where Hong Kong comes in again. Ultimately this could accelerate the internationalisation of the currency and potentially drive the growth of Hong Kong as an RMB pricing centre, which is one of our three main objectives.

As you can see, Hong Kong is uniquely and ideally positioned to provide value to the Belt and Road projects. With investments of this scale and involving countries with various stakeholders and different political, economic, and regulatory priorities, all sides prefer a neutral, trusted venue. Hong Kong has earned the trust of both Mainland China and the international investment community because of our experience, institutions, regulations, and international standards required of a financial centre that could facilitate such large-scale investment projects. We can leverage these advantages to provide a platform to collate resources, raise funds, and diversify risk for Belt and Road projects while fulfilling our own objective of growing into a wealth management centre, risk management centre, and RMB pricing centre.

Hong Kong enjoyed much success over the past 20 years, and I'm confident even more exciting opportunities lie ahead. China's continued economic development along with the Belt and Road initiative are two great catalysts that will help us continue to grow and stay relevant to both our Mainland and international customers over the next 20 years and beyond.

30 June 2017

09 | Innovative wolves protect the pack, and lead the way forward

If there's a single word that turns up the most in business conversations, panel discussions, speeches and lectures, that word might be "innovation". Our generation of technological advancements from the internet to smartphones to mobile applications and crypto-currencies is defined by companies and individuals who can innovate to meet needs that customers didn't even know they had, or find creative, new ways to topple much larger competitors.

While the word tends to be used in conjunction with tech startups, innovation is also a core capability of HKEX. It's deep in our DNA, and not new or trendy to us. In many ways the Hong Kong financial market is at a disadvantage compared to other global financial centres, because we are a very small territory with an economy almost entirely based on services and logistics. We aren't known for manufacturing anymore, and have few natural resources. Instead, we need to rely on our own creativity, work ethic, and ability to innovate. We need to outsmart our competitors to stay ahead.

The introduction of the H-share regime in 1993 is a great example of innovation that drove our market's success for 20-plus years, and Stock Connect is a more recent example. Our teams worked for years to overcome what many felt was "mission impossible": finding a way to open up the Mainland market as much as possible while minimising the risks of an uncontrolled flow of capital. No exchange had ever done anything like Stock Connect before, and it has injected new energy into our market while opening a channel to the Mainland for international investors via Hong Kong.

Innovation is a core part of our company's culture, and we're now focused on innovating in three key areas: connectivity, our listing regime and technology.

Connectivity with Mainland China is a defining characteristic of our market, and we're laser-focused on building on the strong foundation we've created. Stock Connect took years to create, because connecting two very different markets, with two different regulatory structures, market operations, and investor bases isn't easy. The groundbreaking design of the Connect scheme blazed a trail that helped lay the groundwork for the launch of Bond Connect earlier this year. With Stock Connect flourishing, we are working to improve the scheme by adding new products, such as ETFs and listed bonds, and extending it to the primary market so investors on both sides of the border can invest in each other's IPOs.

Our listing regime is also ripe for innovation. We've been seeking comment from the market on plans to allow a broader range of companies to list, provided they can ensure adequate investor safeguards are in place. Those companies could include high-growth companies and those with non-standard share structures, both new elements for our market. We watch economic trends as closely as everyone else, and want to ensure our market is dynamic, diverse, and reflects the exciting, high-growth sectors in our globalised economy. So we need to innovate to build a path for these companies, which give the market more choice while maintaining the high regulatory and governance standards that our market is known for.

Third, we are witnessing tremendous advancements in technology, particularly in Fintech, and we're continually looking at ways to deploy innovative tech in our market to improve businesses processes, market efficiency and transparency. It's too early to pin things down or make any announcements, but we are encouraged and inspired by some technologies such as artificial intelligence (AI) and blockchain. AI could potentially be effective in assisting the Listing Department's review of unstructured data to enhance compliance monitoring, or blockchain might make sense as part of the clearing and settlement process. The cloud is also becoming ubiquitous, and there may be areas for us to utilise the power and capacity of cloud computing to test data and processes. We are first looking at how these technologies might help us on processes that aren't mission critical to ensure they are secure. Then we can look at their effectiveness before possibly rolling them out on a larger scale.

As you can see, innovation doesn't happen overnight; it takes time. I've heard some say we aren't moving fast enough, or we aren't aggressive enough. Others might say we are moving too fast, marginalising those who need more time to adapt. **The reality is that the Exchange is unlike other businesses, and innovation for us involves a number of unique considerations**.

Imagine a pack of wolves, slowly meandering through trees on the side of a cold and snowy mountain. Wolves move in packs, because it's the best way for the entire pack to stay safe and find food to survive. There are some similarities to the different functions at HKEX.

At the front of the pack are the young, dynamic wolves that break down barriers and find the best routes. They symbolise our market development efforts. Behind them are more vulnerable wolves that might be older and weaker, or young cubs who need protection. Some are slower and some faster, and some need more help than others to keep up. Following them are the big, strong wolves — the ones tasked with protection, making sure the pack is moving along and doesn't lose its way. At the back of the pack is the leader, who runs back and forth checking on the pack, making sure nobody is left behind and determining the overall direction.

The Exchange plays multiple roles in the market, just like the different wolves in the pack. On the one hand we are the young wolves in the front, but we are also the rear-guard wolves and leader wolf. We need to determine the direction and lead the pack, but at the same time we need to ensure each member of the wolf pack, or market, can keep up and is comfortable with the pace. The pack won't be successful unless they work together.

Once we identify the best path, we need to be determined to push forward because it's vital to our future survival. The Connect scheme is a great example, because it's a critical piece of market infrastructure that gives us a strong competitive advantage and creates value for Hong Kong, the Mainland and our stakeholders. But on the other hand, when we embark on any controversial reform, like enhancements to the listing regime, we need to consider the interest and pace of all market players. Then we move a little more slowly in order to reach consensus. If that means we lose a big deal in the meantime, that's the price we pay for keeping the pack together.

These are exciting times, but amid the new opportunities we will not lose sight of our broader responsibility to the market as a whole. We will continue to secure our long-term sustainability, reinforce our standards and ensure sufficient safeguards are in place to protect investors. We have come a long way over the years, and our desire to innovate is as strong as ever. We are well positioned to seize new opportunities, and by moving forward together I'm confident that we will have a bright future.

8 December 2017

Chapter 2

Equity Market and Market Regulation

The annual International Derivatives Conference of the Futures Industry Association in the United States.

10 | Clear waters or murky waters — Thoughts from Boca

This March, I was invited to join Futures Industry Association's International Derivatives Conference in Boca Raton, Florida. The conference has been the flagship event for the global derivatives industry for over 20 years and has been gaining momentum and prominence each year. While I'm sure Boca's sunshine and beach have contributed to its popularity, the more important driver is the rapidly changing industry landscape brought by new regulations. This urges regulators, exchanges, and participants to think together: what does it mean?

Unsurprisingly, this year's key theme was the tightening of regulations in the capital market. For those unfamiliar with these regulations, they have been developed in the aftermath of the financial crisis in the West and are aimed at bringing more transparency and oversight to the trading and clearing of financial products. Up until now these products have been largely traded and cleared in the opaque Over-The-Counter (OTC) market. This is further exacerbated by the broker-dealer model in the West, where the regulators and exchanges have little visibility into client accounts within broker-dealers. Therefore, the key focus of the new regulations is on moving the relatively complicated financial products from the OTC market to on-exchange/central clearing houses, and on protecting the interest of investors.

What surprised me in this year's conference, though, is the emphasis that ALL exchanges put on Asia, in particular China. I was honored to be invited to join a CEO Vision Panel, together with the heads of CME Group, Eurex, ICE, Nasdaq, NYSE Liffe and SGX. Nearly all of our global peers plan to make China a key part of their future and nobody wanted to be left out in this "rush to the East". It is yet another reminder that Hong Kong is not the only one seeking opportunities in China and Hong Kong is in a very unique position to capture the opportunities as China opens up.

However, when international exchanges and investors rush to China, they are not only facing the "wall" of capital account controls, but also face an entirely different market infrastructure and regulatory landscape. The former is well known; the latter is much less understood. The former also sometimes gives rise to the perception that due to closed capital controls, China's market structures are backward and under-developed. If that's the conventional wisdom, people are in for a big surprise.

Unlike its western counterparts, China has developed a completely transparent "see-through" model for trading and clearing — regulators and exchanges have visibility down to the individual investor level, such as how many shares are held and how much margin is paid, all in real time. There are 120 million investor accounts open at six China exchanges connected directly with the custodian banks and clearing houses. There is very little flexibility, if any, for broker-dealers to move funds across client accounts, let alone embezzle funds or engage in other malpractice. This is the result of major market structure reforms undertaken by China in the wake of undesirable market practices a decade or so ago. So today, everything is transparent and clean. If the western market is the murky water filled with catfish and carp, then the China market is close to distilled water.

Has China got it right? Has the West got it wrong?

There is a Chinese saying, "When the water is too clean, there's no fish." The problem in China is that the water is too clean. While investors' money is safe, there is no incentive for financial institutions to innovate as they have been reduced to mere execution agents. Confidence in the market structure might be high, but confidence in the market generating meaningful returns is low and the regulator becomes the ultimate responsible party for the market, which is clearly unhealthy for the market's long-term development. So we can't say that China has got it entirely right.

In western markets regulation was light and the water was muddy. There was a lot of mud and arguably too much innovation. The catfish and carp thrived in it. However, the water eventually got so muddy that the whole market lost sight of what was really going on and the market headed into one of the biggest financial crises of all time. So we can't say that the West got it right either.

We in Hong Kong are in the middle of this tug of war. It is both a curse and a blessing. On one hand, for Mainland and international markets to access each other, huge efforts and investments need to be made to build the "channel" between these two pools of water; on the other hand, once we build the channel, Hong Kong will become the indisputable gateway where the East meets the West. We are uniquely positioned to participate in this historic development.

Recognising how huge the opportunity is, we want to strengthen our own capabilities to take maximum advantage of it — for ourselves at HKEx and for Hong Kong. Investing in our infrastructure, such as our data centre, lays the foundation. Aligning trading hours was another step towards this goal. Building a solid risk management system is another key aspect — we not only need sound management for today's risks, we also need to be prepared for the risks that come with the bigger opportunities of tomorrow. Finally, we need to start designing a compatible model for mutual market access which plays to the strengths of each side.

As I heard repeatedly in Boca, the world wants to come to China, and I know China wants just as much to come out to the world. Hong Kong is right at the centre of this momentous trend and our success tomorrow depends on our finding the right solution for both sides today. With Hong Kong people's entrepreneurship and diligence, I have every confidence that we'll be able to achieve it.

2 April 2013

11 | Hong Kong prepares to seize another historic opportunity after learning from H shares

This year marks the 20th anniversary of H-share listings in Hong Kong and we have kicked off a series of events to commemorate this major milestone. While two decades have passed since Tsingtao Brewery became the first H-share company to list, we are still enjoying the fruits of the H-share regime today. As history is a good guide to the future, I'd like to share some thoughts with you about the past, the present and the future of Hong Kong's capital market.

Nowadays we take H-share listings, or in a broader sense, Mainland company listings in Hong Kong, for granted. After all, Mainland companies including H shares have accounted for around two-thirds of our turnover since 2007 and are already an integral part of our capital market. Without H shares, Hong Kong would not be the top-tier global financial market it is today. H-share listings have drawn international investors looking to tap China's growth, and that in turn has attracted international issuers from around the world and abundant financial talent to our city. But few remember the magnitude of the obstacles that were faced when establishing the H-share listing regime 20 years ago.

Let's step back to those days for a moment. In 1993, China was still very much a planned economy. There were few private enterprises and the concept of modern corporate governance was still new to Mainland China; the domestic capital market had just started with no securities enforcement to speak of. To put it simply, the basic building blocks for a functioning capital market did not exist. So when suggestions were made for Chinese companies to float shares in Hong Kong, there was a lot of skepticism. China was a square pipe and Hong Kong a round pipe, and there didn't seem to be a way to get them to connect.

However, for both sides the opportunity was too big to ignore. China was accelerating its "Open Door" policy and its enterprises, especially state-owned ones (SOEs), eagerly sought capital from overseas. Hong Kong was in the right place and had the right tools to capitalise on this opportunity with its mature capital market, proximity to the Mainland

A ceremony celebrating 20 years of H-share listings.

both culturally and geographically, and investor confidence in its system. But how could the square pipe be connected to the round one? Hong Kong, at this critical juncture, proved its innovative and enterprising spirit by developing the right solution at the right time. It teamed up with the Mainland to build the H-share regime, whereby the existing Hong Kong Listing Rules were reinforced by the addition of a special chapter applying only to China enterprises. The H-share regime was the "adaptor" that connected the square pipe with the round pipe.

This solution was not without risk or controversy. One major debate was over whether Mainland companies should comply with the higher standards of Hong Kong's Main Board Listing Rules or whether they should list on a separate board with lower standards. The leaders of both sides at the time agreed that Mainland companies were aiming to join the international market and they'd better play by the highest-standard rules from day one. It took vision and courage to make the right decision — and that set off a wave of Mainland company listings in Hong Kong and drove the long-term prosperity of the Hong Kong market. In return, Hong Kong contributed to the Mainland's successful transformation towards a market economy.

Today, a large proportion of big SOEs have already come to Hong Kong and Mainland listings are becoming a mature story. Some people are even concerned the stream of H shares will run dry. I am optimistic for two reasons. Firstly, the pipeline of H-share companies remains strong, with the recent relaxation of CSRC criteria for allowing domestic companies to list in Hong Kong, and the conversion of Mainland B shares to H shares. Secondly, more significant growth will come from the floating of more shares of existing H-share companies. Currently, the majority of the shares of H-share companies are held by government authorities. Combined, these shares are worth trillions, and we expect them to be released gradually. When that happens, it will significantly uplift the size and liquidity of our market.

Apart from the above "incremental growth" from the existing H-share regime, Hong Kong is facing yet another historic opportunity which will likely drive its growth in the next decade. But tapping that growth will mean making tough decisions, just as our predecessors did 20 years ago.

In order to understand this new opportunity, we have to look at what's changed — and what hasn't — since 1993.

What's the same? The Mainland's desire and need to open up to international markets remain strong. With its trade and economy more globally integrated, its need for an open capital market and financial sector is greater than ever. At the same time, thanks to the "One Country, Two Systems" policy, Hong Kong maintains its edge in financial markets, system and international access. In short, Hong Kong is still highly and uniquely relevant to Mainland China.

What's different? China has evolved from "capital hungry" to "capital abundant", hence it is transforming from a capital importer to a capital exporter. In addition to providing Mainland issuers with access to international capital, Hong Kong's future role is to provide Mainland investors with access to international issuers and products and provide international investors with access to onshore Mainland issuers and products.

Can Hong Kong make this change and meet the Mainland's needs? Can we adapt to this new development and equip ourselves with necessary infrastructure, system and talent? I remain confident. I believe in Hong Kong people's capacity to innovate and take bold steps. With these characteristics, Hong Kong will be able to continue to reinvent itself.

Our predecessors took the chance 20 years ago and succeeded. Now it's our turn. Let's work together to create a new regime that can connect the Hong Kong and Mainland markets and achieve another 20 years of prosperity. I'll share more thoughts on that in the coming months.

2 September 2013

12 | Voices on investor protection

There has been a lot said and written about investor protection, share structures and voting rights in Hong Kong in recent weeks. I've been listening with great interest. The advocates of certain points of view are extremely loud, while some quieter voices are nonetheless trying to be heard as well. Whenever I try to focus on the issues I can't get the voices out of my head.

The other night I was trying to sleep, and as I dozed off I kept hearing these voices arguing endlessly. This is how my dream went.

The first voice I hear is that of Mr. Tradition. He has prospered with the market system we have in Hong Kong, and he doesn't see the need to change. "Hong Kong's system has worked extremely well for a long time," he says, "so why change it now? Our reputation for investor protection is what makes our market so successful. We have clear rules in Hong Kong and they apply to any company that wants to list. We are continually ranked as one of the top financial centres in the world and were first in IPOs only recently. We clearly don't have any problems attracting issuers and we have not made exceptions for anyone. If it isn't broken, why fix it?" And Mr. Tradition sits down heavily, shaking his head.

Now Mr. Innovation bursts in. He's a young man with spikey hair; he talks fast and excitedly. "Give us a break, Mr. T! What's wrong with different share structures? Most other exchanges in the world permit them, it's just you Hong Kong stick-in-the-muds who can't accept change. Look at the technology companies listing in the US — most of the biggest ones, like Google and Facebook, protect the founder's position with special voting rights. People invest in these companies based on the founder's vision, track record and reputation. The founder has the long-term interest of the company in mind, and that's better than a bunch of hedge funds arbitraging the shares or corporate raiders buying into companies thinking they know how to run them! Look at what happened to Apple! Steve

Jobs got kicked out of the company on perfect corporate governance processes but Apple almost went bust before Jobs was asked to come back — and then he recreated one of the greatest companies on earth!"

Mr. Innovation is out of breath, but now Mr. Disclosure speaks up in a steadier tone. "Calm down, Mr. Innovation. The issue is not who is better, innovative founders or activist investors. It is about disclosure. The regulator's job is to ensure full and fair disclosure and to penalise abuses in disclosure. Don't forget, investors will simply price these companies at a discount to reflect the less-than-equal voting rights of their shares. Let the market decide what the right price is when founders ask for a special voting structure. This system works well in the US and elsewhere, and certainly isn't destroying value or ignoring investor protection. It's time for the Hong Kong market to modernise."

"I do want to point out, though," Mr. Disclosure continues, "that the disclosure regime works well in the US largely because of the large sophisticated institutional investor base there and the aggressive litigious culture of class actions. As such, the US system provides important deterrent forces that can offset the negative impact of the different weight in share rights. Hong Kong must get comfortable that there will be enough checks and balances to keep the founders motivated, but at the same time, honest and prudent should Hong Kong consider similar changes. If you ask me, a more gradual approach is better than a wholesale adoption of the US system."

"Wait a minute," I hear another voice. "You guys keep talking about protecting investors. Did anyone actually speak to our investors to find out what they really want?" "Great idea!" everyone agreed. They first brought in Mr. Big Investor. He said, "I don't care too much whether a company is listed in Hong Kong or New York since I can invest anywhere. I only care whether it's a good company. I don't like disproportionate voting rights, but if you must have them, I know how to value them." Then Mrs. Small Investor comes in; she feels really torn: "I can't invest in the US market, so if it is a great company, don't take that opportunity away from me, please! But I really don't like companies with special rights. It isn't fair. I want the regulators to look after my interests."

Then I hear another voice. It's Mrs. Practical. "Boys and girls!" she says, "Let's get real! People in Hong Kong have always taken a practical approach and made bold moves. We took a chance on H shares and Red Chips — we even took a chance on private enterprises — and we have been very successful. Let's just get on with it. If we miss out on the next wave of big listings from China, just think what we're all going to lose! It's not just the stock exchange and the SFC losing their trading fees and levies, the government will lose its stamp duty, the brokers will lose hundreds of millions in commissions, and the investing public will lose the opportunity to invest billions of dollars in fast-growing and iconic companies! Hong Kong can't afford to miss out on all of that!"

Oh dear, somebody doesn't like this at all! It's Mr. Righteous — clearly agitated now that money has been mentioned. "What do you mean, you, you….!" His voice was rising. "It's perfectly simple — one share, one vote and that's the end of it! How dare you suggest the founder is so special! The founder grows old, don't forget — would you let him entrench himself and extract benefits from the company forever? Would you sell Hong Kong's soul just to win one or two big listings? What about our hard-won reputation? Why should we learn from the Americans? Look at what they did to the world in the financial crisis with the so-called financial innovation of Wall Street. If anyone doesn't like what we've got in Hong Kong, they should just pack up and leave..." "And one more thing," Mr. Righteous continues: "Why is the Exchange even considering this? Is it because the Chinese government asked them to?"

I can feel people becoming really uncomfortable, but no one wants to disagree openly with Mr. Righteous, because, well, he is always so righteous. But Ms. Future, who has been listening to her music all along, takes off her earphones. "Don't make it personal," she says to Mr. Righteous. "The world is changing, China is changing and so should Hong Kong," Ms. Future goes on. "Hong Kong missed the technology revolution a decade ago. Looking into the future, there's a wave of new economy companies, particularly in the internet space, that may fundamentally reshape the entire economic landscape of China over the next decade. This could be Hong Kong's chance to claim true global leadership: combining China with technology and the new economy. It's OK for you," she looks pointedly at Mr. Righteous, "you've already made it, but think about my generation in Hong Kong."

"But why does this future of yours have to have special rights for founders?" Mr. Righteous objects.

"If the only way to secure the listing of these companies is to allow special rights for founders, so be it," Ms. Future retorts. "You have no right to deny us the opportunity. These innovative companies are growing so fast they are threatening —

and could possibly overtake — the traditional businesses which are listed on our exchange. Do we want to pass on these companies, and plant our flag firmly in the past?"

Ms. Future is clearly getting annoyed. By this time, I have already begun to sweat in my dream…

"OK, guys, let's not get carried away!" I hear that familiar voice calling for calm. Thank goodness, it's Mr. Process, finally speaking with that very deliberate manner he is known for. "The issue is not about who is right and who is wrong." Mr. Process goes on, "the issue is not whether a particular share structure is good or bad for the markets. The issue is not about who can create or destroy value, founders or activist hedge funds. People can make a case for both sides and many sides. The issue is not about whether Hong Kong should embrace tomorrow or stay in yesterday. We all want to be part of tomorrow."

By now, everyone has sat down to listen to Mr. Process more carefully. "It is all about due process." he says. "Hong Kong's Listing Rules are clear and, if there is a need to change them, we should do it via due process. If we chop and change our regulations to fit whoever comes along we will lose all credibility. What is due process? Well, it means that if a company is asking for something narrow, modest and balanced that can be reasonably dealt with within the letter or the spirit of the overall listing regime as it currently stands, waivers or permissions can be allowed. That is what the Listing Committee and the SFC do all the time. We should also consider whether any discretion we exercise can be articulated clearly as a precedent. This is important because Hong Kong adheres to the rule of law and the regulators need to draw a clear line for future listing applicants seeking similar treatment to follow and carefully articulate a clear rationale for that line."

"If what is asked for is beyond this narrow space of discretion permitted under the rules, however," Mr. Process continues, "then such significant changes to the rules and policies should be adopted only after proper consultation with the community so that they will stand the test of time."

Well, where does this leave us then? I was asking in my dream. "Why don't we call Mr. Solution?" someone suggested. "Great idea," everyone agreed.

And then I woke up!

In real life, there isn't a Mr. Solution who can put the right decision together for us. We have to make the decision ourselves, drawing on the wisdom of the community as a whole. We need to look objectively at the issues and not be swayed by emotional arguments or be distracted by specific circumstances of any given company or issue. In the end, we should take responsibility for doing what is right and best for Hong Kong, not just what is safe and easy.

I went back to my office, completely awake from my dream, and began finishing up my blog. I then began to hear another voice in my head, a very clear voice: "Charles, people are already complaining that HKEx has a 'vested financial interest' in this debate. I know you think that such criticism is totally unfounded, but isn't it a good idea to stay out of the controversy and be silent?" I reflected on this carefully and decided still to proceed with the blog mainly for the following three reasons:

Firstly, yes, I am the Chief Executive of HKEx and part of my job is to promote and protect the interests of HKEx's shareholders. However, as enshrined in our charter, in the event of a conflict, *public interest* is put ahead of shareholder interest at HKEx. It is in this context of the broader *public interest* of Hong Kong that I chose to make my contribution to this important debate;

Secondly, decisions in relation to individual companies or broader policy are not decisions of mine or indeed that of the board of HKEx. They are determined by the Listing Committee, on which I am a small voice, and ultimately by the SFC. The Listing Committee consists of members who are among the best and brightest minds in Hong Kong's financial community. They devote their vast experience, wisdom and an incredible amount of time to public service. Their decision process and the SFC oversight are motivated and driven by consideration of the best interests of Hong Kong;

Finally, I am not using my blog to change any minds or advocate any particular position. I am simply trying to promote an honest, balanced and respectful debate on an important issue of public interest. Nobody should be made to feel shy, guilty or afraid about expressing their views whether he or she is an individual investor or the Chief Executive of a large institution like HKEx, as long as we all do so with honesty, openness and the best interests of Hong Kong in our hearts.

25 September 2013

13 | Taking the debate forward: Should Hong Kong act on non-standard shareholding structures?

Since I published my "dream blog" a few weeks ago, a lot has happened. Now that the dust has settled and people have moved on, it's time to figure out the next steps. In this blog, I want to ask a few more questions that are yet to be addressed in order to deal with future challenges. Just to avoid any misunderstanding, I want to make clear that these are my personal views for the purpose of furthering this debate.

1. It seems the debate on investor protection has died down and people may have already moved on. Why are you bringing it up again?

The debate may have subsided, but the issue has not been resolved. The Hong Kong market has still not reached a consensus. Some think that we've won a moral victory by maintaining the one-share-one-vote principle, while others are disappointed because they believe Hong Kong isn't adapting to change fast enough. The two sides are as entrenched as ever.

What's lost in the debate is whether Hong Kong should embrace new, innovative companies and, if so, how. Losing one or two listing candidates is not a big deal for Hong Kong; but losing a generation of companies from China's new economy is. And losing it without a proper debate is even more unacceptable. We can't ignore this question just because raising it might ruffle feathers or make us uncomfortable; all of us have the responsibility to find an answer. It is in this spirit that I've decided to come forward and share my thoughts as a way to kick off a broader debate that leads towards a solution.

2. Why do innovative companies deserve special consideration when it comes to governance?

Innovative companies are distinct in two ways. First, their success largely comes from the founder's unique vision and plan, rather than other factors that drive business success in traditional companies. The vision and ideas of these founders are core assets of these companies. Protecting the founders and allowing them to deliver on their vision is usually in line with shareholders' interests. The founders deem these companies "their baby" and are generally more focused on their long-term development. In many ways, the business is all they have, unlike other subsequent investors who can more readily exit.

Secondly, these founders tend to start with nothing and usually had to rely on outside funding when they started out. By the time they consider a public listing, the founders' shareholding may have been diluted from rounds and rounds of financing. Therefore, they have a legitimate fear of being removed from the board at the whim of a short-term activist outside investor.

Arguably, because these founders are so vital to their companies, protecting them is also a form of investor protection. In fact, most international markets are willing to allow shares with differentiated voting rights.

3. Is the conflict between protecting shareholders and giving founders these special concessions a truly irreconcilable one?

Not if the system is designed properly. The key is to find the right balance between the concessions allowed to founders and the strength and effectiveness of the counter-measures available to public shareholders in the event of disagreement or conflict. Founders can build great companies, but they can also destroy great companies and their interests are not always aligned with public investors. I believe the stronger the checks and balances are, such as in the US markets, the greater the concessions that can be allowed for founders. In a less institutionalised and less litigious market such as Hong Kong, such concessions, if given, would need to be moderate and come with checks and balances for use in the event of abuse or true conflict.

4. What options are there between doing nothing and adopting dual class shares?

On one extreme is the status quo. This would mean we do nothing. Hong Kong can maintain the moral high ground and the purity of its investor protection principle. On the flip side, this option comes at a cost. First, we have to ask ourselves whether our moral high ground is really so high, or whether we are actually intervening too much in the relations between issuers and investors. Secondly, we could lose the chance to embrace the future and all the benefits it would bring in the long run. Hong Kong lost out in the last technology revolution in the US and many believe we can't afford to lose out in the next one, especially since so many of the large future new economy companies are likely to come from China and would otherwise consider Hong Kong their top choice.

At the other extreme is dual class shares. While this is a big departure from our current system, it's used in other markets in the world. The US, in particular, adopts a primarily disclosure-based approach, while in Hong Kong we supplement this with vetting by regulators. There have been more voices recently calling for Hong Kong to adopt a more disclosure-based approach. But this is a broader debate with many aspects that would need to be considered. For example, the US has a whole range of other key differences in its market structure that we don't have in Hong Kong.

In between the status quo and allowing dual class shares are a wide range of possibilities that could be considered. The main differentiating factor is whether founders have the right to nominate a minority or a majority of board directors.

A minority nomination right, if adopted, would preserve the proportionate voting power of shareholders, albeit within certain limits, while founders are given some leeway to influence the company without fear of being kicked off the board at the whim of activist shareholders or of being stuck with a CEO who has no roots or credibility within the company. What is trickier is how to structure founders' influence on senior management appointments without encroaching on the fundamental power of directors to appoint the leadership on behalf of shareholders.

A majority nomination right, if adopted, would entrench the founder and his or her team. The key here is to ensure a proper check and balance mechanism is in place. Shareholders should be able to veto nominations at the shareholders' meeting. If the nomination right is more moderately structured so that it would be forfeited in appropriate circumstances, e.g. after shareholders vote down the founder's nomination, say, one or two times within a given period, it would ensure that the founders will exercise this right very judiciously. In the event of real conflict, shareholders could effectively take back the nomination rights by voting down the founder's nominees on successive occasions.

5. How do we ensure due process if the market decides to adopt any of these possible changes?

If the market decides to keep the status quo, I hope that it would be a proactive decision we reaffirm after thoughtfully debating the issues and concluding that it is the best course of action. We should not become victims of inaction out of fear or inertia.

If the market decides to choose to consider other policy changes, we should adopt a decision process appropriate to the scale of the proposed change. Slight changes may be made within the existing discretion of the regulators, although a soft consultation with practitioners may be helpful sometimes to ensure a robust solution. More significant changes normally require a full market consultation, while major changes may require legislation and the involvement of the entire community.

In reality policy considerations are complex with many issues on each side. However, the points I am trying to make really are simple — that we should have a proper debate before making a decision, that the decision should be made in accordance with due process, and that we should make the decision proactively.

6. Does a partnership mechanism help?

There has been a lot of news coverage of partnerships lately, but frankly I am not sure how this fits in the context of listed companies. Partnerships are not new; in fact, anybody can form one. It's simply an agreement among people, enforced by the partners themselves. Certain shareholders, directors or senior management members of a particular listed company are free to organise themselves into a partnership for whatever motive, but in the eyes of regulators, you are a shareholder, director, or manager or all three; whether or not you are a "partner" is of little relevance.

Therefore, the partnership issue is a bit of a red herring. Any special rights that are generated from the regulatory debate on weighted share structures will attach to a person in their role as a shareholder, director or manager, respectively. If management forms a partnership among themselves that acts in concert in terms of management decisions, this would be a significant factor for the company's operations and clear disclosure will need to be made.

Overall, the regulator will tend to look through any partnership to the underlying structure and persons beneath. The regulators will not be able to "codify" the partnership concept in the context of governance structures of listed companies. In other words, you can't mix oil with water.

7. If we decide to make changes, how can we limit who qualifies to receive special consideration?

The companies that qualify for special consideration would depend on the reasons behind the change. If it is because of the uniqueness of founders and innovative companies, then we would have to carefully define these terms to limit their application. They shouldn't apply to everyone.

We could also potentially consider imposing a minimum market capitalisation or public float to ensure an adequate level of large and sophisticated international institutional holdings to maintain a healthy counterbalance of power. There could also be a requirement that the founder or members of the founding team maintain a minimum shareholding within the company to ensure there is a sufficient alignment of interests with public shareholders. There could be other approaches, too — one of the benefits of having a proper debate is to flush out new and better ideas.

So what about a partnership mechanism? As I said, it has little relevance here. If the partners fulfil the criteria above, they would be eligible for the concessions.

8. As the Chief Executive of HKEx, you have clear business interests involved here. Are you worried about being criticised for raising this issue again?

I am not worried because I think the subject is a matter of public interest and I feel that I have a duty to make sure that such an important question is properly considered. No doubt some people will criticise me, but I am not paid to be comfortable.

What is the public interest that HKEx is required to promote and protect? Public interests encompass first and foremost the respect of the rule of law and the due process, which represent Hong Kong's core value. Public interests include maintaining an open, fair, and orderly market for the protection of investors and other users. Last but not least, public interests also include seeking the sound development of our markets and maintaining and promoting the long-term competitiveness of Hong Kong as an international financial centre.

As market operators and regulators — and as leaders — we have the responsibility to act for the betterment of Hong Kong. This is not about one listing candidate or fees earned from a listing here or there, it is about choosing a future path and all the responsibility that entails. We need to have broad perspectives and base our decisions on sound judgment. And we need a debate that focuses on the merits of the arguments, and not the person giving his or her opinion.

Lastly, just to be clear: what I say in this blog reflects my personal view only. Here I don't represent the views of the HKEx Board; nor do I represent the views of the Listing Committee, nor am I speaking even in the capacity of a member of the Listing Committee. It is the Committee which has the primary role, subject to the oversight of the SFC, in deciding whether to take matters like this forward.

I hope this blog is taken in the spirit in which it is intended, which is to promote an objective debate that takes a serious look at the benefits and risks of the choices before us and leads to action. Let us have the confidence to embrace this responsibility and do what's right for the long-term interests of Hong Kong.

24 October 2013

14 | Clearing the air on circuit breakers

It's good to be back after the Chinese New Year break. At HKEx we are excited to begin the Year of the Horse.

We are starting the year after a strong finish to the Year of the Snake. Our markets showed great resilience over the last year, finishing second in IPO funds raised and setting turnover records in ETFs and several derivatives contracts. Our subsidiary the London Metal Exchange also saw record trading volumes. But as I've argued before, we can't be complacent. We're in a fast-paced industry with nimble competitors, so we need to remain focused on ensuring our market is as competitive as it can be while maintaining high risk management and investor protection standards.

I've been asked about a variety of topics over the past few weeks, many of them related to market microstructure. The question of circuit breakers, in particular, has drawn quite a bit of interest from the market, and people have been forthcoming in expressing their opinions, which I think is a good thing. I would like to share some of my thoughts on the subject.

First of all, what are circuit breakers? A circuit breaker is a mechanism whereby in case of extreme and abrupt price volatility that moves prices beyond a pre-set threshold in a very short space of time, a pause is triggered in the trading of the security or group of securities concerned, or of the market as a whole. The objective is to give the market a chance for reflection so as to avoid a panic reaction to a price movement that is unrelated to a change in fundamentals — for example, a faulty algorithm, or a "fat finger"order. During the pause, depending on the design of the circuit breaker, trading may or may not continue within certain limits; following the pause normal trading resumes, and the circuit breaker trigger is reset to a new level.

Our market has discussed circuit breakers from time to time before, with the conclusion being that they weren't necessary. Some people say that circuit breakers are not suitable for our market and amount to market intervention, therefore are a waste of time to consider. But I don't think that previous decisions necessarily apply to evolving market conditions today. The Exchange has to maintain a fair and orderly market — this is in fact our statutory duty — and we have to stay alert continuously for changes that may require new mechanisms to maintain market integrity.

In normal market conditions, investors find the right price for a security of their own accord without intervention from the market operator. But in recent years the proliferation of computers has changed the way trading is conducted in almost every market, including our own. Trading has become much faster than it was a decade ago, and it's often done automatically via algorithms that have a chance, even if a remote chance, of generating erroneous orders which could create an overreaction and threaten market integrity. Of course, this overreaction would eventually correct itself, but perhaps not before considerable turmoil which could impact confidence in the market. We have seen such cases in overseas markets, and we have to ask ourselves whether they could occur here in Hong Kong.

Some market participants have told us that they are concerned about the adequacy of the measures to safeguard the Hong Kong market against disorderliness caused by human and machine error. Indeed, this area is one that the International Organisation of Securities Commissions (IOSCO) has asked markets to review, and the SFC's new electronic trading regulations which became effective early this year push in the same direction. I think it is timely for a renewed debate on circuit breakers to determine if they are now needed in the Hong Kong market.

So that's why we're looking at circuit breakers again. I would also like to clarify some misperceptions that I have heard expressed about circuit breakers by some market participants.

One perception is that the Exchange wants to introduce circuit breakers in order to align Hong Kong with the Mainland exchanges. However, aligning with the Mainland is not the objective. The objective is first to consider whether there is a need for circuit breakers in Hong Kong, and if there is, to develop a circuit breaker mechanism that is suited to our market's specific needs.

Another misperception is that circuit breakers mean strict daily price limits as in the Mainland and some other Asian markets. However, this is not necessarily the case. Many overseas markets have circuit breakers with thresholds that are not fixed but adjust dynamically with market movements, allowing price discovery to take place more normally. The key point is that if we are to introduce circuit breakers here, we'll have to find a model that suits Hong Kong best.

A third perception is that circuit breakers are the same as trading suspensions. Again, that is not necessarily the case. There are circuit breaker models used by other exchanges that provide a short "cooling off" period (usually of a few minutes) during which trading continues to take place subject to certain conditions such as a limit on the price at which orders may be executed.

In my view, any introduction of circuit breakers in Hong Kong would have to meet the following conditions.

(1) They should mitigate the risk of extreme and abrupt market volatility caused by non-fundamental factors (such as faulty algorithms);

(2) They should safeguard the integrity of the market;

(3) They should be flexible enough to allow for fundamentals-driven price movement; and

(4) They should be straightforward enough for our market participants to understand and implement.

This is not an easy list of conditions to meet, and we certainly need extensive discussion and debate in the market to arrive at a conclusion.

As yet the discussion is only beginning. At the Exchange we have not come to any conclusion as to whether there is indeed a need for some kind of circuit breaker for our market and, if so, what kind. Any decision would only be made after we have sufficiently consulted the market and the various views are thoroughly considered. Some people may feel passionately one way or the other; but until we are able to provide a forum through the consultation process where all views and opinions can be heard, debated and carefully considered, we will not be able to make the best decision for our markets. So please be patient and ready to contribute your views as well as be willing to hear other people's views.

I should also say that studying circuit breakers is not the only item on our strategic agenda this year. In addition to other new business initiatives, we are looking at ways to strengthen our established businesses.

We recognise that any new measures will not be without challenges as market participants may have different interests at stake. We will therefore consider participants' views carefully. Where we decide to proceed with new initiatives, we will consult the market comprehensively; if the initiative is supported by the market, we'll implement it in an appropriate manner, giving the market time to prepare and adjust. We will also do our best to explain any new measures, so that they are well understood by Exchange Participants and investors.

When we consult on these strategic issues, I encourage everybody in the market to share their views with us. By working together, we can further enhance Hong Kong's competitiveness and continue our momentum in 2014.

May you have a very healthy, happy and prosperous Year of the Horse!

13 February 2014

15 | At last! Kicking off a pre-qualifier on weighted voting rights

Last year I launched a discussion in this blog about weighted voting rights shares in Hong Kong. I tried to present a number of different positions and possibilities on these pages, and said that whether we decide to make changes to Hong Kong's listing regime or maintain the status quo, it should be done proactively after a thorough and comprehensive debate in the market.

Well, the day for that debate has finally come! On Friday, HKEx announced a Concept Paper on weighted voting rights and is seeking market opinions.

It has sure taken a long time for this Concept Paper to be released! Almost 15 months since we first began to consider these issues!

Let's all be frank, it would have been nice if this could have been released earlier. But the time it has taken also means that the final paper fully reflected the broad spectrum of views held by the key stakeholders after months of debates behind closed doors. The deliberateness speaks to the soundness of Hong Kong's system and our steadfast belief in the merits of due process. This is something we should be proud of, even though we might sometimes wish things moved along faster. In the end, it's always better late than never.

I am not speaking here on behalf of the Listing Committee which oversees this consultation process, but I do want to offer a few personal observations.

The Concept Paper has two main questions: first, whether weighted voting right structures should be permissible for companies listed or seeking to list on the Exchange; if the answer is yes, then the second part digs a bit deeper into under which circumstances they might be allowed.

It's important to be clear about what the Concept Paper is, and what it is not. It is not a traditional consultation that seeks feedback on specific proposals for market reforms. It does not contain or advance any specific rule amendment proposal. It is also not an initiative to attract a particular issuer or group of issuers. Rather, the paper is designed to promote a focused and coherent discussion on the concept of a weighted voting rights structure in Hong Kong and determine whether our adherence to "one share, one vote" should be universally required or whether we could accommodate more flexibility under appropriate circumstances.

I am a soccer fan and sometimes like to speak in soccer terms. Thirty-two teams went to Brazil for the World Cup last month and they first played in the qualifying round, where the wins and scores did not count toward the final result. The qualifying round did help group and rank the winning teams so the best teams may meet in the final. This Concept Paper will do the same thing for the contest of views on weighted voting rights — it will help the regulators "rank" the ideas and the sentiment on the issue according to their importance to the community. Then, if there is a community will to proceed further, a meaningful proposal can be put forward as a next step. But we need that qualifying round to decide whether to proceed at all.

Now that we have a structured forum for debate, I consider my part completed. The stage is now yours, be you a broker, a lawyer, an accountant, a listed company, an investor, or a common citizen. I encourage everyone to share his or her views with us. Hong Kong is a big and diverse international market, and along with that comes a multitude of opinions. Our Listing Rules should "reflect currently acceptable standards in the market place." The market needs to be regulated, but it also needs to be developed. That means all of us have a responsibility to take a closer look at how our market is evolving, assess our competitiveness, and consider the uniqueness of our market in terms of opportunities and risks to determine if any changes or flexibility are required. By doing so, we can work together to ensure Hong Kong remains a leading and competitive financial centre in the world.

It is now time for you to speak up on a critical issue in Hong Kong and contribute to a discussion that will shape our market for years to come. I look forward to an open and constructive debate.

31 August 2014

16 | What happened? Taking a closer look at the A-share market

The extreme volatility in the Mainland's A-share market has drawn international attention over the past few weeks, and raised some specific questions about how the A-share market functions and broader questions over the pace of internationalisation on the Mainland. Now that the market appears to have stabilised and everybody has had time to catch their breath, people have begun trying to make sense of what happened and why. The most commonly asked questions include:

(1) How did the rapid bull-run come about? What fuelled it?

(2) Why did the bull-run end so quickly and why was the decline so steep?

(3) Should the government have intervened? Should it have intervened earlier? Maybe later? Did it intervene enough or too much? How has the intervention impacted China, the capital markets, and international sentiment? What would have happened had the government not intervened?

(4) How did various stakeholders contribute to the boom and bust? What else contributed? Short sellers? Weak regulation? Leverage? OTC financing? The internet? Domestic speculators?

(5) In what ways has the Chinese market structure itself played a role in this crisis?

There is a lot to digest here. Identifying the root causes and getting the right answers could take months, if not years, of effort and study, particularly as more precise and extensive data becomes available over time. In this blog, I want to focus on China's market structure and leave the other questions to be addressed later once the dust is more settled. With Stock Connect providing more opportunities for funds to flow across the border, a greater understanding of the A-share market is an important starting point to better make sense of what happened.

Many people are already aware that the Chinese stock market is, in essence, a retail investor-dominated market. Most of the direct participants are retail, so the market is uniquely organised to help meet their specific needs while many policy decisions are made with a strong orientation toward their interests. But understanding how China's market structure developed around retail investors and assessing whether it's been effective at protecting them is helpful, particularly to market operators such as HKEx, which has worked hard to develop mutual market access with the Mainland.

What are the key features of the Mainland market?

In a nutshell, the Chinese stock market is the world's only completely "see through" market where almost all of its approximately 200 million investors directly participate in the exchange markets through their own unique account number, without much involvement from the intermediary layers of brokers, dealers or institutional fund managers. More specifically, the Chinese stock market has the following key features:

Individual accounts of retail investors are all directly opened with the Shanghai and Shenzhen exchanges, China Clear and the custodian banks. In contrast, international and Hong Kong investors typically maintain their accounts with their brokers, who in turn are participants of the exchanges. Typically, only participants are able to trade directly through the exchanges in international markets.

Investor assets such as cash, securities, margin, reserve fund contributions, or lent and borrowed stocks are all centrally maintained, controlled and managed by the central market structure operators, such the two exchanges, clearing houses, state margin protection, and stock lending and borrowing corporations. In contrast, international investors' assets are typically maintained and managed by their broker/dealers while stock lending, borrowing and financing are also handled by brokers. Brokers then put up margins and contributions to the central counterparty on a net basis.

Trading, clearing and settlement of all securities, though channelled through brokers, are matched and cleared directly and centrally at the exchange and clearing houses on a gross basis. In contrast, international investors typically trade and clear through their brokers, who can internalise matching and only net clear with the central counterparty.

These unique market structures, with central control over all investor assets and market access, allows the Chinese regulators to essentially regulate through software and hardware so that prohibited conduct and practices will not be able to be executed directly through system.

Beyond market structures, the Chinese stock markets also embrace many regulatory prerogatives with a strong bias toward retail investors, such as constant tight control over primary market fundraising by corporates, strict electronic auction control of IPO subscriptions, daily 10 per cent price movement and eligibility hurdles for retail investors to trade in the futures markets.

Why does China's market work this way?

The international markets have developed and evolved for well over a century, with the broker-dealers initially acting as the primary intermediary or mini exchange for their own investing customers and the stock issuance corporates. The broker-dealers then formed exchanges in which only they were members, and therefore qualified to deal and trade on behalf of the investing public. Large institutional investors emerged over the more recent decades and became the primary intermediaries to deal with the broker-dealers and the exchanges.

China started its stock market development essentially along a similar path, although the concept of issuers, brokers, exchanges, securities regulators and investors (largely individuals) were all "born" about the same time at the beginning of the 1990s. The market was initially structured with a bias toward financing state-owned enterprises; but, given the large number of retail investors, regulatory orientation has decisively begun to shift to how to comprehensively protect retail investors.

In the early 2000s, China encountered significant stress with the broker-dealer system as there was a lot of abuse of investor assets, with some intermediaries speculating using investor funds in their custody. China took decisive action against the wrongdoers, but also tried to completely eliminate the chance of further wrongdoing. The fear of a retail investor backlash and a strong sense of accountability for their interests led to reforms that started in 2004 and resulted in the system China uses today. It wasn't originally planned like this, but China changed course partway through.

Is China's system better or worse than the international system?

This is not such an easy question. There are pros and cons to both systems.

First, there's no doubt that China's market structure is more transparent because regulators can see everything that happens on the exchanges. Nothing escapes the regulatory eye. With the single identification number, regulators can see right through the system to quickly identify wrongdoing and take immediate action. On the flip side, this "see through" system only applies to on-exchange activities, and has failed to reach the off-exchange market. Internet lender-fuelled leverage, which some speculate contributed greatly to this round of boom-and-bust in China, is relatively new and perhaps caught the regulators off guard. Several investigations have begun in this area and the jury is still out.

Second, China is the "safest" market in the world today from a customer asset perspective. The system is structured in such a way that investors' assets are left with a central custodian, making them completely off-limits to broker-dealers who may otherwise use the assets to leverage or monetise them. Again, though, there is a downside: the broker-dealer community is deprived of access to a very important pool of resources they could otherwise use to conduct and grow their businesses. Furthermore, broker-dealers in international markets differentiate themselves by learning about their clients, offering appropriate products and managing risk on behalf of investors, which leads to a differentiated and more sophisticated investor base. China lacks this diversification, which could provide an important counterweight to wild swings in the market when most of the investors are pulling in the same direction.

Finally, I have mentioned before that China's market is the most egalitarian, in the sense of widespread equal accessibility to participate in price discovery. China is one of the only countries in the world where ordinary folks crowd into trading halls to watch market data on the big screen as if they are watching a sporting match. The flipside is retail investors are making investment decisions that do not usually go through institutional investors and brokers, who could provide professional and rational investment analysis.

There are many unintended consequences from this retail-focused regulatory regime and I will list only a few:

- As many regulatory and risk management requirements are already pre-built into the system, there is an inherent absence of an institutionalised, compliance and risk management culture among China's investing public;
- Restrictions on primary market issuances results in an imbalance in the A-share market between issuance supply and demand. Secondary market valuation is less

about anticipated corporate earnings and more about anticipated secondary market liquidity;

- The strict controls over the timing, pace, pricing and auctioning of IPOs deprives the A-share market of its natural ability to properly price IPOs.

What does this mean for Hong Kong and the Stock Connect programme?

There's no doubt that the Mainland regulators will learn from this experience, just as Hong Kong and international regulators have learned from past crisis of their own. The last several weeks have shown how volatile the Mainland market can be, but its pace towards greater global integration will remain unchanged. If anything, the recent volatility underscores the urgent need to improve the market, introduce more institutional participation and further open up the market to foreign investors. In this regard, the Stock Connect scheme is a controlled, safe, yet flexible way for the Mainland to continue its liberalisation. It is now more important for China than ever.

While a natural instinct may be to pull back from the Chinese market for a while, the country's long-term roadmap towards internationalisation is set. As its market becomes stronger and more robust over time, our unique position will bring benefits to Hong Kong. As bridge builders, it is our job to build new bridges, establishing greater connectivity and creating more investment choices. We helped build the Stock Connect bridge; whether and when to cross the bridge is up to investors.

The differences in how the Mainland and Hong Kong responded to the stock market plunge were in stark contrast, and demonstrated to global and Mainland investors alike that our market in Hong Kong is mature and operates in a fair and orderly manner. Our deeply-rooted compliance culture along with our stringent regulatory regime gives investors confidence. Compared to the US or European markets, we have much more direct retail participation, but we remain an international and largely institutional market with no capital controls. We have been an open market for a very long time and have experienced many different crises over the years which have taught our regulators, intermediaries and investors to remain calm in times of market turbulence. These form the foundation of our success.

We intend to utilise these advantages to provide the Mainland with a secure, trusted environment with which to experiment as it grows and matures, while we pursue transformative growth of our markets in Hong Kong.

China is a huge and yet still young market. It will go through some more growing pains and tough times. But there's no doubt that with patience and time, China's market is on its way to becoming the global centre of gravity in the Asian time zone and will help secure the brightest future for Hong Kong.

16 July 2015

17 | Putting the pieces together for an ideal market

We are hard at work on our new Strategic Plan, but have come across some questions about two of our suggested initiatives: the Third Board and Primary Equity Connect. Both are important components of our roadmap to build Hong Kong into an ideal market, and they can only be achieved with regulatory guidance. Speaking purely in my personal capacity, I want to examine these initiatives in more detail, explain some of the logic behind my thinking and provide some perspective to stimulate a constructive debate on whether these initiatives are sound, necessary and realistic.

Why are we talking about a Third Board?

The Hong Kong IPO market has been a great success story over the last 20 years and our listing regime is considered to be world class with its unique track record of bringing Chinese companies into the international norms of sound corporate governance and investor protection. Despite the success, we are still very sensitive to periodic calls for greater efforts to attract promising, emerging companies to list; to complaints that some companies become "bad apples" shortly after listing; and to occasional frustration that some "zombie" or "shell" companies are allowed to stay on the market for far too long.

I think we all agree that these issues should be tackled in our effort to create an ideal market, which is one that efficiently allocates capital by allowing good companies to list and raise funds and poor companies to be quickly removed from the market. The question is what is the best way to achieve it.

Some argue we should make the initial listing criteria even more stringent and build a higher wall to keep bad companies out. However, a side effect of this higher wall is it may keep companies with great potential out, too. If a higher wall could guarantee that bad companies are kept out, maybe this would be a worthwhile trade off — but it's not the case. China has arguably the highest wall possible with hundreds of companies waiting for years to be listed, and even it can't eliminate the many failing companies that cause substantial investor losses each year. The reason is simple: a few companies that meet those higher standards at listing might still become poor quality companies later.

In my view, instead of focusing on the entrance criteria, we should be focusing on the exit instead. That's why I think we should be looking at a much more robust post-listing regulatory regime and a more efficient delisting process, a system which wouldn't allow "zombie" companies or "shell" companies to stay for too long.

So why hasn't this happened yet? In fact the market considered reforms to the post-listing regulatory regime many years ago, but the plans never came to fruition. Hong Kong rightly values its tradition of rule of law and due process, particularly when it comes to material regulatory changes that might have far reaching impacts on stakeholders, especially existing issuers and investors. There are also a lot of entrenched interests (especially those "shell" companies), which means any potential regulatory change would likely require a very lengthy consultation period. Even after that process, the market may still be unable to reach a consensus, as we found out previously.

It is not clear to me whether the markets, including the regulators, the exchange, our participants and investors, are prepared to take another hard look at this thorny issue again today. So if we are not able or don't want to touch the existing regime but still want to do something, what are the alternatives? It is in this context that we begin to explore the idea of a Third Board as an option to consider.

A Third Board would require no changes to our existing regulatory regime and could contain three important features that would differentiate it from the GEM board: (i) a lower entry threshold which will allow more emerging companies to list in Hong Kong; (ii) more aggressive and robust continuous listing obligations for issuers to ensure investor protection; and (iii) a much more effective and efficient delisting procedure, which will allow the market to "flush out" bad companies quickly to protect the reputation of our markets. If lower entrance hurdles and a faster exit process give rise to concerns about investor protection, particularly regarding retail investors, regulators could consider limiting participation to professionals. A Third Board, if successfully launched and smoothly operated over time, may eventually create market pressure on our other boards, leading to further improvements in those markets as well.

I am not suggesting that a Third Board is a preferred option over changes to GEM. This decision is ultimately for the regulator and market to make. If market participants agree a Third Board is an option worth exploring, we can then take guidance from our regulators and work together to design details of the new market such as the admission criteria, post-listing regulatory regime, investor eligibility and other rules. Of course if the market has other solutions, we're open to hearing those, too.

As a market operator and the frontline regulator of issuers, it is our duty to continue looking at ways to facilitate economic development by helping issuers raise capital and strengthening investor protection through enhancing market quality. We continue to think about how we can fulfill this mission, and I believe we need to have the courage to take action now.

Primary Equity Connect

Now to the Primary Equity Connect. This is our plan to expand Stock Connect, which has connected the secondary markets in Shanghai and Hong Kong, to the primary market. In other words, it would allow Mainland investors to participate in Hong Kong IPOs and vice versa.

There are very good reasons for both the Mainland and Hong Kong to embrace a Primary Equity Connect. The Shanghai and Shenzhen exchanges both have their sights set on building international boards to welcome foreign listings and investors, however we expect this to take considerable time, if it happens at all. The urgent need for the Mainland is to diversify its investor base, including attracting more institutional participation, and to allow Mainland investors to have greater access to high-quality international investment assets so they can diversify their holdings. Both of these key objectives can be met by opening up the primary markets on both sides of the boundary to investors from the other side.

For Hong Kong, it would strengthen our attractiveness as a global listing destination to enable international companies to access China's vast domestic savings. It would also reinforce our role as the primary access point to welcome international investors to directly participate in new issuances from China.

So how difficult would it be to get this done? There has been some skepticism about the feasibility of introducing a Primary Connect, with some people believing it would take substantial regulatory changes and is therefore unlikely in the short term. While it's true regulators would need to cooperate and some changes would be required, it's not as far-fetched as it might seem.

For Southbound, which means Chinese domestic investors buying primary shares offered by companies already listed or newly listed in Hong Kong, there should be no regulatory changes needed in Hong Kong as our city is already open to international (including Mainland China) investors who wish to participate in IPOs in the city. However, approval from the Mainland regulator would be required. While it is far from certain that such approval would be forthcoming, there are good reasons to believe that Mainland regulators could eventually look at this as an effective, efficient and low risk way to help China direct its massive domestic savings into more diversified global allocation through the Connect programme.

For the Northbound channel, which means international investors participating in offerings of primary shares in Mainland China, I do not believe that there should be any material Hong Kong regulatory issues or concerns if shares are offered to professional investors only. Chinese issuers in these instances are primarily interested in attracting international institutional investors to their domestic offering and see no need to tap Hong Kong's retail investors. However, I recognise that different regulatory considerations would arise should Hong Kong retail investors be permitted to participate.

The Primary Equity Connect is an example of the best kind of initiative, because it brings real value to the Mainland's development and liberalisation while also bringing value to Hong Kong as the connector and platform that can make it happen. Still, this is ultimately a decision for regulators; while there would be some regulatory hurdles, we believe it's in Hong Kong and the Mainland's best interest for us to overcome them.

The Third Board, too, is an idea to help us build an ideal market, and ensure we are ready for future opportunities. Both of these initiatives are in the early stages and we are always open to a vibrant debate and feedback from the market, including alternatives that make us better. Only by working together can continue building Hong Kong and ensuring its position for another generation of success.

2 March 2016

18 | Addressing Mainland investors' questions about how Hong Kong's market works

With the announcement of Shenzhen Connect, we've begun to see a substantial increase in the Southbound trading of Hong Kong shares by Mainland investors, particularly after Mainland insurance companies got the green light to access the Connect programme. With growing interest in the Hong Kong market, we have also begun to notice that Mainland investment commentaries have become increasingly focused on one particular area of the Hong Kong markets: the easy access to refinancing through instruments like deeply discounted rights issues and share consolidations, which happen occasionally. Citing individual cases where some Mainland retail investors felt "trapped" in some of these transactions, the commentators cried foul and asked why Hong Kong regulators don't "weed out" such practices all together.

We clearly recognise the continued need to enhance our market quality. In light of the increasing pace of regulatory actions to strengthen the oversight of the secondary markets, I wrote a blog directly in Chinese to offer our Mainland investors some perspectives on how the Hong Kong markets operate, particularly some of the philosophical differences in market regulations between the Mainland and Hong Kong. The following is a summary of the key underlying messages, but not a direct English translation of the blog which was written to answer some typical questions from a Chinese retail investor.

Should the regulators be smart enough to identify the bad guys and take decisive actions to weed them out?

As much as the regulators want to "weed out" the bad guys, the regulator cannot possibly tell who is good or bad until a violation has taken place. Hong Kong regulators typically don't pass their own judgment on good companies against bad companies in terms of investment. All it cares about is whether a market player has committed any legal or regulatory breaches and if so, take action. Since our system is based on due process and rule of law under which one is presumed innocent until proven guilty, it will take time, resources and efforts to prosecute and convict offenders in Hong Kong. While this means bad guys may not always be effectively caught, it is extremely unlikely that an innocent person would be wrongfully convicted.

Why can't regulators in Hong Kong do more to prevent bad acts from happening?

There are two fundamentally different approaches in regulating a market. The first is to set up a system of strict pre-approvals so that regulators can try to screen out bad actors before they act. This is a great approach in theory, but in reality it does not always work and it has significant negative side effects. Such an approach might be appropriate in a market that is dominated by retail investors like the Mainland; Hong Kong, however, which is primarily an institutional investor market, follows a different practice: it assumes that people have good intentions and will follow the rules, but monitors the market closely and will punish those caught doing something wrong. An excessively stringent system of prior approvals would impact the freedom of all players and undermine normal market activities.

So instead, with a focus on compliance and enforcement, we have Listing Rules that contain mandatory disclosure requirements and give shareholders the right to participate in the decision-making process for major transactions. Regulators in Hong Kong can't dictate the business decisions of listed firms, but they do ensure shareholders have a say. This is a fundamental difference between the approaches in the Mainland and Hong Kong.

Some market instruments in Hong Kong, such as deeply discounted rights issues and share consolidations, can be easily abused. Why don't Hong Kong regulators ban them or make them much more difficult to use?

Again, philosophically, rights issues, placements, share consolidations, and other re-financing tools are not inherently problematic and are merely neutral tools available to companies that allow them to raise funds more easily, conveniently, and efficiently at lower costs in the secondary market. But we know that some bad guys may use these tools for nefarious purposes; lots have been done to clean them out and more is still needed. In doing so we need to strike the right balance between providing the right tools to support normal market operations and doing our best to follow-up on those who use the tools for the wrong reasons. Think of it like a kitchen knife: it's great for chopping meat and seafood, but it can also be used for much worse things. But we don't ban sharp kitchen knives because of what they *might* be used for. Regulators in Hong Kong work very hard and very diligently to try and maintain this balance.

What can Mainland retail investors do to protect themselves?

The first thing Mainland investors need to know is that the Hong Kong markets are different from the Mainland markets and they need to understand the differences and act accordingly. They need to take responsibility for their investment decisions and not hold out unrealistic hope that someone else, like the government or regulators, might come to their rescue should they make a bad investment decision.

They should therefore focus on companies that are well known, well traded and well covered by the analyst community. They should seek advice and counsel from investment professionals. If they choose to invest in some "special situation" companies where they believe they have a "tip" or some other advantage, they need to exercise extreme caution. Shareholder approvals are typically required for major transactions and in some cases the controlling shareholders are not allowed to vote. So Mainland investors must pay attention to these events and exercise their rights to reject transactions they don't believe are in their best interests. They should also stay vigilant, regularly review their investments and make timely decisions.

Finally, the differences in the two markets are why investor eligibility requirements and eligible stock lists were introduced with Stock Connect in 2014 and why some restrictions remain in place for the upcoming Shenzhen Connect. We feel over time, as Mainland investors become more comfortable and more familiar trading in Hong Kong, these restrictions will be relaxed.

11 September 2016

19 | The difference between a traffic cop and the sheriff

My friend, Ashley Alder, the CEO of the Securities and Futures Commission (SFC), gave a speech recently about the regulator's renewed emphasis on policing the market through a "front-loaded" approach. The SFC's decision to recalibrate how it monitors the market and listing approval process has sparked some discussions about the listing regulatory system we have in Hong Kong, and the roles of both the SFC and HKEX.

For most market professionals, the roles and functions of both regulators are quite clear and without controversy, but I want to tackle a few questions that have popped up from other market players and friends from the media who might be a bit confused about who is responsible for what and whether this means the current system does not work well anymore.

Let me use an analogy here: imagine HKEX is like a city's traffic cop and the SFC is the sheriff. Both have the same overall goal of keeping traffic moving and ensuring the city's roads are safe, but there are many differences in their roles and the tools they use to carry out their duties.

The traffic cops get their power from the traffic rules (i.e., the Listing Rules): they are tasked with making sure driver's licenses (i.e., listing approvals) are given properly, traffic is flowing in good order, drivers are following the rules and traffic lights and speed signs are in the right places.

The sheriff's department gets its power from the law (i.e., the Securities and Futures Ordinance or the SFO, and the Securities and Futures (Stock Market Listing) Rules or the SMLR), and has a much broader scope and mandate: it's responsible for safety, enforcing laws, punishing criminals, surveillance and preventing as much crime as possible. The sheriff's department also supervises the traffic cops and can, under special circumstances, intervene in individual traffic violations or deny the issuance of driver's licenses.

This may sound simple and clear, but what to make of the recent changes that the sheriff's department is considering? Three questions come to mind:

1. Does this mean that the traffic cop has not done a good enough job so the sheriff needs to intervene?

I don't think so. As said earlier, the jobs of the traffic cops and the sheriff are different, and their mandates and approaches are also different. Most importantly, their tools and weapons are different. Standard gear for the traffic cop is a whistle, a night stick and an alcohol detection device. Those tools are effective to control the traffic, but are not terribly useful when it comes to apprehending suspects or detering violations. The sheriff's department is, however, equipped with big flashlights, handcuffs, tear gas and big guns. Those heavy weapons do not work well with the general traffic but they are highly effective in capturing and deterring criminals.

By the same token, the focus of work is different between the two departments. The traffic cop's primary objectives are to keep the traffic safe (i.e., market regulation) and keep the traffic moving (i.e., market development). If a driver is caught driving erratically, the traffic cops can pull them over and investigate, and even give them a ticket if necessary. But all along, the traffic cops need to ensure minimal disruption to the traffic flow as they conduct their work, and they won't block the entire road causing inconvenience to good road users just because they need to catch a person who jumps the yellow light or speeds. Meanwhile, if the traffic cops discover that a driver is a wanted criminal and a threat to the public, then the case is immediately turned over to the sheriff's department which will catch the fugitive even if that means setting up road checkpoints and bringing the entire city's traffic to a grinding halt.

2. Will stricter policing minimise or eliminate traffic violations and other criminal activities on the road?

Yes and no.

Greater vigilance and more aggressive patrolling will help detect, deter and minimise violations; if the traffic cops identify trends of certain driving behaviour or misconduct, the traffic cop should proactively enhance its policing by increasing patrols, intensify checkups and put up more speed limits and no stopping zones. The traffic cops should also refer more cases to the sheriff's department for more effective crackdowns. But can the traffic cop weed out all bad drivers or criminals through the vetting system or when granting licenses (i.e., listing approvals)? The answer is unfortunately that it would be very difficult.

Imagine an applicant for a driver's license successfully passing both the written and road tests. That would qualify him for a license based on the rules, and the traffic cop would likely issue one, despite the fact that he may become a careless or reckless driver, or even turn into a criminal on the road in the future. For sure, if the traffic cops detect a pattern in the breaches of traffic regulations, they can heighten the bar for the driving test before issuing a driver's license. But even if the traffic cops do that, it's still not possible to weed out all potential bad drivers. Also, a change in the driving test usually requires a market consultation to minimise the impact to other innocent citizens and the changes adopted should not be used retroactively to those who already secured a license under the old rules.

In contrast to the traffic cop, the sheriff has broader powers in this regard. He can deny the issuance of a license or directly reject the application even if the applicant has successfully passed or is able to pass both the written and road tests based on the traffic rules, as long as the sheriff has sufficient reason to believe that it is in the public interest to do so. In the same spirit, if a driver was stopped by the traffic cop but passed the alcohol test and without breaching any other traffic rules, the rules don't give the traffic cops a lot of power to detain the driver even if the traffic cop feels some suspicion. In this instance, the traffic cop should refer the case to the sheriff who could intervene and decide to detain the driver if he has sufficient cause to do so, for example, in the public interest. Either way, more aggressive policing on the road could help deter bad drivers from using the road.

3. How should the public be expected to deal with the two different enforcement agencies which end up operating in parallel?

In order not to confuse the traffic and to allocate regulatory resources efficiently, the traffic cop and the sheriff have historically reached an understanding that the city's traffic should be administered by the traffic cop under the oversight of the sheriff, who has generally worked behind the scenes through the traffic cop.

This means traffic cops administer the written and road tests to grant driver's licenses (i.e., eligibility and suitability tests), and the sheriff conducts deep background checks of the driver license applicants (i.e., public interest test) based on a wider network of intelligence. Ideally everything is done in one test to minimise additional burdens to the applicants, and although the sheriff can certainly either "front-load" or "back-load" the background test, he may want to avoid making applicants go through another written or road test.

Similarly, the sheriff's department has the power to get onto the highways with its big flashlights and heavy weapons whenever it deems necessary and appropriate, but doesn't do it frequently to avoid unduly disrupting the general traffic. It'd also be better to avoid taking the traffic cops' tools like whistles and sticks away and begin to police the traffic directly without the traffic cops in order to avoid confusion and chaos on the highway.

We have discussed earlier that the sheriff is the senior authority and has broad power under the law. Therefore, if the sheriff believes the prevailing market situation warrants he can certainly decide when and how to use his broad powers but coordination with the traffic cop and communication with the general public about the change would still be important in order to avoid confusion on the roads.

I know that the analogy of the traffic cops and sheriff is not the most precise description of the relationship between HKEX and the SFC, but I thought it could provide some interesting perspectives. On the subject of HKEX's role in this space, I want to point out that contrary to what some commentators suggest, HKEX, as part of the Hong Kong Inc., has every incentive to ensure high listing standards and a quality market. It's not worth getting a few extra dollars in revenue and allowing bad actors onto our market, which hurts the overall market reputation and diminishes our business.

In fact, although HKEX is a listed company, it's not an ordinary business organisation because, under the SFO, we must put the public interest ahead of our commercial interest, something we take very seriously. And from a commercial perspective, it is in our best long-term business interest to ensure quality companies that attract investors and liquidity list in Hong Kong. Half of our Independent Non-executive Directors are appointed by the Hong Kong government, our Chairman is appointed by the government and the Chief Executive is approved by the SFC to ensure the public interest is the top priority.

The regulatory system in Hong Kong has generally served Hong Kong well over the past two decades. Our city has grown from a regional market to a major player in global finance, often finishing among the first in the world in IPO funds raised over the last eight years. Nonetheless the fast evolution of our market and the welcoming of new market players and investors from different backgrounds and cultures have created new challenges, and we need to ensure our regulatory regime is keeping pace.

There is a perception recently of market quality problems. Even if the companies causing the problems just form a very small fraction of our overall market in terms of number, market cap and turnover, we take it seriously. We don't want a few bad apples to spoil the entire market. We will continue to make the best use of our tools to ensure our market runs smoothly and fulfil our duty as a traffic cop.

Our market has proven to be very resilient through massive transformation all around us, and I am confident that we will continue to thrive by working together to address outstanding market issues and ensure our market is high quality, robust, and safe for everyone.

23 July 2017

20 | Questions and thoughts on positioning Hong Kong for the future

There are still a couple of weeks to go to submit your feedback regarding our Concept Paper on the creation of a New Board in Hong Kong aimed at attracting new economy companies. I won't re-hash the entire paper here (you can read it or watch our webcast), but in a nutshell it proposes the establishment of a New Board with two segments: New Board PREMIUM would have roughly the same listing criteria as the Main Board today but allow companies with weighted voting rights (WVR), while New Board PRO would target early-stage pre-profit companies with light touch initial listing requirements and be accessible to professional investors only.

This is a Concept Paper by definition, so we are looking for thought leadership from the market and are open to any outcome. Meanwhile, a few issues have cropped up persistently over the last few weeks in my discussions with friends in the market, so I wanted to clarify our thoughts on them here.

1. Why would HKEX want to undertake such a major reform of the listing regime by creating the New Board?

We have a very good business model at the moment. We have traditionally had Chinese issuers and international investors, which is a great match. With Shanghai and Shenzhen Connect, we are now welcoming more Chinese investors with the goal of attracting more international businesses to list. With the four customers — Chinese issuers, foreign investors, Chinese investors and overseas issuers — we are in an advantageous position and it has driven our growth.

However, that isn't enough. Hong Kong has to "know its clients" and what they want, particularly future clients. If you look at the make-up of our market, it is overwhelmingly in traditional industries such as property and finance. But we're in an unprecedented period of creativity and entrepreneurship that has created new industries, new sectors and new ways of business and life that we have never seen before. Hong Kong has to find ways to inject that new energy into our market and change its DNA. These creative and new economy companies represent the future and we have to make Hong Kong a welcoming home for these companies and their investors from China and everywhere else.

Our present listing regime isn't agile enough to cope with the demands of this new age. This global relay race is in the first leg, and we are lagging behind; the good news is that we can still catch up if we do well in the second and third legs. After all, among global markets, only Hong Kong stands a chance at servicing all four of our customers well, which is a unique advantage of our market.

2. GEM didn't work. Why are you confident launching another new board will?

There are a lot of opinions out there on GEM, but whether it's perceived as successful or not shouldn't prevent us from continuing to develop our market. If something didn't work out before, it doesn't mean we should give up and go home. If we never try, we will never be successful. But if we do try, we have a chance.

Some also wondered why we are at it again after our last consultation on WVR did not come to a fruitful conclusion almost three years ago. The answer is: things have changed a lot in the last three years; the world has moved on — so has Mainland China and so has Hong Kong. Becoming a new economy market is too important an aspiration for us to shy away from reopening a debate no matter what the outcome might ultimately be this time around.

3. Hong Kong doesn't really have a history of technology companies. How could the New Board succeed without a broader ecosystem?

We agree that Hong Kong still has some ways to go to be competitive in the new economy, but our market is competitively positioned given our unique connection between the world and China. We already have some of China's largest and best new economy companies — some which are world leaders — on our markets. We have a good foundation but we have to begin earnestly building upon it rather than finding excuses why it can't be done.

In that spirit, we plan to launch a completely new venture called the HKEX Private Market in 2018 to provide early stage companies and their investors with a share registration and transfer platform based on blockchain technology so they can conduct pre-IPO financing and other activities on an off-exchange venue not under the regulatory remit of the Securities and Futures Ordinance. The Private Market will serve as a "nursery" for early stage companies before they are ready to enter public markets. We are hopeful that the Private Market will help foster a welcoming and supportive ecosystem in Hong Kong for new, early-stage startups and their investors and form a critical mass for success.

4. What exactly is a "new economy" company? How will you be able to determine which companies qualify and which ones don't?

It's obvious that some companies will be considered "new economy", such as those in technology, e-commerce, internet software and services, biotech and similar industries. It's a broad concept and we expect it to evolve over time. The key principle is to identify companies whose businesses are in sectors where *people* rather than *investment capital* is the key, and creativity, innovation, technology, intellectual property, and new ways of commerce in totality are the primary drivers for its growth and business successes. Our plan is to work out a definition that is principle-based with feedback from the market.

If it's anything like our listing suitability requirements, I anticipate that the overwhelming majority of the time it will be clear whether a company qualifies as "new economy". It might be a bit murky in a small number of cases so we'd need to deliberate further. I'm sure we may not get it completely right all of the time, but that's part of the learning process and I think the process will be refined and improved as we go along. It's certainly no reason not to attempt this at all.

5. What if we just accept secondary listings from companies with WVR already listed in the United States? Wouldn't that be easier?

The Concept Paper already proposes having New Board PREMIUM welcome secondary listings of US-listed firms with WVR structures, however there's no reason to restrict WVR to *only* companies already listed on the US market. Hong Kong should be confident in its listing regime, as we have proven over many years that we have a robust and successful market. We should have confidence in our own vetting abilities, and not simply defer to the US market to make decisions on our behalf. We were confident enough to pioneer the H-share listing regime 20-plus years ago and start the mutual market access programme with the Mainland three years ago. There is no reason why we can't accomplish what we set out to do without having to defer to or seek validation from someone else.

6. Why not just make changes to the Main Board to accommodate companies with WVR?

This is definitely something that can be considered, but I want to shed some light on why we proposed a New Board outside of the Main Board.

Our Main Board is the premier board in our market with rules and regulatory oversight that are clear to everyone. Any changes to this model, which has proven to be successful over the years, would require an in-depth consultation and would be a more controversial endeavour.

Imagine if a family wants to add a full-set of new, modern smart appliances to its existing kitchen. You could do it, but it would involve some big changes, be complicated and messy. Rather than cause such an inconvenience to the family, we propose building a kitchen extension so as not to disturb the family's daily life. Once the new kitchen is done, we can enjoy having both or we can tear down the wall between them and have one large kitchen. But this is a decision that can wait for another day.

The important thing is to introduce a key new dynamic to our market with as little disruption as possible, but we are open to the best ideas for making it work.

7. New Board PRO is high risk, so why does it have a light touch initial listing regime? Isn't that counter-intuitive? Could New Board PRO soon become a market of "listed shells"?

In an ideal world, we all wish that great companies could be allowed onto capital markets early on so that investors could benefit from their spectacular success while at the same time poor quality companies are barred from our markets through high hurdles and aggressive regulatory scrutiny. We do not, however, live in that ideal world. Nobody can tell in advance which companies will be great and which ones will turn out to be lemons. All great companies started as risky outfits in garages and basements and for every home run, there are dozens if not hundreds of strike outs.

If we do not want to miss the grand slam home runs, we have to make it easy for early-stage companies to access capital without making them spend limited valuable resources to work through a rigorous vetting and regulatory process. The only way to protect investors against such inherently high risk companies is either through full disclosure or, if still not sufficient, restricting access to professional investors only. This is a trade off that we all have to learn how to best assess.

Could New Board PRO potentially become another market of "listed shells"? No, it will be just the opposite. As listing on New Board PRO will be much easier, there would not be demand for the listing status or "shell" which would drain away much of its value as a tradeable commodity. We would also be focusing on the exit and looking at a more efficient delisting process, a system which wouldn't allow "zombie" companies or "shell" companies to stay for too long. So we think easy-in and easy-out is the way to go for New Board PRO.

8. Is it fair to only allow professional investors to access New Board PRO, depriving public investors of the right to invest in future new economy stars?

There is no straight-forward answer to this question. It is a regulatory decision reflecting the market regulators' philosophy and self-defined purpose. It is also a matter of how the regulators see the balance between investors' freedom of choice and their need for protection.

Regulators of different markets adopt widely different approaches. A developed market like the US is a buyer-beware market, with the regulator stressing full disclosure and investors all treated equally. That means public investors do not enjoy more protective measures than professionals. On the flip side, in the retail-dominated Mainland market, Chinese regulators see themselves as having a mission to protect average people, so they have laid down elaborate mandatory measures including strict initial listing and post-listing criteria to protect individual investors. Hong Kong is a hybrid, in-between the US and the Mainland. It is a highly developed market but at the same time a market with a strong Chinese element. This is why we designed the current proposal for New Board PRO, restricting retail investors' access to the higher risk market in order to ensure strong investor protection.

The bigger question we need to ask ourselves is this: what level of risk are we comfortable with as a market? If we break down our proposals, it spans a spectrum, as shown in the illustration below, from our private market, which is at the conception stage, is very high risk and involves very little or no regulation, to our New Board Premium which is a heavily-regulated board with stringent listing criteria and profits requirements, and is open to the public.

We can put ourselves anywhere along this spectrum: the further to the left we go, the more risk there is and the lighter touch that is required. For every iconic company developed this way, we might have many failures. If we aren't comfortable with that, we can move along the spectrum towards the right, and slowly raise the barrier to entry and thus lower the risk.

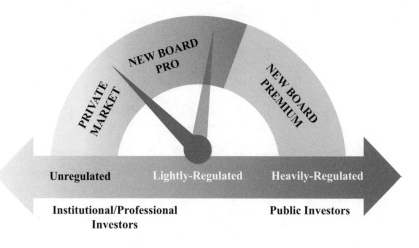

359

There are pros and cons to each of the above approaches, and I think this is the framework in which we should be discussing how we want to position our market going forward. The proposals in the Concept Paper are a start, but the opinions from the market will be vital to developing a more diverse market that can serve the best interests of Hong Kong and remain globally competitive.

These are just a few of the questions that have come up, and I'm sure there are many more. We encourage everyone in the market to speak up and tell us what you think. The future of Hong Kong's financial market depends on your participation.

We have spent many years considering how to strengthen Hong Kong as a financial centre while adhering to the city's core values of fairness and investor protection. We have also considered many diverse points of view from different segments of the market. No solution is easy and without challenges, but doing nothing is not cost-free or risk-free either. Inaction does not eliminate risks and costs; it simply passes them on to the next generation. It is our duty to act and proactively make a decision in the best interest of the future of Hong Kong. That's why it's more important than ever that we work together and overcome obstacles to build a market that is well positioned to capture the next generation of opportunities.

1 August 2017

21 | Shaping the future of our IPO market

Today the Hong Kong market has decided to take a big step forward and secure our relevancy as a premier global capital formation centre. Following an extensive market consultation, we have reached a clear consensus that Hong Kong must broaden its listing regime and proactively embrace the new economy.

We have truly come a long way. Think back to just four years ago, in 2013: the thought of welcoming pre-revenue companies or those with weighted voting rights would've been almost unthinkable. But as the years have passed, and more exciting new internet and e-commerce stars have emerged in China alongside the further growth of US tech giants, we began to ask ourselves if we were doing enough to compete.

I wrote a blog in 2013 that shared multiple perspectives on reform, with one key takeaway: whatever we do, it should be a pro-active decision. We shouldn't hold on to the status quo because of fear of change; if we were to decide to maintain the status quo, it should be because we felt, in our hearts, that it was the right way forward. For us to do that, we needed to look honestly at the question of our competitiveness and have a thorough debate about the benefits and consequences of different actions or inactions. We shouldn't be swayed by emotional arguments or be distracted by the circumstances of any one particular company.

Our market has evolved a lot over the past four years, and the time was right for us to issue a Concept Paper on listing reform. We received so many responses with so many ideas, covering a wide spectrum of possibilities to enhance our competitiveness. While there was some variation in precisely how to move forward, the market was unequivocal in its view that we need to change; we need to welcome companies to our market that better reflect the drivers of the new global economy, companies that are transforming existing industries and shaping new ones. As a global financial centre, our stakeholders have told us they don't want to miss this chance.

Our initial proposals included the creation of a New Board PREMIUM and New Board PRO, each with different listing and investor eligibility requirements depending on the risk associated with each board. The market told us clearly that all new listings should be held to the same high vetting standards as all other listed companies in Hong Kong, and questioned why mom and pop investors would be excluded from certain potential high-growth segments of the market. They felt it would be simpler and less complicated to include companies that are pre-revenue, pre-profit or employ weighted voting rights (WVR) to be included on the Main Board.

We therefore propose creating two new chapters to the Main Board to accommodate new economy companies that use WVR and pre-revenue companies, which will broaden our issuer base and diversify our capital market.

In formulating our proposals, our goal was to look at a number of ways we could open the door a little bit wider to welcome a broader diversity of companies to our market. Accepting WVR with adequate protections is one component of that plan. Whatever views we may have with respect to WVR, the competitive reality is Hong Kong can't afford to rule out these companies solely on the basis of their use of WVR.

Another new chapter allowing pre-revenue companies would initially apply only to the biotech sector because it has some particularly unique characteristics. Biotech companies make up a majority of pre-revenue listings globally and the sector is strictly regulated under a regime that sets external milestones on development. This provides a bit of a measuring stick for investors, and gives them an idea of how to judge companies that do not have traditional indicators of performance, like revenue or profits, as their products are not yet approved to be sold in the market.

If pre-revenue companies or those with WVR list in our market, it would trigger additional investor protection safeguards.

With respect to investor protection in the context of WVR, it is important to note the distinction between two different questions: (i) how to protect minority investors against possible misconduct by controlling shareholders and (ii) how controlling shareholders obtained their controlling power. Our proposed WVR rules do not alter any of the measures in the current listing rules related to the first question; the proposed WVR rules only address the second question by allowing founders of "new economy" companies to acquire controlling powers through contractual arrangement rather than the conventional model that only looks at the investment of capital. There are obviously additional measures to ensure that investors are protected against misuse of the powers acquired. Those protections include prohibiting the issuance of new WVR shares after listing, limiting who can hold WVR shares and the transferring of those shares, and much more.

Admittedly, coming up with a precise definition of "new economy" is a challenge, which is why we are proposing a requirement that the Listing Department put the factors that determine eligibility into guidance letters.

We are also creating a new route to secondary listings in Hong Kong to attract companies from emerging and innovative sectors. We are aware that many successful new economy companies already listed in the US and UK would benefit from these reforms.

These proposals would re-structure our market with the Main Board catering to established and high-growth companies with large market capitalisations, and GEM serving as a board for small and medium-sized firms. With stricter rules and heightened listing criteria, we expect GEM to attract quality issuers that meet our high standards.

Hong Kong is already a leading global IPO centre, and with these reforms we will be presenting an even more compelling case to new economy companies when they choose their listing venue. In fact, we've seen a marked increase in the number of listing enquiries we've received of late.

In closing, I want to give my heartfelt thanks to the entire market for considering these issues honestly and openly. The Hong Kong Government and the Securities and Futures Commission have also done a deep dive into the possibilities for reform and have guided us to find solutions that work. They deserve our sincere respect and gratitude.

We have come a long way, but we are not at the finish line yet. We will consult the market on Listing Rule amendments in the first quarter of next year, so we will be calling on the market once again to share your thoughts and suggestions.

When our market comes together as a collective to push forward positive change, there is nothing we can't accomplish. I am confident that together, we will put Hong Kong in a position to capture the next generation of opportunities and ensure our long-term prosperity.

15 December 2017

22 | Hong Kong well-positioned for the listing of Mainland unicorns

This article is only available in Chinese. (It is on p.106.)

23 | Dawn of a new era

We have been on a remarkable journey in the Hong Kong financial market over the past several years, and a brand new era will begin tomorrow when new rules take effect that will open the door to innovative companies that use non-standard share structures and pre-revenue biotech companies.

The changes are reflected in new passages in our Listing Rules, but their significance is so much greater: they culminate in the single most transformative change to our market in a quarter century, dating back to the introduction of the H-share regime back in 1993.

On the eve of this change, I wanted to share some thoughts with you.

Building a consensus takes time

First, Hong Kong has a long history of punching above its weight, competing hard, and winning. We are a small market, so it's vital that we stay relevant, agile, competitive, and always ahead of the curve. This is in our DNA, and we've proven it again by taking a very difficult—and sometimes divisive—issue and working out a solution that is broadly accepted across our market.

It was just five short years ago that the thought of welcoming WVR companies would not have been taken seriously. We had just lost a major listed firm to the United States, and we needed to examine whether we were doing everything we could to be competitive. Without the broader market consensus to move forward, we could only begin by debating all sides of the issue through a virtual reality of a dream blog*.

In the intervening years, we saw the continued rise of new economy companies,

---

* Refer to article #12 "Voices on investor protection" of this chapter.

many from our own backyard in Mainland China. We issued two concept papers and a consultation paper, heard from market participants, investors and other stakeholders, and saw the debate play out dramatically on television and in newspapers. I also put in my two cents over the years, blogging several more times to argue that Hong Kong needs to position itself as a leading venue for the next generation of technology companies, and that the market should study the issue carefully and make a proactive decision. Hong Kong is simply too important for us to avoid grappling with these questions.

With all sides having had their say, the Hong Kong government, SFC, HKEX and market participants joined hands to work out exactly what the new rules should look like. Without the leadership of the government, the support of the SFC and valuable contribution from all sides, today simply wouldn't be possible. It's a strong signal that we can cooperate, overcome differences, reach a consensus and find the right path forward together. It's what this city does best.

Staying relevant, staying competitive

There is no question that Hong Kong is under tremendous competitive pressure — capital markets around the world are in a fierce battle to attract listed companies and investors. We are not immune from these dynamics in Hong Kong.

The explosive growth of the new economy, particularly in China, has changed the economic landscape. Innovative companies are reshaping industry after industry, and many of today's economic champions are deeply rooted in new technologies. Our competitors abroad, particularly in New York, have been very effective at identifying trends and attracting these lucrative firms to their markets.

We have been slow to adjust, but it's not too late. We are at the precipice of a gold rush of new economy companies from China, and it's vital that we position ourselves to benefit from this development. In fact, Hong Kong has a long history of leveraging China's growth to develop our own market. A large share of our success over the past 30 years has been due to China's rapid growth — as China becomes more prosperous, Hong Kong benefits. It's a virtuous cycle that I don't expect to change anytime soon.

China is developing and internationalising faster than ever before. Mainland regulators recently announced the introduction of China Depository Receipts (CDR), which will attract more new economy firms to list on the Mainland. CDRs give these

The new listing regime becomes effective at the end of April 2018.

companies more choice, and will likely substantially accelerate the development of the new economy in China. This is good news for Hong Kong, because more companies seeking listings enlarges the pie for everyone. In relative terms, the Mainland may benefit much more from the introduction of CDRs and other reforms, but Hong Kong is still much better off than had the Mainland taken no action at all. The more China opens, the more its economy grows, and the more prosperous the country becomes, the better off we are in Hong Kong.

I have no doubt that many firms will choose Hong Kong as their listing destination because of our many competitive advantages. We have now expanded our Listing Rules to welcome new economy companies, like New York, but we are close to the Mainland and more connected to China in terms of culture, language and trading habits. Compared to the Mainland exchanges, we are much more international and market-oriented; our regulatory framework is basically based on disclosure and operates according to market principles. We are both international while having a deep understanding of China's national conditions. We have the best of both worlds, which gives our market a special appeal.

Protecting investors

When we looked at ways to enhance our competitiveness and open our market up to a broader array of companies, we had to find ways to ensure that Hong Kong retained its reputation as a market that treats investors fairly and protects them to the greatest extent possible from potential malfeasance. I want to address this issue directly, because it's one that speaks to the core of who we are.

The rule changes taking effect tomorrow do not diminish protection for investors in any way. We have not changed the current regime at all in terms of investor protection — rather, we have changed how we look at how controlling shareholders obtain their controlling position.

Here's how it works: we've traditionally recognised a controlling shareholder's position by looking at the financial capital he or she has contributed to the company. In other words, the shareholder's control is commensurate with how much he or she has injected into the company. This is easily understood and how we've operated our market for a long time.

Under the new rules, we are recognising that there are other ways for shareholders to control a company beyond simple capital injections. For instance, a shareholder's human capital, such as intellectual property, new business models, or founder's vision can be considered as acceptable means of acquiring control. So we are not changing how minority shareholders are being protected, we are simply opening the door to allow different ways for controlling shareholders to establish control.

Once that happens, those controlling shareholders will be regulated in the same way as we currently regulate controlling shareholders. On top of that, we are introducing additional safeguards to protect investors against the potential misuse of power.

The biotech revolution

Among all the changes to our Listing Rules, it's the biotech chapter that I'm most excited about.

We are on the precipice of an explosion in the development of new drugs and technologies that have the potential to transform human development on a global scale. We are seeing breakthroughs in science that could solve real problems like disease, infant mortality, a sustainable food supply, and potentially the ability to drastically improve our life expectancy. There is currently tremendous energy and investment in new pharmaceuticals, treatments and service models that could revolutionise our lives and those of our children, helping us to live healthier, longer.

This goes far beyond financial markets or Listing Rules. This is about human development, and we want to deploy our capabilities and advantages as a deeply liquid global financial centre to push this industry forward and provide the funding they need, so the rest of us can enjoy the benefits of their hard work and important breakthroughs.

In the United States, we've seen the baby boomers grow into retirement, which is putting new demands on the healthcare system. In a couple of decades, China's middle class — now the largest in the world — will also be seeking new treatments that help them handle pain, improve hearing or eyesight, or move around efficiently so they can enjoy precious time with their families. China's regulatory authorities have also recognised these demographics, and are keen on overhauling the drug approval mechanism and accelerating the research and development process for innovative drugs. These market trends will drive significant development and a massive need for capital in the coming years.

We are already seeing innovative companies in China using the Internet to transform the way healthcare is delivered, and we want Hong Kong to be a platform for these companies to grow and develop products and services that benefit all of us. Capital markets have an important role to play, and Hong Kong is perfectly positioned to be at the forefront of this global biotech revolution.

Of course, with great promise comes great risk. A pharmaceutical firm won't generate any revenue until it successfully goes through a long process of clinical tests and regulatory approvals. Biotech companies require huge investments into research and development, have extremely long life cycles, and a low success rate. For every home run, there might be a dozen strikeouts — or many more. While this underlines the risks of investing in pre-revenue companies, the good news is biotech is strictly regulated by national pharmaceutical regulatory authorities in the US, Europe and China, which is one reason that biotech made sense for our market. Each stage of development for a biotech firm has clear and explicit regulatory standards, which investors can reference when making investment decisions.

The inherent risk in biotech means it's more important than ever that investors do their homework. When we set a suitable threshold for listing, we did so according to industry standards while incorporating special disclosure requirements so investors can make informed decisions. However, there's no way we can guarantee only successful companies will list. The best we can do is to strengthen our knowledge about the sector, exercise judgment in our vetting process and require sufficient and appropriate disclosure. We are close to finalising appointments to a new Biotech Advisory Panel and have invited experts and others with deep experience in the sector to provide guidance to our Listing Department. We are learning from other markets in how they handle biotech firms, and will improve as we go along.

I'd like to reiterate that biotech investment is not for the faint-hearted investor. Investors should be aware that one clinical test failure could be a major setback for a listed pharmaceutical firm and instantly evaporate much of its market capitalisation.

The future

Tomorrow is the first day of our new journey, but we are not finished trying to improve our listing regime and enhancing our competitiveness. Innovative and biotech companies are new to Hong Kong, so we will be taking a very humble approach and learning as we go along. We will be agile, adopt some practices used elsewhere, and continually fine tune the disclosure standards and post-listing regulatory systems until we get it right. We will also carefully monitor the new rules and listen to feedback from the market. In short, if change is needed we'll do it.

We need to keep communication lines open, we need to be honest with each other, and we need everybody to pitch in to make it work. We are all invested in the future success of Hong Kong.

I don't expect companies to choose to list in Hong Kong simply because we now allow WVR or pre-revenue biotech firms. I expect companies to come to Hong Kong because we have a deeply liquid market, because we have a world-class legal system, because our rules and regulatory structures are transparent and accountable, and because we operate according to well-understood international standards. In addition, we are on the doorstep of the world's second largest — and arguably most dynamic — global economy with innovative trading connections with Shanghai and Shenzhen that are bringing even more Mainland investor participation to our market. This is why companies will continue to choose Hong Kong. The new rules simply mean we no longer need to automatically turn away exciting new economy firms, and we no longer need to deny Hong Kong investors from benefitting from these firms' successes.

Tomorrow is a new dawn for the Hong Kong market. It marks the start of a new era. Now we must get to work, in the greatest spirit of Hong Kong, to compete hard and solidify our standing as one of the brightest and successful financial centres in the world.

29 April 2018

24 | Bracing for the ups and downs of biotech

There is a lot of excitement in our market over the new rules we introduced earlier this month to welcome new economy and pre-revenue biotech companies. We've already had several companies express interest in listing under the new rules, and we've heard a lot of positive things from investors who are excited about being able to invest in some of today's most dynamic companies right here in Hong Kong.

While it's exciting to have access to the new economy, some of these companies come from sectors that are not as familiar to Hong Kong investors. We have a lot of retail investors in our market and a strong regulatory regime that strives to protect investors as much as possible — but there are still inherent risks when investing in any company. Biotech, in particular, is one field where companies could become a wild success or flame out into nothing, quickly.

So why did we want to open our market to risky biotech companies?

Pre-revenue companies tend to be high risk and lack the usual metrics of revenue or profit that investors use as yardsticks for valuation. This is why our Listing Rules have not allowed such companies to list in the past. However, biotech companies are something of a special case, since the very nature of the strict and lengthy approval processes for their products means that they are unable to produce revenues at precisely the time when they need capital the most.

On the other hand, since developing a new medication is a lengthy process that requires substantial research and development phases, multiple clinical tests, and ultimately an approval process by national healthcare regulators, biotech companies follow a path with clear

milestones that can help investors judge their stage of development and level of risk.

I mentioned in my previous blog that there is growing demand for new healthcare products and services to meet the needs of China's middle class, which numbers some 400 million people. Because of their potential to change people's lives, heal ailments, cure diseases, and allow us to live healthier, happier and longer, it is important that we connect the capital market with the biotech industry so emerging companies in need of investment have access to capital.

What are the particular risks involved in investing in biotech?

As I mentioned, developing a new medication does not happen overnight. It could take years, and most drugs never make it. Because these products involve the health and safety of the public, they are strictly regulated by governments. So even if a company passes multiple clinical trials, there's no guarantee a problem won't emerge in the last mile. One poor test result could lead to a dramatic decline in a company's market cap.

We tend to see a few big risks that investors should keep in mind. First, investors can become overly enthusiastic about biotech, especially when it's newly available on public markets, and get ahead of themselves. It's true that many of these companies are working on products that could literally change the world, so their potential can be limitless. This generates a lot of enthusiasm and excitement that often glosses over the potential risks. Second, we've seen some unsophisticated investors enter this space with confidence, thinking it's like other sectors, when it has unique characteristics and a much higher degree of risk. Third, biotech is more susceptible to insider trading. The reality is the life sciences and biotech sectors are extremely complex and require very specific expertise to understand. Public information, especially during the research and testing stages, is scarce; it is much different from a car company or luxury goods brand, with huge returns from dealing on inside information even more tempting than usual.

If the risks are so high, why not just set the entry bar higher?

As the operator of a major global stock exchange, we knew we wanted to open our market to these companies — despite the risks — because they need funding and investors are eager to tap their vast potential. But as a responsible regulator, we had to ask ourselves: "How mature should a company be before we allow it to list publicly? Where do we set the bar?"

If we allowed early-stage companies to list, we would be exposing investors to a higher degree of risk. But if we allowed only late-stage companies, they would already be more established and have fewer funding needs. It's a spectrum, and we needed to set the bar somewhere in the middle, somewhere it makes sense for Hong Kong — we want to ensure companies that need funding can get it, while the risk is kept as manageable as possible for investors.

Did we get it right? I don't know. We spent a lot of time on it, we examined the experiences overseas, we looked at the needs of the industry, and we consulted with a series of experts as we drafted the rules. This is new to us, too, so we are in regular communication with the market, have hired people more familiar with the sector, and recently formed the Biotech Advisory Panel to help us clarify or answer questions that pertain to a specific biotech issuer, or more generally to the biotech industry as a whole.

What are the main risk mitigation measures for investors?

With the advice of experts in the sector, we have set a range of entry criteria that biotech companies must meet before they can list. These include a minimum market value, the need to have reached certain milestones in their research and development, and to have received investment from sophisticated investors. The most important measure, however, is ensuring relevant, accurate and adequate disclosure.

If a company meets these criteria, then they will be allowed to list. We cannot pick and choose listing candidates depending on our whims. As a regulator and market operator, we are never going to be as smart as the market and, while we will exercise due care in our review of listing applications, it is up to investors to determine value and whether or not to invest.

What's the role of the Biotech Advisory Panel?

The Biotech Advisory Panel doesn't have any authority to approve or reject listings, but members can advise the Listing Department and Listing Committee on certain areas that may need to be clarified or expanded on in a company's draft prospectus, for example. In other words, the Panel will play the role of a passive advisor to provide guidance on information disclosure; it will not be vetting listing documents or play a gatekeeping role.

If the Exchange can't pick and choose the optimal listing candidates, then why not control the number of applicants at the beginning to limit speculative activity?

It's been a long road leading to this new Biotech Chapter of our Listing Rules and we are about to receive the first listings of pre-revenue biotech companies here in Hong Kong. While we hope for a steady and gradual development of the market, there is of course a risk that investors get carried away, leading to a market boom and bust. This is something that has happened in other markets, like the US, Europe and Taiwan. This can obviously be painful for investors, so it's important we learn from their experience.

That said, Hong Kong has always operated an open and fair regulatory regime, adhering to the rule of law. Therefore, while we can set the listing criteria, clarify the rules, and require more stringent disclosure requirements for biotech, it's up to market to set the pace and decide the inherent value of any given firm. Nobody can entirely eliminate risk in this sector, not even regulators. So it's a "buyer beware" market.

So what should investors do?

With the first wave of these companies set to list soon, investors must be cautious and exercise good judgment when trading in these shares. Because of the complexity of the industry, investors should have good professional knowledge and be willing to take on substantial risk if they plan to jump in. It's not for the faint-hearted investor. So do your homework, make sure you have a diversified portfolio, and seek advice if you're not sure whether to invest in a specific company. If you want exposure to the sector but don't fully understand its inner workings, a fund or ETF might be a more suitable option.

It will take a few years before we see a deep pool of talent, investors, and analysts in our biotech ecosystem. Our goal of a healthy and robust ecosystem could be threatened if a large number of investors jump in early and one or two companies ultimately fail. So let's be patient, be prudent, and do our best to make this work over the long term.

Biotech is a burgeoning industry globally that is drawing a lot of attention and excitement. I am delighted that Hong Kong can play a critical role by providing needed funding to these firms so they can continue on their quest to create new medicines, treatments or services that could improve real human lives. With the support and understanding of our market, I have no doubt that we can make Hong Kong a leading fundraising centre for the best and brightest biotech companies in the world.

23 May 2018

Chapter 3

Connectivity and Mutual Market

The launch of Shanghai-Hong Kong Stock Connect on 17 November 2014.

25 | Shanghai-Hong Kong Stock Connect — A big step in a long journey

We've been busy preparing for the launch of Shanghai-Hong Kong Stock Connect since it was announced last month. We had a packed house of market participants on Monday to learn more about how the scheme works, and will hold more seminars in the weeks ahead. I want to use today's blog post to set out the fundamentals of the scheme, so people can return to it at any time if they have questions.

What is Shanghai-Hong Kong Stock Connect?

Shanghai-Hong Kong Stock Connect is a mutual market access programme that will allow investors in Hong Kong and Mainland China to trade and settle shares listed on the other market via the exchange and clearing house in their local market.

Mainland investors who meet certain eligibility criteria will be able to trade eligible shares listed in Hong Kong, namely constituent stocks in the Hang Seng Composite LargeCap Index and Hang Seng Composite MidCap Index, and all H shares not included in these indices but that have corresponding A shares listed in Shanghai. It also allows Hong Kong and international investors to trade eligible shares listed in Shanghai, namely constituent stocks of the SSE 180 Index and SSE 380 Index, and all Shanghai Stock Exchange (SSE)-listed shares not included in these indices but that have corresponding shares listed and traded in Hong Kong.

Here is how the order routing and execution works:

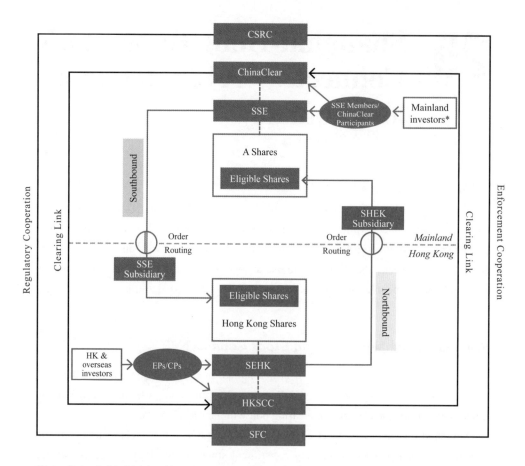

*Note: Only eligible Mainland investors can participate in Southbound trading.

What makes this scheme unique?

The scheme is a very significant breakthrough in the opening of China's capital market. The programme was structurally designed to achieve the maximum benefits of a free and open trading market while maintaining adequate risk management parameters to ensure smooth and sustained governmental policy support. More specifically, the programme requires very limited changes to the existing legal and regulatory systems and market structures in each market, but allows investors to trade across the markets with minimum incremental costs or inconvenience. The following highlights the key features of the programme.

(1) Order routing in gross for maximum price discovery

Orders through Shanghai-Hong Kong Stock Connect will be routed to the other market for matching and execution. This means that price discovery will continue to take place in Shanghai for A shares as a single liquidity pool and similarly in Hong Kong for Hong Kong shares. The *gross* of all buy and sell orders under the scheme will go through the boundary, enhancing market driven liquidity.

(2) Clearing and settlement in net for minimum cross-boundary fund flows

Clearing and settlement will take place in the local market where investors are based and only the *netting* of the buy and sell orders in one market will be cleared and settled between the two clearing houses at the end of each trading day. In other words, for Northbound trades, China Securities Depository and Clearing Corporation Limited (ChinaClear) will act as the host central counterparty and Hong Kong Securities and Clearing Company Limited (HKSCC) will be a clearing participant of ChinaClear. HKSCC will take up the settlement obligations of its clearing participants in respect of Northbound trades and settle the trades directly with ChinaClear in the Mainland, while HKSCC clearing participants will settle the Northbound trades with HKSCC in Hong Kong.

The opposite will happen for Southbound trades. HKSCC will be the host central counterparty and ChinaClear will be its Clearing Agency Participant. ChinaClear will take up the settlement obligations of its clearing participants in respect of the Southbound trades and settle the trades with HKSCC in Hong Kong, while ChinaClear participants will settle Southbound trades with ChinaClear in the Mainland.

Also, when Hong Kong and international investors invest in A shares, ChinaClear will impose its own risk management measures onto HKSCC. HKSCC will, in turn, impose those risk management measures on Hong Kong clearing participants.

(3) All Renminbi (RMB) conversions happen in Hong Kong

All trades under the programme will be done in RMB. Mainland investors will use RMB to invest in Hong Kong stocks, with the exchange to Hong Kong Dollars taking place in Hong Kong by ChinaClear. Hong Kong and overseas investors must also use RMB to purchase stocks in Shanghai, with HKSCC paying RMB to ChinaClear. Unlike the QFII and QDII schemes, the RMB exchange will happen offshore, so it will not impact the onshore RMB exchange rate or China's foreign exchange reserves.

The expected outflow of RMB from the Mainland and the expected inflow of RMB from Hong Kong will facilitate a healthy and significant offshore investment ecosystem for the RMB and help accelerate the internationalisation of the currency.

(4) Closed loop fund flow supporting prudent risk management

The design of the programme is such that fund and securities flows are insulated in the closed loop of the two settlement systems. Once Hong Kong and international investors sell their A shares or Mainland investors sell H shares, the money flows back to their home market bank accounts. Funds from the sale of shares cannot be used for speculation in other asset classes, such as real estate in the destination market. This design ensures that the programme is used exclusively for the purposes of investment activities between the two markets and cannot be used for other purposes as such money laundering and hot money asset speculation.

(5) "Home market" rules apply

When we drive in different countries we need to follow the rules of that country, and Shanghai-Hong Kong Connect is no different. Investors must abide by the rules in the market in which they're investing.

Specifically, this can be reflected in the following three ways:

Firstly, listed companies remain subject to the listing regulations in the market in which they are listed and investors from the other market will no longer be able to rely on their own home market regulators for investor protection other than through mutual enforcement assistance programmes by the two regulators.

Secondly, trading and clearing participants are still subject to the rules and laws of the markets where they operate.

Thirdly, the trading and clearing rules and practices of the home market where the trades are executed will not be altered because of the programme. However, where there are differences, the relevant regulators may require their local participants to comply with certain local rules as well.

(6) Interests aligned

Both sides' exchange and clearing company have an equal vested interest in the scheme as all trading and clearing revenues are shared equally, no matter where the majority of the trades may be executed. In fact, the entire scheme was designed symmetrically. Both sides have the same rights, responsibilities and obligations. If one side benefits, the other side benefits. If one side has a concern, it will be a concern for the other side as well. It is completely reciprocal.

This design also means regulators have a huge incentive to cooperate and ensure that their home market investors are adequately protected in the destination market and their home market trading is protected against market misconduct from participants based in the originating market.

7)   Quotas for a smooth and stable launch

Orders routed and executed through the Trading Links will be subject to a maximum cross-boundary investment quota, together with a daily quota that will be monitored in "real time" and used on a "first come, first served" basis. The quotas are in place to pace out the initial launch smoothly to avoid undue market volatility in the short term.

In the initial phase, Southbound trading of SEHK Securities will be subject to a daily quota of RMB 10.5 billion and an aggregate quota of RMB 250 billion, while Northbound trading of SSE Securities will be subject to a daily quota of RMB 13 billion and an aggregate quota of RMB 300 billion. As I mentioned earlier, calculations will be based on "netting" of buy and sell orders, which means the quota should be able to support much larger trading volumes.

(8) Scalable in size, scope and markets

Quotas are only one of the initial restrictions on the scheme, with the others being investor eligibility for Mainland investors and eligible stocks. These requirements are also in place to ensure prudent risk management at launch. Assuming a successful launch and smooth running of the programme, and subject to regulatory approvals, these restrictions are expected to be phased out over time.

What does this mean to Hong Kong and Mainland China?

Some commentators have characterised this programme as a government initiative to boost the domestic A-share market or a supportive policy in favour of Hong Kong. While the announcement of the programme did provide short-term positive momentum in market sentiments both in Shanghai and Hong Kong, the significance of the programme is much broader and deeper. In many ways, it is a major step in the journey towards China's financial market opening and the internationalisation of the RMB.

The Mainland benefits from the scheme by opening to a much broader international and institutional investor base while enabling greater diversity in the portfolios of Mainland investors. It creates a mutually accessible, comprehensive, closed-loop, scalable, and controllable scheme for the very first time that lets the market take on a more vital role and creates additional room for the further opening of China's capital account.

Hong Kong will also secure significant strategic benefits from the programme in consolidating its position as the international market of choice for the Mainland to go international and the China market of choice to access China. It also enhances our position as the world's primary offshore RMB centre, increases liquidity and broadens our investor base. It even paves the way for possible mutual market access across other asset classes and connecting with other markets in the future.

We first welcomed H shares to Hong Kong in 1993 in a scheme that, with hard work and patience, contributed massively to Hong Kong's success over the past 20 years. If we nurture and grow Shanghai-Hong Kong Stock Connect in the same way, I'm confident it will have an even more transformative impact on Hong Kong and Mainland China.

For now, our first priority is to help prepare the market thoroughly to ensure a smooth launch later this fall. I look forward to continuing to keep you posted on all key developments in this important journey.

8 May 2014

26 | Bridging the differences between Hong Kong and Shanghai

We've been very busy over the past couple of months preparing for Shanghai-Hong Kong Stock Connect. Our team has held a number of briefings and seminars for market participants, media, and investors, and we're delighted that the response has been positive and many are excitedly waiting for the scheme to begin.

We have also received a strong response from brokerages. Many have signed up to participate once the scheme starts, and we have scheduled market readiness tests for later this month. Once they are complete, we will publish a full list of eligible brokers on our website.

That strong response also extends to the investor community. Many people have told us they are excited about mutual market access and what the scheme represents in terms of the long-term development of China's financial market. Some investors have also raised some good questions, namely pertaining to certain constraints under the scheme—things like investor protection, the quota system, pre-trade checking, and investor and stock eligibility.

Before I turn to these topics, it's important we talk about why the scheme was developed. The Chinese market will open over time, but it will be a long and challenging process. If we wait for the Mainland market to align with Hong Kong's, we could be waiting a decade or longer. So we had to figure out a way to connect the markets while respecting the differences between the current regulatory regimes in the Mainland and Hong Kong. We needed to build a bridge connecting the two markets without fundamentally changing them.

However, it's not perfect. While we have managed to find a solution to most of the challenges of aligning two very different markets, some of the differences were so significant that our solutions will inevitably constrain the market. I know that these constraints aren't ideal, but considering the magnitude of the scheme and what it represents, in terms of progressing China's financial market opening and enhancing Hong Kong's role as a global financial hub, I believe that it is important to move forward rather than let such an important opportunity pass us by.

The constraints of the scheme at launch are necessary to get it off the ground, but I'm confident that over time, once people become more familiar with the scheme and the risks involved with cross-boundary investments, these constraints may eventually be removed.

Three areas of constraint are quotas, holiday arrangements, and trading mechanisms. I'd like to explain why these constraints are in place and what they are meant to accomplish.

Will investors be protected by their home market when they invest across the boundary?

One of the main principles of Shanghai-Hong Kong Stock Connect is "home market rules". That means investors must obey the rules of the market they are investing in, which is no different than if investors buy shares in London or New York. They will also be protected by the regulators of the market they are investing in. For example, the CSRC will be in charge of regulating listed companies in the A-share market, but it cannot regulate listed companies in Hong Kong for Mainland investors. So when Mainland investors buy shares in Hong Kong through the scheme, they have to obey the rules of the Hong Kong stock market and will be protected by the SFC. The same also works in reverse.

In terms of investor protection, it's important that investors taking part in the scheme understand the risks. Regulators can protect investors from losses from unlawful activities and can punish market malpractice. But investors also need to be clear about the rules and regulations of the market they are investing in.

I should particularly note that the CSRC and SFC deserve our gratitude and respect for their leadership and foresight in developing Shanghai-Hong Kong Stock Connect. They took on a lot of additional regulatory responsibility and challenges so that investors from both sides are afforded the opportunities and convenience of mutual market access.

Why are quotas imposed on the scheme? Will they be removed in the future?

The quota system is a key component of the scheme designed to minimise any potential unforeseen risk of excessive capital flows. It is in place for the initial stage to regulate the pace of capital flow and ensure the scheme is rolled out smoothly. I believe the quotas may be expanded over time depending on market conditions, or removed altogether.

Is the current quota allocation unfair?

Not at all. Unlike the QFII system, which grants quotas to individual institutions, quotas under Shanghai-Hong Kong Stock Connect are granted on a "first come, first served" basis. Quota allocation is based on the timing of buy orders, regardless of size or other factors. Orders submitted through the scheme will be subject to a throttle control mechanism, which works the same way regular orders are handled in the Hong Kong market. In other words, there are no inherent advantages for any particular broker or investor.

In order to prevent "placeholder" orders, we will strictly monitor buy orders to minimise the price gap between the orders and the most recent transaction prices. Furthermore, cross-boundary investors will not be able to modify orders after they are issued. Instead, the investor will need to cancel the first order and issue a new one, thus re-entering the queue.

Why is Shanghai-Hong Kong Stock Connect closed over so many holidays?

The scheme will only be in operation on days where trading and clearing arrangements are open in both Shanghai and Hong Kong. This is to reduce the burden, in terms of cost and staffing, that market participants would have to bear to execute trades across the boundary when the local market is closed for a holiday. Nevertheless, this is something we will continue to look at with market participants and the banking sector. If there is demand for cross-boundary trading and settlement during holidays, and it's feasible, it's something that can be considered in the future.

How should investors manage risk if one market is closed for a holiday while the other remains open?

The holiday schedules for Mainland China and Hong Kong are posted in advance. It is investors' responsibility to be aware of holiday arrangements and take appropriate risk management measures in accordance with their own needs.

How will pre-trade checking work in Hong Kong to comply with Mainland requirements?

Pre-trade checking is an example of a major difference between the two markets that we had to overcome.

Currently, Mainland investors are only allowed to sell A shares that are available in their stock accounts at the end of the previous day. The pre-trade checking structure at every trading account in the mainland market is easily able enforce this rule.

However, investors in Hong Kong trade under the T+2 system, whereby investors can execute their trade at a time of their choosing; they don't need to transfer the stocks to their selling brokers until two days after the trade is executed. It is also a common practice for institutional investors, in particular, to engage a custodian to perform share custody and transfer services.

Under Shanghai-Hong Kong Connect, CCASS will need to require international investors to instruct their custodian to transfer the shares to their selling broker before 7:30am on the trading day so CCASS can confirm the shares are with the selling broker before the market opens. This facilitates HKEx and CCASS to trade and settle with SSE and ChinaClear in compliance with the Mainland trading rules.

If the relevant stocks have not been transferred appropriately and in time, the institutional investor will be unable to sell the shares on that day. The same principle applies to retail investors, i.e. the retail investor will have to ensure that the shares are with his/her broker at the start of the trading day if he/she wants to execute trades in the A-share market that day.

This trading practice will inevitably add an administrative burden to international investors who will need time to become accustomed to these trading rules. Despite significant efforts, we will not be in a position to offer investors alternative solutions at the time of launch that will resolve this issue. After the launch, however, we will begin to allocate resources to develop possible solutions to help investors minimise the inconveniences such a practice may cause.

10 August 2014

27 | The beginning of a long journey

We are just a couple of days away from the launch of Shanghai-Hong Kong Stock Connect. More than a year of discussions and months of intensive market preparations all come down to the opening gong on Monday morning, when funds will flow directly between the Hong Kong and Shanghai markets for the very first time.

People on both sides of the boundary are excited about the launch, and are eager to see how much is invested northbound and southbound, how quickly quotas are used up, if international investors can get used to pre-trade checking requirements, and more.

The newspapers and media reports are awash with predictions from market analysts. Some expect there to be substantially more funds flowing north to Shanghai, while others predict Mainland investors will flood Hong Kong's market. Some think the quota is too small and will be used up quickly, while others might think initial trading will be light. With such a wide spectrum of predictions, some people will be right and some will be wrong.

The most important thing for us is to get this historic "train" on the tracks and to depart the station safely. Indeed, whether the initial trains are sold out with large crowds left on the platform or the train departs with some empty seats may not be as important. What matters to us is that this is a long term scheme and its success will be measured in years, not days or weeks. The immediate achievement is the infrastructure itself, which connects such vastly different and disparate markets and systems.

I've also heard a lot of people speculate over who gains more from the scheme; some think the Mainland is the major beneficiary because its market is opening and its currency is becoming more international. Others say it's a gift to Hong Kong that will enhance liquidity and bring more investors to our market. I think both of these perspectives miss the point — the question we *should* be asking is, are we better off with Shanghai-Hong Kong Stock Connect or without it? The answer is easy. Instead of counting marbles to see which side has more, we should be confident that we found a solution that benefits both sides.

The regulators, exchanges, and clearing houses on both sides of the boundary have worked extremely hard over the last seven months to iron out as many kinks in the scheme as possible prior to launch day. Just this week, the HKMA announced that the RMB 20,000 daily exchange limit would be lifted, and yesterday the capital gains tax issues were settled. This should give investors some certainty and confidence.

Still, there may be hiccups in the early going. Our team has worked hard and is prepared for a number of situations, but as with any scheme of this magnitude, there is always the possibility of something unanticipated cropping up. If this happens, we will work to fix the issues as best as we can. Shanghai-Hong Kong Stock Connect will keep evolving and getting better over time, so we shouldn't be too elated if everything is great on the first day, or too disappointed if it doesn't meet our expectations. It is a long road, and we are just getting started.

I want to sincerely thank everyone in the market for your support. It hasn't been easy coming this far, but we're ready to launch. Best of luck to you all. This is a journey we'll be on together for a long time. See you on Monday!

15 November 2014

28 | Taking stock and looking forward

The first week of Shanghai-Hong Kong Stock Connect is now in the books. Ordinarily, one week would be too soon to assess the performance of a major new market infrastructure project such as this. But given the substantial market focus and interest, I thought it might be helpful if I answered a few common questions we've received to date.

Are you satisfied with the performance in the first week? Or are you disappointed?

The most important measures of success for any major market infrastructure project are safety, stability and smooth operations. On that score, I am completely satisfied. I am therefore grateful to our colleagues, our partners and market participants.

Anticipating the flow of funds across the boundary was a bit trickier for us. We all shared in the collective enthusiasm prior to launch, so when we see the volume drop we're obviously a bit disappointed. But even if the market had rallied substantially in the first few days, it wouldn't have been sustainable in our view. Short term rallies usually happen when there are clear short-term trading profits to be gained; in this case, it's been seven months since the stock connect was announced, leaving plenty of time for A- and H-share prices to converge and for the scheme to be priced into the market. So there wasn't much of a "rush" to get into the scheme.

As I've said before, how much traffic travels across the bridge is a secondary consideration for us. As bridge builders, our biggest concern is that the bridge is built, it is safe and sound, and it is ready for traffic. As people become more familiar with it and have more reasons to cross, traffic will inevitably increase. Overall, this is a long-term scheme and we shouldn't get too hung up on the initial numbers.

What is your assessment of Northbound trading?

On the first day when the Northbound quota was used up, we saw a lot of swapping from the QFII scheme. As for the declining volume for the rest of the week, some have speculated the late clarification on taxes or challenges with pre-trade checking may have been factors. In other words, many participants are still going through their internal onboarding processes while others might be waiting for us to finish our central pre-checking enhancement on delivery. With the pending holiday season and year-end processes, we expect the whole market to be a bit quiet in the coming weeks.

Why has Southbound trading been so light? Are Mainland investors just not interested in Hong Kong stocks?

This is probably the most common question I've received since the scheme started. Southbound trading has been less than most market people anticipated. When we take a closer look at the reasons behind it, we've come up with a few potential factors:

(1) Absence of short-term arbitrage gains. Unlike 2007, the valuation gap between the A-share and H-share markets is no longer there and some of the gap is actually the other way around.

(2) Mainland institutions are not yet permitted to invest offshore. This issue is being resolved as the CSRC is already preparing the approval process. Over time, these institutions are expected to provide Mainland investors with more research coverage, product offerings, and promotional efforts to encourage more Southbound trading.

(3) The RMB 500,000 investor eligibility requirement. Investors with an account balance below RMB 500,000 contribute significantly to A share market turnover. Whether we are able to lower or eliminate this barrier is something we will work on with our regulators.

(4) Absence of small-cap eligible stocks. Mainland retail investors favour small cap stocks which were not included in the list of eligible stocks for Shanghai-Hong Kong Stock Connect. While more stocks in both markets will likely be included in the scheme over time, we need to give the regulators more time to get used to the new joint regulatory and enforcement regime to protect investors against market misconduct, which tends to have a disproportional impact on small-cap stocks.

(5) Lack of familiarity with the Hong Kong market. The biggest factor is still

Mainland investors' knowledge and familiarity with the Hong Kong market. The A-share market is a very unique market and A-share investors are used to an investment and regulatory environment that is very different from Hong Kong and other international markets. So it's understandable some of them may dip a toe in the water before deciding to jump in. The broker-dealer community will continue doing its part to promote the scheme and we fully expect more Mainland investors to gain confidence and become more active over time.

Did you launch stock connect too soon? Should you have waited longer to give the market more time to prepare?

This is really a chicken-and-egg question. The impending launch of the scheme is what allowed us to get the tax situation resolved and the market infrastructure upgraded. If we had called for this prior to the scheme being announced or a firm launch date, it would have been very difficult for market players to get on board and commit to making the necessary investments without knowing when or even if the scheme would ever happen.

I spoke to the media last week and said the launch of Shanghai-Hong Kong Stock Connect is like a baby being born. Maybe the parents haven't decorated the nursery yet or purchased all the baby clothes, but that can be done later. When the baby is ready to be born, we can't make it wait. So let's welcome it into the world and then make sure we take good care of it and help it grow.

What is HKEx doing to help promote the scheme so we see increased volumes in both directions?

Continuing with my analogy, the baby has been born so now we can look at ways to make sure it is as healthy and happy as possible. Northbound, we are working hard on pre-trade checking to find a solution that works for institutional investors. We hope this can be done by May next year, but there's a possibility we can even get it done sooner on a trial basis if the industry is willing to work with us more closely.

Southbound, we will work with regulators to allow more investors to take part while working with the broker community to help promote the scheme, understand Hong Kong better, and get more research going. Perhaps we can consider expanding the scheme to smaller stocks initially through index products like ETFs. This will all help.

Do you think the slow take-up of the Stock Connect will impact other links or "bridges" to the Mainland market?

I truly don't know the answer to this question. The slower take-up may give people pause to think about the next move, or it might make people less worried about the risks of financial stability that are typically created by fast and excessive cross-boundary fund flows. If I had to guess, the reaction is more likely to be the latter as the first week of the Stock Connect gave a positive sign that the market has been calm and investors have been rational and sophisticated.

Do you still think you're on the right path? How will this end up?

Without a doubt we are on the right path. In all of this punditry and early obsession with the numbers, we should not lose sight of what the scheme represents. It's an historic opening of the Mainland's market that will grow and get better as China accelerates its strategic goal of reallocating its massive domestic national wealth beyond its own domestic markets. We have a strong conviction that Southbound flows will arrive *en masse* over time. What happens this week, this year, or next year will probably be forgotten in the future when we look back at the importance of the scheme to Mainland China and Hong Kong.

When the H-share listing regime began in 1993, few companies in the Mainland used the scheme years after it was announced. It wasn't until the late 1990s that more companies listed in Hong Kong as H shares, and it really began to take off in the early 2000s. Now, we couldn't imagine not having H-share companies listed in Hong Kong.

We have built the first bridge for mutual market connectivity. Now, as bridge builders, we will start looking for the next place to build a bridge and continue our mission as we help unlock China's potential and bring more benefits to Hong Kong, securing our role as a trusted partner to China and the world.

23 November 2014

29 | From Shanghai-Hong Kong Stock Connect to establishing a "Mutual Market"

Shanghai-Hong Kong Stock Connect has just passed its two-month anniversary, and continues to grow steadily. The scheme is operating smoothly, with Southbound turnover gradually picking up towards the Northbound level. While we continue to tweak and improve the existing Stock Connect scheme, we are looking forward to where we will go next.

As we work towards a link with Shenzhen Stock Exchange, we are already thinking ahead about how Mainland capital markets will further open up in the Stock Connect era. Discussing the development of an Asian wealth management centre at the Asian Financial Forum yesterday, Chairman Xiao Gang of the CSRC stressed the long-term impact that the internationalisation of Chinese markets will have on the established global market order. In my blog today I would like to discuss the possible implications of the Connect programme and how that could impact the way we prepare ourselves for the accelerated pace of opening of China's capital markets. More specifically, I would like to raise a few questions:

(1) What is the backdrop of China's capital market liberalisations against which the Stock Connect programme was conceptualised?

(2) What is the essence of the Stock Connect model that makes it work — the emergence of a "Mutual Market"?

(3) What are the key features of the "Mutual Market" embodied by Stock Connect that makes it unique and valuable?

(4) How does Hong Kong move from Stock Connect to a more comprehensive "Mutual Market" encompassing other products and asset classes?

1. What is the backdrop of China's capital market liberalisations for the Stock Connect programme?

China today is closely connected globally in terms of its economic fundamentals; its financial system and capital markets are, however, still largely closed. As the second largest economy in the world, its currency is still not freely convertible. There is wide consensus both inside and outside of China that this is unsustainable and that China will inevitably accelerate the pace of opening up its capital markets. As China treads the path of capital market liberalisation, it will naturally have a tremendous impact on its citizens' livelihoods and entail enormous risk and responsibility for the country's leaders.

On its surface, opening up is simple enough: either international investors are allowed in, Chinese investors are allowed out, or both. In reality, this is easier said than done. Given the gulf between China's legal and regulatory standards and those of the rest of the world, it will take years before international investors are able to arrive in large numbers. Setting Chinese capital completely free runs contrary to the instinct of control and raises concerns of capital flight. Besides, Chinese investors have never had any meaningful exposure to the outside world and are highly unlikely to venture out lightly in any event.

It is against this backdrop that the key conceptual framework of Stock Connect was designed, such that it could capture the consensus and support of policymakers and regulators, and which ultimately led to the historic breakthrough.

2. What is the essence of the Stock Connect model that makes it work — the workings of a "Mutual Market"?

The greatest significance of Shanghai-Hong Kong Stock Connect is that it provides China with an interim model for opening up before it is completely ready for the large-scale arrivals of international investors and departures of Chinese domestic investors. The interim model works like a "Mutual Market" whereby investors on each side of the boundary are able to trade products of the other market within their home time zone, relying on their home market infrastructures. With the joint oversight of the two regulators, capital flows from China and international markets are able to congregate and interact with each other in this "Mutual Market", facilitating the gradual convergence of the Mainland and international markets.

This model has the potential to be extended to other products, including equity derivatives, commodities, fixed income and currencies. Those Chinese capital markets products that are of interest to international investors can be placed in this "Mutual Market", while those international products that are needed by Chinese investors can be added, thereby enabling Chinese investors to diversify their investments and hedge against international price risks. In this way, a swift and large-scale opening up of China's markets can be achieved, giving China greater international market and pricing influence without waiting for lengthy and extensive legal and regulatory reforms to be carried out first. And with the currency of transaction being the Renminbi (RMB), this "Mutual Market" would further accelerate the internationalisation of China's currency, helping to realise yet another goal.

3. What are the key features of this "Mutual Market"?

(1) **Close regulatory cooperation between the two markets.** Building on the Stock Connect model, the joint regulatory model would import the best features of both markets. The trust and confidence developed and the model of regulatory collaboration and enforcement assistance will be the foundation for the expansion

of the "Mutual Market".

(2) Gross order routing ensure maximum price discovery and revenue sharing ensures alignment of interests. Orders from each market are electronically routed in gross into one matching engine in the target market, ensuring one price discovery point. Shared economic interests allow the two exchanges to aggressively strive for success in both markets.

(3) Clearing and settlement in home market first before net settling between the two CCPs. Such a clearing model minimises cross-border fund flows and helps preserve the integrity and independence of the clearing houses in each market.

(4) Home market norms and practices for investors. Investors in each market can continue to rely on existing infrastructure, including brokers, trading platforms and clearing systems, while continuing to benefit from protection from their domestic regulator.

(5) Expanded fundraising market for issuers. Companies seeking capital would benefit from a larger pool of investors and enhanced liquidity.

(6) Enhanced risk management tools for capital markets users. For companies and investors, they would be able to access a wider number of derivatives to help manage market and economic risks.

(7) Greater incentives for intermediaries. Expands the pool of international products that Chinese intermediaries are able to offer their clients, fostering a natural process of international integration of the Mainland brokerage market.

The Stock Connect model has in essence created a new "Mutual Market" between the Mainland, Hong Kong and international capital markets. Through this two-way, scalable, closed-loop structure, international capital and products have been brought to China's doorstep for Chinese investors and vice versa, all taking place in an environment where risks are transparent and easily controlled.

4. How does Hong Kong move from Stock Connect to a more comprehensive "Mutual Market" encompassing other products and asset classes?

Shanghai-Hong Kong Stock Connect is really just the opening act for the "Mutual Market", with its potential far from fully exploited. In addition to the cash equities that are available now, there are many more products that could be included in this model. However, in order to move forward, we need to have some sort of common ground with policymakers and regulators about what we do and how. Perhaps I can offer a few initial thoughts on this below.

Building on the equities link

With Shanghai-Hong Kong Stock Connect already established, there is a clear template for a Shenzhen-Hong Kong link in cash equities. The only things we need to consider are when to roll it out and whether we make some incremental improvements to the rules and trading mechanisms, and add a few more products.

However, if cash equities are the trunk roads supporting capital flows between the Mainland and international markets, equity derivatives are the junctions and slip roads that help regulate the traffic. Equity index futures help investors in both markets hedge their risks, and would be a helpful tool to smooth the transition to greater A-share inclusion within global market indices, such as the MSCI Emerging Market index.

In simple terms, there are two ways in which equity index futures could be incorporated into the "Mutual Market": via cross-border licensing of index benchmarks or through the more comprehensive method of allowing cross-border trading. We are in close discussion with our partners and regulators to explore the best way forward.

Addressing urgent needs in the Mainland commodities markets

From an objective standpoint, if you ignore all the structural impediments, all China would need to do to internationalise its commodities markets would be to invite foreign investors in. However, given the realities of a closed capital account and wide differences in the legal and regulatory framework between China and the outside world, it will still take some considerable time before the Mainland commodities markets can be fully internationalised on its home soil.

Notwithstanding the fact that China is today the largest consumer and producer across a number of commodity classes, it is frustrated that it still feels like a "price taker" rather than a "price setter". If China truly intends to secure its rightful place in influencing international commodity prices, it will need to take bold action and allow its capital and/ or products to go international and exercise its influence in the international markets.

The crux of the issue is how to do all this? And should it be done alone or in partnership? In partnership with whom? And can partnership in the commodities area be a lasting win-win? Commodities derivatives are very different from equities and we need to carefully consider with the Mainland commodities exchanges how we can achieve an opening up of the market in a mutually beneficial way. I hope that we will be able to reach a suitable solution soon, and am convinced that the "Mutual Market" model will be the key to resolving this.

Liberalisation in fixed income and currencies is already steaming ahead

Reforms have been accelerating in China's interest rate and foreign exchange markets. However, as fundamental pillars within the capital markets, any shifts in policy in these areas are likely to have significant reverberations around the entire market. Given the complexities, the pace of reform needs to be measured in order to properly control the risks.

Against this backdrop, Hong Kong's role as a testing ground for the offshore RMB market is all the more important. To fully leverage Hong Kong's function in this regard, we need to broaden and deepen the offering of RMB interest rate and currency derivatives here. Initially, this can be through the licensing of Mainland benchmarks, gradually adding Mainland liquidity to the mix.

Rapid developments in the primary market present new opportunities and challenges

Primary capital raising is one of the most basic services provided by the capital markets to the real economy. Does the "Mutual Market" have a role in this? Hong Kong has long served as the primary offshore capital formation centre for China; with the accelerating reforms in the Mainland (particularly around registration-based IPO reform) coupled with the massive pools of domestic capital chasing primary issues of equity, there could be profound implications for the Hong Kong stock market going forward.

We need to carefully analyse the changes and the options available to corporate issuers, and consider how best to maintain our competitive advantages in this new environment:

- Mainland issuers often choose to list domestically because investors are familiar with their company and are willing to grant them a premium valuation due to substantial demand for new issues. However, this market has historically been constrained by uncertainties created by regulatory interference in the market, preventing issuers from having certainty about the timing or size of capital raising;

- The main driver for Mainland issuers selecting Hong Kong is that we have an open and market-oriented capital raising regime, with access to international institutional investors. The downsides have traditionally been lower valuations compared with the Mainland, as Hong Kong IPOs are not available to be subscribed by domestic investors, and there are limitations on full fungibility with the Mainland;

- International issuers considering Hong Kong and the Mainland for listings primarily wish to attract Mainland investors, but often find themselves unable to cope with the legal and market structures in the PRC.

The factors most likely to affect the Hong Kong IPO markets are the Mainland's enormous liquidity and the accelerating reforms in the domestic capital markets. Correctly understanding and anticipating these changes are crucial to developing mutually beneficial win-win market solutions for the future.

In summary, I believe that the "Mutual Market" model can become a new template for China's capital markets internationalisation, at least in the short to medium term. I've only raised many questions here and it will take our collective wisdom and efforts to find the right way forward. Without doubt, however, Hong Kong's "One Country, Two Systems" model provides strong underpinnings for us as the best "Mutual Market" partner for the Mainland.

So is the "Mutual Market" model the best option for Hong Kong? Can we be a part of this common marketplace without sacrificing or compromising our core values and standards? I believe so, but this is something that Hong Kong people need to debate carefully and decide.

20 January 2015

30 | Improving the bridge we built

We've had a very busy first quarter of 2015 at HKEx with the launch of a consultation paper on a closing auction session and volatility control mechanism, our review of submissions to our Concept Paper on weighted voting rights, and many other pressing issues. But along with these tasks, one of our top priorities has been to continue listening to the market and improving Stock Connect. We've built our first bridge to the Mainland, and now we want to make sure traffic can flow efficiently; fortunately, we're starting to see some real tangible results. With the strong market performance over the last few days, I thought now would be a good time for a quick reflection on what has transpired in the last several weeks and what is in our pipeline.

First, we're very pleased to see the increase in Southbound trading, which set another record today (2 April) with turnover of almost HK$6 billion. It was the third day this week that a new record was set, which indicates that Mainland investors are becoming more interested and comfortable trading Hong Kong-listed shares. In fact, the market capitalisation of companies listed on our exchange also hit a new record today at HK$27.3 trillion, while on Tuesday (31 March), stock market turnover reached HK$149.6 billion, the highest since January 2008.

The market will no doubt ebb and flow along with changing investor sentiment, but we should all be confident that we are heading in the right direction overall. Several enhancements have been rolled out in recent weeks to improve Stock Connect, as we promised to do when the scheme launched in November last year. Let me list them here.

First, we are pleased to see that the China Securities Regulatory Commission came out last week with the long anticipated announcement approving Mainland mutual funds to invest in Hong Kong shares. While formally allowing new mutual funds to invest in Hong Kong, its most important significance is that it clears the way for existing mutual funds to begin the process of amending their original investment charters to allow them to use

Stock Connect to invest in Hong Kong. This is a great start, but there is more to be done. Since the launch of Shanghai-Hong Kong Stock Connect, we've organised more than 50 investor education seminars for Mainland investors, and just recently introduced a range of market data promotions aimed at meeting the needs of Mainland investors interested in Hong Kong stock market data. We will continue to roll out additional investor education programmes to help familiarise Mainland individual investors with the Hong Kong market, and have some exciting initiatives coming soon.

Secondly, we are also making it easier for international and Hong Kong investors to participate in Northbound trading. On Monday (30 March), we introduced a mechanism that largely resolves international investors' concerns over the Mainland's pre-trade checking requirement, which made some institutional investors reluctant or unable to trade through Stock Connect. Now, investors will only be required to transfer shares they are passing to their broker for settlement after their sell orders are executed, which is similar to how they settle trades of Hong Kong-listed stocks.

Thirdly, an issue that frequently came up was beneficial ownership in A shares held through the nominee structure established under Stock Connect. Under the existing structure, Hong Kong Securities Clearing Company Limited (HKSCC) holds the shares for investors as a nominee holder. Initially, some investors were worried that they would not have proprietary rights over the A shares held through HKSCC. Funds with fiduciary obligations in the US and Europe need to show that they've got good ownership title, and they need to be able to enforce their shareholder rights. We have since clarified the role of HKSCC under the Hong Kong and Mainland China legal and regulatory frameworks and have engaged with the international fund industry, investors and regulators to answer questions and explain the nominee structure. Most investors now accept that they do have beneficial ownership under both Hong Kong law and Mainland law, which eases those concerns. Also, HKSCC recently amended its rules to enable it to provide certificates to A-share investors as evidence of their beneficial ownership, and even to help investors take legal action in the Mainland where necessary.

When we launched Shanghai-Hong Kong Stock Connect, we said it would be tweaked and refined over time to make trading through the scheme more convenient and efficient. While the changes I've outlined here have all helped, we can do more. We are now working alongside our counterparts in Shenzhen on the launch of a link with the Shenzhen Stock Exchange, which will open up even more investment opportunities and bring further improvements to the Stock Connect model.

Finally, it's important to keep Stock Connect in perspective. It's easy to get excited about record-breaking days or new links with Mainland exchanges, but these are all part of a much more profound structural change: the gradual but accelerating opening of Mainland China's financial markets. As I've written before, our goal is to work with Mainland authorities to create a multi-asset class Mutual Market that connects Chinese investors with international products, and international investors with Chinese products. The Shanghai and Shenzhen links bring equity products into the Mutual Market, and we aim to have equity derivatives included soon as well. Over time, we believe we can create a space that connects Mainland products with international investors and Mainland investors with international products across multiple asset classes including equities, equity derivatives, commodities, fixed income, and currencies. If international investors want to invest in Mainland products, they'll be able to do it here, and if Mainland investors are interested in international products, HKEx will have those too. This will be a win-win situation for all.

As I have said often, we are bridge builders and the bridges we build will be here for decades, not just months or years. We are now building the first bridges and will continue to improve the access roads, the direction signs and service stations, which will make it easier for people to cross. I am convinced that the traffic will increase significantly over time as investors on both sides of the border become more comfortable and more willing to venture out of their home markets now that new bridges are available, especially as the Mainland is accelerating the opening up of its markets. If China's capital markets one day get included in key global indices, hundreds of billions of dollars will need to be invested into China; and if China's national wealth is one day truly invested, not just sitting in bank accounts earning next to nothing, trillions will be deployed to capital markets both onshore and offshore.

I often get asked: will the bridge you have built become obsolete and useless once China fully opens up? I never really understand this question. If China truly opens up, it will just mean that they allow more bridges to be built and more tunnels to be opened. Yes, some people may then decide to move to the other side altogether, but a lot more will be just as happy to join the party by travelling freely back and forth. As long as our bridges are not more difficult to cross, why would they become obsolete? Especially if we build our bridges at the best locations with the highest traffic flows. Don't forget, the bridge we just built is actually toll-free. Investors on both sides just pay their normal exchange levies and brokerage commissions. There is no separate charge for people to use Stock Connect.

In short, this week's market performance and positive investor sentiment is an encouraging sign that we're on the right path and we should continue pursuing our vision. But ultimately this is just a start, and there is much more to come.

2 April 2015

31 | New records for Stock Connect, and new questions for Southbound trading

The Hong Kong securities market has been very active since the Easter Holidays ended, with the Hang Seng Index rising sharply for consecutive days. We've set a number of new records on back-to-back trading days, including market capitalisation, securities turnover, Stock Connect turnover, and many more. But amid all this excitement, both Mainland and Hong Kong investors have been asking me some key questions. Southbound investors are wondering if now is a good time to get into the market and how they can join if the quota is filled. Meanwhile, Hong Kong investors have asked if our local market will become more volatile once quota restrictions are relaxed.

I don't have a crystal ball to tell me which direction the market will move; your guess is as good as mine. But as one of the builders of the Stock Connect bridge, I want to say we should all take a deep breath after these record setting days and remember we're on a long road.

Take a deep breath

Stock Connect is a bridge that was built to stay open, so there's no need for Mainland or international investors to rush to cross. Everyone knows that if you travel over the Christmas holidays, you'll see plenty of crowds and potentially delays at the airport. Likewise, the Stock Connect bridge is crowded with traffic now, so investors can be prudent about when they feel is the right time to cross.

We made an enormous investment in this bridge, and we are continuing to work with brokers, information vendors, and technology vendors to beef up our investment to provide a reliable trading environment for investors.

The scheme will evolve over time

When Stock Connect launched, we introduced aggregate and daily quotas to ensure the scheme got off the ground in a smooth and stable way. Over the past two days, the Southbound quota has been fully used up before the market closed. This has prevented some Mainland investors from getting into the market when they wanted to, and they are urging authorities to quickly expand the quotas.

I would like to remind our friends in the Mainland: don't be too anxious. You should be patient. Regulators are closely and cautiously monitoring market developments and will consider expanding the quotas at the appropriate time. I am confident about this for two reasons: first, despite the active market, the aggregate amount going through Stock Connect is very small. The aggregate Southbound net flow of funds, as defined by the quota, has only amounted to RMB 47.9 billion*, which represents just 19 per cent of the total. RMB 115.7 billion has been used of the Northbound quota, or 39 per cent. To put this in perspective, the A-share market has been recording turnover of over RMB 1 trillion since late last year, regularly setting new records. The fund flows under Stock Connect are thus a tiny trickle compared with turnover in the A share market.

Second, Stock Connect has been designed as a closed-loop, transparent, and manageable scheme in which all transactions are clearly recorded and reflected in the exchange and clearing systems of both markets. There have not been any disorderly fund flows and after sell orders are executed, capital is routed back to the investor's home market. The design of this system allows enormous amounts of trading to take place without substantial fund flows actually having to cross the border every day, and also ensures that regulators in both markets can sufficiently monitor and manage the risk of cross-border transactions.

Endless opportunities, but not without risk

With the start of mutual market access, international investors who emphasise value investing are having their first historic encounter with Mainland investors, who are

---

* Aggregate quota consumption up to 8 April 2015.

primarily individuals. This will have a profound impact on both the Mainland and Hong Kong securities markets. These two groups of investors are totally different, and there will be certain chemical reactions as they meet, leading the development of China's capital market into a new era.

Out of this new era will come many new opportunities for local Hong Kong investors, but they won't come without risk. Stock Connect is the first time many Mainland investors are able to invest overseas, while their participation in Hong Kong will bring enormous opportunities and inject new vitality into our local securities market. At the same time, their differences in investment values, risk awareness and regulatory cultures will bring new challenges and risks to Hong Kong investors, particularly retail ones. Staying calm and exercising caution in a more active market will be a challenge to each investor in Hong Kong.

As the market operator and regulator, we know that the ever-increasing trading volume means that we are shouldering more and larger responsibilities. We must continue to ensure the reliability of our trading systems, constantly monitor the market and take appropriate risk management measures when necessary to ensure the market operates in a smooth and orderly way.

In closing, Stock Connect is more than just another simple investment channel. Its real, lasting meaning will be in enabling Chinese investors to internationalise their asset allocation in order to grow their wealth over the long term. Stock Connect is here to stay.

9 April 2015

32 | Stock Connect: The bigger picture

One year ago my colleagues at HKEx were busy getting ready for the start of a ground-breaking new programme called Shanghai-Hong Kong Stock Connect. We had the rules in place, the regulators had established cross-border regulatory cooperation, and technical preparations were ready. Our primary goal at the time was to ensure the programme launched smoothly. Evaluating early trading volumes was secondary, because Stock Connect was going to be a long-term component of our market's infrastructure.

It's amazing how fast a year can go by.

Here we are at the one year anniversary, and already people have been asking me to look back on Stock Connect and evaluate the scheme after its first year. Personally I do *not* like to spend too much time reflecting on past milestones; I like to look forward, and we're already working on a number of other initiatives. But given the significance of the anniversary, I'd like to share a few thoughts about what Stock Connect has meant to Hong Kong, and particularly what it means for our future.

First, there are the raw statistics. Total trade value going North into the Mainland was at RMB 1.475 trillion from the start of Stock Connect to the end of October, using up RMB 142 billion, or 47 per cent, of the aggregate quota. There was a record trading day on 6 July this year, when Northbound trading turnover hit RMB 23.4 billion. Southbound total trade value was HK$721 billion up to the end of October, using up RMB 89 billion, or 36 per cent of the quota. Our biggest trading day Southbound happened on 9 April, with turnover of HK$26.1 billion. It is worth noting that — in the context of high Mainland market volatility — the past three quarters were the Hong Kong stock market's best in terms of volume in the past five years. Perhaps more importantly, the Stock Connect mechanism performed without any operational lapses — notwithstanding the many operational differences between the two markets — and Hong Kong smoothly weathered the Mainland market turbulence and the authorities' subsequent rescue measures. This is indeed a tribute to the maturity of the Hong Kong market, its regulators, investors and intermediaries, and augurs well for Hong Kong's longer-term potential to handle Mainland and international investment flows.

There is still a lot of room for improvement in Stock Connect, in terms of new products, expanded quotas, enlarged stock eligibility criteria, and more. We are also excited about Shenzhen Connect, which will open up another Mainland market for international investors and strengthen Mainland links with Hong Kong.

But the focus shouldn't be on aggregate quota usage or trading turnover, because Stock Connect is much bigger than that: it's a catalyst, and a model, for the future. It shows us what's possible when we innovate and exploit the great advantages Hong Kong has under "One Country, Two Systems", and the benefits we can earn by working alongside Mainland China as it grows and internationalises. It unlocked a door to the Mainland, and in future will likely be seen as the pre-cursor to a new generation of cross-border regulatory cooperation and market connections.

Our vision is to transform HKEx into a global exchange group by facilitating China's opening and bringing benefits to Hong Kong. For the last 20 years, we have been an equity market that is very good at pricing companies, launching IPOs and trading equities. This has been our bread and butter for a long time, and because we have such a deep and liquid market we had a strong value proposition when we began building Shanghai-Hong Kong Stock Connect. This first bridge gave Mainland investors access to the Hong Kong equity market using their own regulations and clearing houses, making going "international" much more comfortable and familiar, and conversely gave international investors access to Mainland China equities under a familiar regulatory and legal framework.

True financial centres, however, go beyond equities. They price more than companies; they also price goods and money. China is the biggest consumer and producer of many bulk commodities, yet is poorly integrated with world commodity markets. As a result, China does not feel that it has the proportionate level of influence over global commodity prices that it deserves. Meanwhile, China's home-grown commodity markets cannot truly reflect global market prices because they are still largely closed off. There's no reason Hong Kong couldn't play a role here, but traditionally we've had no products to trade — we haven't been a commodities hub so have had no strong value proposition. But our 2012 acquisition of the London Metal Exchange (LME), which hosts over 80 per cent of global on-exchange base metals trading, gives us the chance to bring global products into Hong Kong. This is the rationale behind our second, cross-continental bridge — London-Hong Kong Connect: to give traders in the Asian time zone access to LME products denominated in Renminbi (RMB) with Hong Kong's own legal, regulatory, trading and clearing apparatus. Over the longer-term, we envision Chinese players exerting their pricing power more directly through a similar commodities connect scheme with Hong Kong, in a familiar regulatory environment, and have those trades relayed back to London via the London Connect. We are building our value proposition and securing Hong Kong's place as a trusted partner for the Mainland.

The most significant financial centres are able to price money, and that is our long-term objective. With the RMB offshore market expanding, Hong Kong is well set to grow in its role as a major pricing centre for the RMB. We are tackling that directly via our RMB futures product, which has seen an encouraging increase in turnover recently. However, another way for us to promote RMB is via commodities. In the international markets, commodities are nearly all denominated in international currencies. China's interests can be served if many of those commodities could be denominated, traded, and cleared in RMB, something we can make happen in Hong Kong. By combining commodities and currencies, we can potentially fully develop our FICC market and solidify our role as a significant global financial centre.

These plans require building more bridges, creating more connectivity, and more networks. None of this would have been possible — or it would have taken much longer — without Shanghai-Hong Kong Stock Connect as a precedent. It is a vital piece in the bigger picture of China's connections with global markets via Hong Kong. Stock Connect showed what could be done when we innovate, work across boundaries and join hands with regulators, and it laid the foundation and set the stage for the future. That will be Shanghai-Hong Kong Stock Connect's enduring legacy.

5 November 2015

33 | Taking mutual market access to the next level

Shenzhen Connect was announced last week, the culmination of a lot of hard work and effort between regulators and the exchanges in Shenzhen and Hong Kong. It was also a major milestone in our mutual market initiative, giving investors on both sides of the boundary more choices and enhanced access to each others' markets while playing a valuable role as a trusted partner in China's financial liberalisation drive.

For us at HKEX, it marks the completion of our mutual market access plans for China's secondary equity market. The vast majority of A shares traded in the Mainland are now directly available via Stock Connect to global and Hong Kong investors without the need to apply for a quota. The inclusion of 200 ChiNext stocks also gives international investors access to a new universe of technology companies, opening up even more investment possibilities. The abolishment of the aggregate quota and expansion of access, which will include ETFs sometime in the future, proves Stock Connect is scalable and flexible.

So with the launch of the scheme just a few months away, I thought it would be a good time to look more closely at how it will function. I took some questions at our news conference last week and have had friends ask me about Shenzhen Connect as well, so I figured it would be a good idea to publish answers to these below.

1. What are the highlights of Shenzhen Connect? Why is it significant?

Shenzhen Connect provides more access, more flexibility, more products, and more opportunities. It gives international and Hong Kong investors access to more A-share stocks and more sectors, such as technology and healthcare companies listed on the Shenzhen Exchange and ChiNext. With 880 new stocks included as part of the link, most companies traded in Mainland China can now be accessed directly by foreign investors for the first time.

We also abolished the aggregate quota, which took effect at the time of the announcement last week. That means international investors can feel more comfortable investing in China and Mainland investors have more freedom and flexibility to invest in Hong Kong without fears of bumping up against a quota.

Mainland investors now have more choice too, with 100 small cap stocks listed in Hong Kong now eligible for Shenzhen Connect. This, combined with Shenzhen Exchange's marketing and investor education efforts, will likely bring new energy to Hong Kong over time.

Finally, we announced there are plans to include ETFs in Stock Connect. We said from the beginning that Stock Connect was scalable and new products and asset classes could be added over time, and we plan to continue monitoring its development and looking for ways to make it better.

2. Why was the aggregate quota abolished but the daily quota was maintained?

Removing the aggregate quota should give institutional investors more confidence that the A-share market is open for investment on a large scale.

The daily quota, though, will remain in place for risk management considerations but it has effectively doubled. That's because the RMB 10.5 billion quota currently in place for Shanghai Connect has been replicated for Shenzhen. That means Mainland investors have a total RMB 21 billion quota to invest in Hong Kong stocks.

In fact, our experience with Shanghai Connect shows the RMB 10.5 billion daily quota was only completely filled a couple of times in the nearly two years since it started. Investors should therefore not be too concerned about the daily quota. It is just one extra safeguard in case there is an extreme flow of funds in either direction.

3. Why is investor participation in ChiNext stocks restricted?

We've taken a prudent approach to Stock Connect all along to ensure that it succeeds as a stable, successful programme.

The Quadrilateral Agreement signing ceremony for Shenzhen-Hong Kong Stock Connect.

The ChiNext board is a start-up board in Shenzhen with a number of small cap companies that is high risk and high volatility, so we will initially provide access cautiously. In fact, even the Mainland has restrictions in place vis-à-vis ChiNext. For example, Mainland investors who want to participate on ChiNext must sign a risk disclosure statement. In time, it's possible the Hong Kong securities regulator could introduce something similar here. But for now, retail investors who are interested in ChiNext stocks can trade them via institutions in Hong Kong, which creates new business opportunities for market practitioners in the city.

But like everything else in the mutual market access programme that is designed to be manageable, sustainable and scalable, I expect this to evolve over time. Regulators will keep a close eye on how the scheme functions and individual investors may be allowed to trade ChiNext shares eventually, subject to the resolution of related legal and regulatory issues.

4. Do you think the launch of Shenzhen Connect will help narrow the gap between A and H shares?

There was speculation the gap would narrow with the launch of Shanghai Connect, but that didn't happen. That's because the investor makeup and sentiment in both markets are fundamentally different. The Mainland market is mostly made up of retail investors who can be fickle, while Hong Kong is dominated by institutional and value investors. The result is a pricing gap that will probably be sustained in the short term, while long term the pricing will likely converge as stocks become fungible and investors have more options.

I do think Shenzhen Connect is another step towards the internationalisation of the Mainland's A-share market, however, and it's only a matter of time before A shares are included in major global indices.

5. Some investors have said the Shenzhen market is already expensive. Do you think there is much international investor appetite for stocks listed there?

I've read some commentary mentioning the high multiples, volatility, and low correlation with the world's major markets as a concern. Whether there is merit in these arguments is for investors to decide. As bridge builders, our job is to provide more access and more choices for investors to diversify their investments to the greatest extent possible. The stocks listed in Shenzhen represent much of China's new economy and the country's future, so I anticipate interest from foreign investors in terms of diversifying their portfolios and having exposure to high-growth sectors.

6. How do you see the market reaction to the launch announcement?

As I said at the start of Shanghai Connect, this is a long term market facility that can't be judged in one day or even one year. Short term market fluctuations are based on the sentiment of the day, but increasing access to the Mainland market is something that will be sustained and enhanced over the long term. Mutual market access will have a far reaching impact on both markets.

7. So what is the next step? What new Connect schemes are in the pipeline?

We said at the very beginning of Shanghai Connect that the programme could be expanded over time, and the decision to include ETFs is a great example.

We're going to need some time to work out some differences in the two markets, such as trading and settlement arrangements over different holiday schedules as well as the short selling mechanism. The exchanges and clearing houses on both sides are working towards the goal to provide investors with greater market availability and convenience. We hope we'll have some updates soon.

There are a number of ways we can provide greater cross-boundary access for both Mainland and international investors, and we will continue looking at ways to broaden Stock Connect, include new asset classes, and build more bridges so investors will have more opportunities and more choices.

21 August 2016

34 | Mutual Market Access 2.0

We are on the eve of another major milestone for HKEX: Shenzhen Connect. Launch day, set for next Monday, is special for us because it symbolises the beginning of Mutual Market Access 2.0, following Shanghai Connect's launch in 2014. It links the secondary equity markets between Hong Kong, Shanghai and Shenzhen for the first time, making it, when combined, the second largest stock market in the world after the United States.

We have been on the road for the past several weeks introducing international investors to Shenzhen Connect, with more than 50 meetings with investors across Europe and North America. The response has been very positive, with many investors enthusiastic about the opportunities that are being brought with the new cross-border trading scheme. We have also been traveling around Mainland China, with forums in Shenzhen, Shanghai and Beijing to introduce Hong Kong's market and share with Chinese investors some of the benefits of investing here. We are all excited about the long-term prospects of Shenzhen Connect, and the mutual market programme in general.

We've heard a lot of feedback from international investors, and we've been working with our counterparts in Mainland China since the launch of Shanghai Connect to address some of their concerns. For instance, the aggregate quota was abolished in August so institutional investors can trade confidently without fears of potentially bumping up against a limit.

We also introduced a mechanism last year that largely resolves international investors' concerns over the Mainland's pre-trade checking requirement, which made some investors reluctant or unable to trade through Stock Connect. The Special Segregated Account service means investors are only required to transfer shares they are passing to their broker for settlement after their sell orders are executed, which is similar to how they settle trades of Hong Kong-listed stocks. Both of these moves further lowered the barrier to participation.

The launch of Shenzhen-Hong Kong Stock Connect on 5 December 2016.

We have also largely resolved the issue of beneficial ownership in A shares held through the nominee structure established under Stock Connect, alleviating some investors' concerns that they would not have proprietary rights over the A shares held through the Hong Kong Securities Clearing Company (HKSCC). Last year, the China Securities Regulatory Commission clarified the role of HKSCC under the Hong Kong and Mainland China legal and regulatory frameworks, and most investors now accept that they do have beneficial ownership under both Hong Kong law and Mainland law, which eases those concerns.

Over the last year, we have broadened the list of eligible investors to welcome more people to trade through the Connect scheme. Mainland funds began using Stock Connect last year, and they've now become an important component of Southbound trading. Mainland insurance funds were given the green light to use Stock Connect recently as well, and we believe they will become active players over time.

Shenzhen Connect also brings along some notable improvements, including an expanded list of eligible stocks. Shenzhen is known as China's Silicon Valley, and is home to some of the country's most exciting technology and new economy companies. Foreign investors will now have direct access to this exciting frontier for the very first time. About 880 stocks in Shenzhen are eligible for Northbound trading, opening up a plethora of new opportunities. While investors in Mainland China will generally have the same choice of stocks in Hong Kong through both Southbound channels, they will also have access to 100 small cap stocks in the Hang Seng SmallCap Index for the first time exclusively via Shenzhen Connect, giving Mainland investors more investment choices and opportunities.

So what can we expect once the new link begins? As I've said before, I'm less concerned about the immediate market reaction because Stock Connect is a long-term market facility. Its impact will be judged in years, not in days or weeks. We expect Southbound trading under Shenzhen Connect to take some time to grow, as Southbound investors who have been using Shanghai Connect can continue doing so unless they are attracted by the 100 small caps in Hong Kong exclusively available via the Shenzhen Southbound channel. Northbound investors, who are no longer subject to the Aggregate Quota, can invest at their own pace in the Shenzhen market according to their investment strategies.

From a big picture perspective, Shenzhen Connect will lead to the further integration of the Chinese and international markets. While the two markets have big differences in terms of regulation and investment philosophy, the increased interaction between investors, regulators and market participants under Stock Connect will help the markets learn from each other, and hopefully our future marketplace will be the one with the best and most advanced features of both sides.

Stock Connect will also be here for a long time, even as Mainland China continues to open up. While sophisticated investors will have the means to invest directly into China and some Mainland investors have the knowledge and ability to go abroad, Stock Connect will continue to be a reliable platform for investors who seek access to investible assets abroad from the comfort of their home market rules.

Mutual Market Access also has an important meaning for Hong Kong. We have always played a vital role in different stages of China's financial market development. This began in 1993 when the H-share regime started, and Hong Kong facilitated the fundraising needs of Mainland companies as they grew and became global leaders. In the process, Hong Kong cemented its position as a global financial centre. Now as China's needs have evolved, so has our role. With Mutual Market Access, overseas investors can make use of Hong Kong as a convenient access point to Mainland China, while Mainland investors can use Hong Kong as their first stop as they begin to diversify their assets beyond Mainland China's borders. Over time, this influx of capital from the Mainland will make Hong Kong a focal point for international products. By playing this dual role, Hong Kong will grow as a wealth management centre alongside our traditional role as a global capital formation centre.

With the secondary equity markets now connected, we are setting our sights on growing the Connect model. Next, ETFs will be added under Stock Connect, and then we plan to expand into other areas such as primary listings, commodities, bonds and more. Our team is working hard to realise these dreams, which will further strengthen Hong Kong, and I hope we have something else to announce soon.

1 December 2016

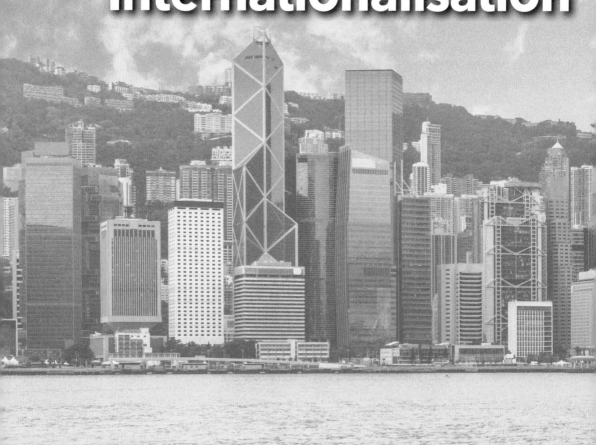

Chapter 4

Fixed Income & Currency Market and RMB Internationalisation

35 | "Raising children" & "Building a nursery"— Additional thoughts on the internationalisation of the RMB (synopsis based on Chinese original)

The offshore Renmnibi (RMB) market in Hong Kong has seen rapid development over the past 18 months. Since July 2010, RMB deposits in Hong Kong have increased from around RMB 100 billion to RMB 620 billion, and the total value of trades settled in RMB has increased by nearly 15 times. This has spearheaded the acceleration of the internationalisation of the currency. Not only has the pace quickened, the thinking has also broadened and several new initiatives have been put in place.

At the same time, debates about how to proceed with the internationalisation of the RMB inside and outside of the country have intensified. In the last six months there has been an easing in the growth of offshore RMB deposits in Hong Kong, and for the first time, there are expectations of a depreciation of the RMB in the offshore market. This has prompted all kinds of questions centred on the following issues:

(1) **Before the Mainland completes the reform of its top-level mechanisms for interest rates, exchange rates and capital account controls, and in the light of the current balance of trade, is it appropriate or feasible to proceed with the internationalisation of the RMB now?**

(2) **Is the RMB's internationalisation justified at this stage given the risks and costs involved?**

(3) Now that a large part of RMB deposits in Hong Kong have returned to the Mainland, will Hong Kong be able to really help the internationalisation of the RMB? Will the internationalisation of the RMB actually bring real benefits to Hong Kong?

Here, I would like to draw on an analogy involving some popular thinking behind raising children to illustrate my points.

Chinese people have a traditional concept that no matter how hard it is, they must have their own children so that they can depend on their children for support in their old age. The internationalisation of the RMB is, to Mainland China, similar to raising a child: it will experience birth, gradual growth and eventually moving out of the home to become useful as an international currency for trade settlement, pricing, investment and reserves.

It is important to raise this child because Mainland China is now the world's second largest economy and its currency, which is relatively closed, is starting to hinder the economy's further development. **If the Mainland is to achieve its long-term goal of securing greater global economic and political influence, it is important that it let the RMB participate in international currency systems in a more proactive manner. In other words, the Mainland has reached a point where it has to consider raising its own child.**

Only under this big picture can we view the process of RMB internationalisation holistically and treat the frustrations, risks and costs which have to be endured objectively. Below are my brief replies to the three key questions.

1. Before the Mainland completes reform of its top-level mechanisms for interest rates, exchange rates and capital account controls, and in the light of the current trade situation, is it appropriate or feasible to proceed with the internationalisation of the RMB now?

Although the Mainland has yet to complete reformation of its top-level mechanisms for interest rates, exchange rates and capital account controls, reformation of these mechanisms should not be a precondition for the internationalisation of the RMB. Just as there is no absolutely perfect moment to give birth to a child, there will probably never be an absolutely perfect moment to launch the internationalisation of the RMB.

Surely, the long-term absence of top-level mechanisms will ultimately restrain the pace of RMB internationalisation. As the RMB internationalisation "baby" grows, we have to improve the living standards and environment for it. On the other hand, the RMB's internationalisation can help apply pressure and speed up the implementation of necessary reforms of top-level mechanisms.

To answer the question of whether the Mainland's structural global trade surplus will frustrate the internationalisation of the RMB, I think we should look at the Mainland's trade structure from a dynamic and differential perspective. First of all, the dynamic changes between surpluses and deficits and between appreciation and depreciation will be a constant. The ever-changing global trade structure will also create favourable conditions for the internationalisation of the currency. Secondly, the Mainland's trade performance differs across regions. It is absolutely possible for the RMB to become regionalised. The RMB, as a new-born, may not be able to go far at this stage, but it can be nurtured nearby and raised in the vicinity.

2. Is the RMB's internationalisation justified at this stage given the risks and costs involved?

Since the internationalisation of the RMB is important and feasible — just like raising a child is — we have to provide the process with sufficient "nutrients" and "care", which will understandably cost money, take up time and involve risk.

Debates about risks and costs should be conducted in light of the larger framework of the long-term objective of the internationalisation process. Given this objective, some of the risks and costs are short-term; some of them are bearable, and some of them are controllable. Most importantly, once we clearly identify the long-term objective of the internationalisation of the RMB, we have to stay firm and move ahead boldly.

Apart from bearing risks and costs, we have to give the RMB internationalisation process patience and care, not blame it for failing to be a high achiever in the initial stages. The internationalisation of the currency is a long-term process. If we harbour premature and unrealistic expectations of its effectiveness, we will only jeopardise its development.

In short, like raising children, the internationalisation of the RMB will not be without hardship, difficulties and challenges. But we do not refrain from having children because costs and risks are involved. Rather, we endure them because of the happiness, bliss and ultimate benefit of having their support that raising a child will bring.

3. Now that a large part of RMB deposits in Hong Kong have returned to the Mainland, will Hong Kong be able to really help the internationalisation of the RMB? Will the internationalisation of the RMB actually bring real benefits to Hong Kong?

Hong Kong is the Mainland's only natural choice for the internationalisation of the RMB. In the early days of this process, Hong Kong can play the role of a "nursery" to provide a safe and reliable testing ground.

Currently, in the absence of a rich supply of offshore RMB products, the overwhelming majority of offshore RMB deposits immediately return to the Mainland without playing a large role in offshore markets. As a result, some may doubt the functionality of Hong Kong as a "nursery".

Although the majority of RMB deposits in Hong Kong quickly return to the Mainland, the nature of the deposits has changed: they are now owned by non-Mainland citizens and have the freedom to cross the border more freely. A child in a nursery goes home every day, but when he or she is in primary and secondary school they may go home during the weekend only. When the child is in university, he or she only goes home during the summer. After finding a job and joining society, the offspring may be stationed abroad and only come home for holidays or festivals. Today, the RMB in Hong Kong returns directly to the Mainland because there are insufficient offshore RMB products, not because the child is not promising. In the future, when the offshore RMB eco-system is further developed, international confidence in the currency will increase and offshore RMB will stay overseas for a longer period of time. This is why HKEx is working hard to promote the rapid development of RMB equities in Hong Kong. This is also why HKEx will continue to promote the development of the RMB interest rate and currency product markets, as well as the commodity derivative markets, to expand the possible applications of offshore RMB.

Some may also doubt the direct benefits that RMB internationalisation will bring to Hong Kong. My view is that the benefits will be huge both in the short and long terms.

In the short term, the benefits are intangible. Having run a "nursery" for the internationalisation of the RMB, Hong Kong has made a tremendous and irreplaceable contribution to the development of the country at a most critical moment. Naturally, this will enable Hong Kong to receive, in many aspects, tangible and intangible policy support and opportunities from the country.

In the long term, after the "nursery" for the internationalisation of the RMB proves successful and the internationalisation process begins to deepen, Hong Kong may become the "primary school", "secondary school", and even "university" for the internationalisation of the RMB. In this sense, the development of this "nursery" is an enormous business opportunity for Hong Kong.

4. Conclusion

The long-term objective of the internationalisation of the RMB is to rebuild the international monetary structure so that Mainland China can obtain greater global economic and political influence. With this long-term objective clearly in mind, one is more prepared for and has a greater understanding of the permanency of the process and the costs and prices that have to be paid. The process for the RMB's internationalisation is, to the Mainland, similar to raising a child. It's necessary, but returns will not be immediate. It will also cost money, take effort and involve risk. Hong Kong can serve, in the early days, as the "nursery" for the internationalisation of the RMB, facilitate its rapid growth, and become the first stop for the "child" to go out of the country.

In summary, the internationalisation of the RMB may proceed on a larger or smaller scale, but not internationalising the currency is not an option. Its speed may be fast or slow, but it must not be stopped. The atmosphere may be warm or cool, but its healthy development must be ensured.

3 January 2012

36 | OTC Clear: Building out a new asset class

Today marks an important milestone for HKEx, as we have officially launched our over-the-counter (OTC) derivatives clearing house, OTC Clear. The launch may not attract much attention locally or internationally, but it does represent an important step for us in fulfilling our long-term strategy to diversify into more asset classes and develop Hong Kong as a comprehensive financial centre.

We launched OTC Clear in response to changes in both the regulatory and business landscape. From a regulatory perspective, the financial tsunami in 2008 revealed some serious problems in the growing OTC market, such as counterparty risk, a lack of transparency, and uncertainty about the exact size of OTC losses and their distribution. In the aftermath of the crisis, the Group of 20 (G20) industrialised nations decided that considering the size of the OTC market, greater oversight was needed. The G20 mandated that all standardised OTC derivative contracts be cleared by central counterparties in order to reduce systemic risk to the financial system. This presents significant opportunities for exchanges and market infrastructure players such as HKEx.

From the business perspective, OTC Clear fits strategically into our long-term vision of building a leading horizontally complete and vertically integrated exchange in the Asia-time zone. We already have our own clearing houses for our exchange-traded equity and equity derivatives markets, which means all parts of the trading ecosystem — products, trading and settlement — are under one roof. But we are now moving into new markets and new asset classes, like fixed income and currencies. In these markets, a lot of the trading is OTC. Therefore establishing OTC clearing is necessary if we want to serve these markets well.

More importantly, we are positioning ourselves to be the leading exchange and Central Counterparty Clearing House in handling the growing use of the RMB as an investment and settlement currency. Hong Kong has already established itself as the most important

OTC Clear formally commences business on 25 November 2013.

offshore RMB centre in the world. OTC Clear will leverage this advantage by clearing RMB-denominated derivatives contracts, a tremendous growth area, further enhancing our status and differentiating us from many of our competitors.

That being said, it's important to understand that OTC clearing remains a relatively new market and, as I mentioned, there are a number of global regulatory issues that still need to be worked out. Therefore we will start small, testing the water with interest rate swaps covering the Hong Kong Dollar (HKD), RMB, US Dollar (USD) and Euro, and non-deliverable forwards with currency pairs of USD/RMB, USD/Taiwan Dollar, USD/ Korean Won and USD/Indian Rupee.

We are proud to have 12 founding shareholders as part of the business, most of which are leading banks in China and globally. They will also be the first clearing members of OTC Clear. While it is not a public utility given its highly customised products and services, the business is not a "club" either; the shareholders are key members of the trading community for these products, and we hope more members of the community will join as well. We expect OTC Clear to be competitive because it provides international risk management measures and capital efficiency, all based on Hong Kong's sound regulatory environment and market structure.

What is the benefit for Hong Kong? In the short term we are able to fulfill our G20 mandate; over the long-term we will grow this business and begin clearing a broader range of OTC products. That helps keep some business in Hong Kong that might otherwise go elsewhere, and it contributes to making the capital markets and the banking system safer for everyone.

Initiatives like OTC Clear sometimes fly under the radar. But as it turns out, some of the most important work at the exchange is behind the scenes to make sure we're competitive over the long term. OTC Clear is a good example of an initiative being launched as a piece of a much larger puzzle. It is not a standalone item, but a project that completes our vertically integrated business model in the fixed income and currency markets, and serves Hong Kong by bringing transparency and centralised risk management to the important OTC space. While it will take some time to get going, we're optimistic about its long-term potential.

25 November 2013

37 | HKEx: Prepared for greater RMB volatility

The Renminbi (RMB) has been making headlines in recent days because of its decline relative to the US dollar. In fact, the RMB has fallen the most since new foreign exchange rate policies were introduced in 2005, plummeting 1.4 per cent in only three weeks.

For a currency long assumed to be on a one-way path towards further appreciation, this news has naturally sparked a lot of discussion among analysts, investors and the media. Some believe the depreciation may be the result of the People's Bank of China wanting to shake out speculators; others say it may be in preparation of a wider trading band; while some believe China may be choosing to devalue the currency for domestic reasons. In my view, the reasons behind the change are secondary. No matter what happens, the RMB is on a path to greater internationalisation and increasing volatility — and greater risk is a part of that.

And why shouldn't it be? The world has changed, and the RMB is no longer only a domestic currency. Use of the RMB worldwide has grown exponentially, with RMB deposits in Hong Kong skyrocketing from around RMB 60 billion in 2010 to RMB 893 billion in January this year. Trade settlement by banks in Hong Kong has grown from an inconsequential amount in 2009 to RMB 3.8 trillion in 2013, a staggering increase in a very short period of time.

The Bank for International Settlements recently ranked the RMB the ninth most traded currency in the world, one more reason cities like London, Singapore and Taipei are looking to transform themselves into key hubs for offshore RMB complete with a suite of RMB products. We in Hong Kong are well positioned, having the foresight to lay the groundwork to facilitate and benefit from the RMB's internationalisation's earliest developments. Through hard work and strategic investments, infrastructure was put in place in 2011 to facilitate RMB-denominated IPOs, followed by the world's first physically deliverable RMB Currency Futures in 2012. We now have 112 RMB-denominated products listed on HKEx's markets.

Some of our early efforts are starting to bear fruit today. Our RMB Currency Futures contract, for instance, has seen steady growth since it was introduced in 2012. Average daily volumes in February saw a record high of 1,461 contracts (US$146.1 million notional), up 500 per cent from the average daily volume in 2012. We also saw record daily volume of 5,970 contracts traded (US$597 million notional) on 25 February. Meanwhile, open interest has grown more than 600 per cent, hitting a month-end high at 22,636 contracts (US$2.3 billion notional) in January. The open interest's record day high of 23,887 contracts (US$2.4 billion notional) was hit on 14 February. We are seeing increasing need for price transparency and to manage foreign exchange and interest rate risk in light of the RMB's volatility, especially among corporates engaged in business between China and other countries. Futures contracts provide easy access to all investors to address these risk management needs effectively and efficiently.

As Chinese leaders made clear their desire to have the market play a decisive role in China's future in November last year, we expect the RMB's internationalisation to only pick up steam in the months and years ahead. That means RMB volatility is here to stay. We want to ensure we don't surrender our first mover advantage. We need to stay on our toes and make sure we're providing the right risk management tools.

With that in mind, we've been working on a number of new initiatives. Just last week, we announced RMB Currency Futures received approval from the Securities and Futures Commission (SFC) to be included in our After Hours Futures Trading (AHFT) session, starting on 7 April. With RMB Currency Futures contracts available after the end of market close, it will help meet demand from the European and US markets as well.

Secondly, we are looking at extending the maturity of RMB Currency Futures to manage longer-term risk exposures. In the future, we aim to lengthen the maturity even beyond 16 months, introduce RMB Currency Options and further extend AHFT hours. Last week we also announced our first infrastructure footprint in Mainland China, our Mainland Market Data Hub. For now, it is making index and securities market data available to our Mainland clients, but later this year derivatives data will be added as well.

Finally, our RMB readiness has extended into the over-the-counter derivatives space. We launched OTC Clear last year and we expect it to grow over time. It is already clearing single currency interest rate swaps (IRS), single currency basis swaps, non-deliverable IRS and non-deliverable forwards, supporting RMB-related FX and interest rate risk management.

The recent volatility in the currency demands products that can manage exchange rate risk, and our RMB Currency Futures are a key tool for that. But internationalisation of the currency also requires a much wider spectrum of RMB-denominated products outside of Mainland China. We've made a start, but more will be done. We'll stay focused and ensure HKEx Group remains well positioned to serve the needs of investors in this next phase of China's remarkable growth story.

9 March 2014

38 | HKEx expands product portfolio for future growth

As we near the end of 2015, we are already starting to look ahead to what next year will bring. This year was a major milestone in capital market development in Mainland China as we experienced the first full year of Shanghai-Hong Kong Stock Connect. This major market innovation is not only bringing new liquidity to Shanghai and Hong Kong, it is also subtly altering the ecology and investment demands in the two markets.

We just celebrated the anniversary of Shanghai-Hong Kong Stock Connect earlier this month, and I said at the time that we are already moving forward with new initiatives to improve the Stock Connect and prepare for a more widespread opening of the Mainland's financial market, which we believe is inevitable.

Therefore our job is to get ready for these opportunities. This opening will eventually see more Mainland investors invest offshore, and we want to ensure Hong Kong is their first choice. We are creating an environment where Mainland and global investors have more options when it comes to allocating their assets.

That brings me to our announcement of the launch of 34 new stock futures contracts, some of which will begin trading tomorrow (30 November) and the rest on 7 December.

These new futures contracts cover the 34 most actively-traded stocks in Southbound trading under Stock Connect, together accounting for about 60 per cent of Southbound trading turnover. The addition will also bring the total number of stock futures contracts traded in Hong Kong to 71. This gives investors much more choice and flexibility, and come at a time when stock futures trading volume on HKEx rose 91 per cent in the first 10 months of the year from the same period last year.

We are also nearing the debut of the launch of new London Metal Mini Futures contracts for nickel, tin, and lead. The three will expand the range of commodities products offered by HKEx, broaden our base of potential market participants and provide more arbitrage opportunities. The contracts, which are all RMB-denominated, join the first batch of London Metal Mini Futures contracts — London Aluminium Mini Futures, London Copper Mini Futures and London Zinc Mini Futures — that launched in December last year.

Like Stock Connect itself, we believe the launch of these new stock futures, and their commodities counterparts, lay the groundwork for future growth. As more Mainland investors invest Southbound, the risk of volatility will increase alongside the need to manage risk. However, this process will take time; we don't expect immediate results, but we want to ensure we're ready when the demand is there.

Compared to equities, stock futures can create higher capital efficiency for investors. If used properly, they can also be a very effective risk management tool. However, as any kind of derivative can be high-risk, investors should fully understand the contract and have a high risk-tolerance before investing. Along those lines we will launch a number of education sessions with futures brokers to help them understand the new products and explain them to their clients.

These new futures contracts align with our overall vision, which we have been steadily realising since our acquisition of the London Metal Exchange in 2012. We are enriching our product line, providing new opportunities for investors, and ensuring we become China's top international financial centre and the world's best access point to China.

29 November 2015

39 | Risk management and opportunities under RMB volatility

2016 has barely begun, but so much has happened already in the financial markets. Many friends are distraught at the decline in the value of the Renminbi (RMB) in the last week. They have asked me how they can hedge this currency risk, especially as many people had put their savings in RMB in anticipation of the currency appreciating.

The RMB had appreciated for a long time — for almost two decades it was seen as pretty much a one-way bet until only a year or two ago, and I blogged about this at the time, arguing that the RMB rate will be more susceptible to market forces*. Today, with the RMB being included in the International Monetary Fund's Special Drawing Rights (SDR) basket of currencies, we are seeing that two-way volatility is the new norm, and investors need to wear their seatbelts and prepare to manage the exchange rate risk.

This new era of volatility in the RMB brings opportunities to Hong Kong, which is the largest offshore centre for the Chinese currency. When the RMB appreciates, Hong Kong can provide strong RMB-denominated assets to investors who are interested in holding them, and when the currency depreciates, we can provide a variety of debt products because a weakening currency is in borrowers' favour. In fact, a number of companies have recently been reducing their borrowing in US dollars and raising RMB loans, a very sensible market reaction as the RMB depreciates.

With price fluctuations in the offshore market more intense than in the onshore market these days, there is a greater need for exchange rate risk management in the offshore market compared to the onshore market. The contrast also shows that the breadth and depth of the offshore market for RMB-denominated products needs to be strengthened. A strong offshore market for RMB products is especially important at times like this, when a small change of Mainland policy often leads to extreme volatility in the offshore value of the RMB.

With this market demand and our unique advantages, we can develop RMB exchange

---

* Refer to article #37 of this chapter.

rate and interest rate products which will not only help the RMB further internationalise and gain influence, but also build out the RMB ecosystem in Hong Kong.

We took our first step in 2012 with the launch of the RMB currency futures contracts to meet market demand for a tool to manage the RMB's currency risk. We're thrilled the product has already become the most actively-traded on-exchange RMB futures contract in the world. Because of the recent volatility in the RMB, the contracts hit a record open interest level of 29,352 contracts on 11 January, and turnover hit 6,425 contracts with a notional value of US$643 million on 7 January, the second highest ever.

What's the outlook going forward? First, the People's Bank of China (PBOC) has been very clear about its determination to promote market-oriented reform of the RMB exchange rate. It has relaxed controls in recent years, and also widened the trading band, which introduced greater volatility.

Second, more and more countries are holding RMB as the currency internationalises, meaning its exposure to supply and demand of these economies is increasing. With the inclusion in the SDR, central banks around the world are actually required to gradually increase their RMB asset allocation, which will bring further market forces to the determination of the value of the RMB.

These are steps taken by the PBOC as it has undertaken market-oriented exchange rate reforms. It has repeatedly stressed that the RMB is no longer pegged only to the US dollar, but rather a basket of currencies. On 11 December, the China Foreign Exchange Trade System released, for the first time, a new exchange rate index based on the RMB's value against 13 trade-weighted currencies. The PBOC's intention is clear: as an emerging market currency that is internationalising quickly, the PBOC wants the RMB to continue to be relatively stable against the basket of currencies, but has no direct ties to a single currency. That means while the RMB exchange rate will be stable overall, the RMB-USD exchange rate may be volatile, which requires proper risk management.

To help investors meet the new challenges, HKEX intends to roll out a suite of new products to hedge RMB-related risks, including currency-pair products against multiple currencies and precious metals products denominated in dual-currencies. On Friday, we will also host a forum for Hong Kong futures brokers to explain how to manage RMB exchange rate risk in this new era.

This is all a good start, but more will be done. We'll stay focused to better serve the needs of investors and strengthen the position of Hong Kong as a major RMB pricing centre.

12 January 2016

40 | Three steps to developing our fixed income and currency business

HKEX's third annual Renminbi (RMB) Fixed Income and Currency (FIC) Conference kicked off in Hong Kong today with more people attending than ever before to hear about opportunities in the FIC space. When we started the conference in 2014, it was a much different time. China's FIC markets were tiny, the RMB exchange rate was managed in a tight range, one-way appreciation was expected, and nobody had heard of defaults in China. How quickly things have changed!

Since then, global interest in China's FIC market has been growing. The RMB FX market has been opening up, and there are expectations that its bond market — already the second largest in the world — will double or triple in size in the next 10 years. Funding, which has traditionally relied on banks, is shifting to the capital markets, which is creating even more opportunities. On the flip side, Chinese investors are slowly coming out to the global bond markets, which many believe will pick up steam as pension and insurance funds seek greater international diversification. The country is also accelerating its strategy of going offshore through acquisitions, investments, and a willingness to have a stronger voice in the pricing of products.

The RMB, too, has been moving quickly. Its inclusion in the International Monetary Fund's Special Drawing Rights (SDR) basket was an indication that it is becoming a major global reserve currency. There is now greater tolerance in China for volatility in the RMB exchange rate, and the shift to basket pricing is a sign of a major policy shift. Once promoted as a trade and payment currency, the RMB is now becoming an investment asset class of its own.

The RMB is currently at the early stages of being used for payment, ranking number five globally according to SWIFT. But as its use grows, it will develop into a global reserve currency commensurate with the size of China's economy, followed by a global pricing currency.

These are tectonic shifts, and Hong Kong is in a great position to take advantage of these changes. We can make our mark, and build Hong Kong into a comprehensive wealth management centre.

I've said before that for a financial centre to become truly successful, it needs to be able to price companies, goods, and money. Hong Kong is already a leader in equities, and our 2012 acquisition of the London Metal Exchange has given us a solid footing in commodities and we are making good progress. But our FIC business is just at the beginning stages, so we are focused on building out a comprehensive ecosystem for FIC investors.

Our plan at HKEX involves **three key steps**: having the right **platform**, the right **products** for investment and risk management, and **cross-market connectivity**.

On the platform side, we started building OTC Clear in 2013 as a key piece of our infrastructure in anticipation of future clearing by FIC market participants. To take advantage of Hong Kong's unique position, OTC Clear focuses on clearing services for regionally-traded products and particularly on RMB-based derivatives. Over the last year, we have built the capability to provide clearing services for USD/CNH cross currency swaps, which has generated significant interest in the market and is expected to launch shortly, pending regulatory approval. Going forward, we will further expand OTC Clear's offering to cover clearing for wider types of RMB-based FX products.

On the product side, we launched our USD/CNH Currency Futures contract in 2012, which has seen turnover take off since last year and is now the most actively traded RMB futures contract in the world. We are now moving to diversify our product offerings by launching new CNH currency pairs against the Japanese yen, Euro, and Australian dollar, which will begin trading on 30 May to facilitate cross-currency hedging.

There is also tremendous potential for an RMB Currency Index benchmark as the RMB becomes a reserve currency and the market focuses on the relationship between the RMB and global currencies. That's why at the conference today, we announced the first tradable RMB index with Thomson Reuters, establishing a market benchmark tracking the RMB against a basket of world currencies. The China Foreign Exchange Trade System, an arm of the People's Bank of China, introduced the CFETS RMB Index last December indicating the RMB will be measured against a basket of trading partners' currencies rather than just the greenback alone. Our new index will allow market participants to conveniently monitor the RMB's movements. We also plan to introduce futures and options on the index in the future to provide the market with effective risk management tools. A US dollar basket currency index has fulfilled this role for the US dollar since 1973, attracting deep liquidity. We believe the same potential exists for the RMB.

Over time, we plan to launch a full suite of different RMB products, including RMB-USD dual-counter gold futures later this year. The contracts will give gold producers, users and investors a complete solution to manage risks arising from the gaps between the gold spot and futures markets, as well as the difference between the RMB and USD.

On the connectivity side, we are exploring ways to make bond market access more efficient. Our experience with Shanghai-Hong Kong Stock Connect puts us in a good position to try and find cross-border solutions that work for both sides. At the appropriate time we will develop bond index futures and credit default swap products derived from the bond market.

Implementing these three steps will build Hong Kong's FIC space into a mature, comprehensive market that can serve as an offshore wealth management centre for Mainland investors, an offshore pricing centre for the RMB and global asset classes for China, and a comprehensive risk management centre for Mainland investors.

We are still at the beginning of a long road. The FIC market in Hong Kong is small, but has tremendous growth potential. We see that potential at HKEX, and so do the 800 people who turned up today for the RMB FIC Conference. But while the future is bright, we need to be nimble and act quickly to take advantage of our opportunities. I'm confident that over time, we'll develop Hong Kong into a major wealth management centre that serves the needs of both Mainland and international FIC investors.

24 May 2016

41 | Bond Connect: What it means and how it works

Today we held the largest event dedicated to the Renminbi (RMB) fixed income and currency space in Hong Kong, HKEX's 4th annual RMB Fixed Income and Currency (FIC) Conference. This year's was the biggest yet with over 1,000 delegates and top minds coming together to shape Hong Kong's role in the evolving Fixed Income RMB market.

The development of our FIC market is of the priorities in our Strategic Plan, and we've been making good progress. Our flagship product, the USD-CNH futures contract, has had the highest liquidity and distribution among the world's exchanges over the last few years. In January, both volume and open interest broke all-time records, with more than 20,000 contracts traded and open interest of 46,000 lots. This year, we also launched new RMB interest rate and currency derivatives products, and we plan to launch the RMB-USD dual-counter gold futures contract in July and other new FX, rates and credit products in the future. This enhancement of our product suite aims to provide investors with a complete RMB risk management solution to hedge interest rate and FX exposure while holding Chinese bond assets.

While we continue to build out the product side, there was another topic on everyone's mind today: Bond Connect. I was peppered with questions about the breakthrough cross-border scheme, particularly about how it will work and what it means to Hong Kong and the Mainland.

Bond Connect is a pilot scheme that will connect China's interbank bond market with the world, giving international investors "Northbound" access to trade bonds directly on the China Foreign Exchange Trading System (CFETS) for the first time. "Southbound" trading will start later, giving Mainland investors the opportunity to trade in major overseas OTC bond markets.

Bond Connect represents a major breakthrough in the development of the China bond market, which is the world's third largest after the United States and Japan. The market has been growing dramatically by 20 per cent year-on-year, reaching US$9.6 trillion in outstanding size as of March this year. Various factors are driving this growth, including the International Monetary Fund's decision to include the RMB in its Special Drawing Rights (SDR) basket, the RMB's inclusion in emerging market bond indices, funding needs for China's economic transition and strategic projects such as the Belt and Road initiative, green bonds and special project financing.

Bond Connect isn't the first time foreign investors will have access to China's bond market, but we believe it will be the most significant. Right now, foreign investors can buy and sell bonds through the People's Bank of China eligible-institutions scheme in the interbank bond market, known as the China Interbank Bond Market (CIBM) Scheme, as well as through the QFII and RQFII schemes. But Bond Connect will be different, and a much more convenient channel.

For instance, under Bond Connect, there is no quota or the need to stipulate an intended investment amount, which is required under the existing CIBM scheme. Also, while the access rules are the same under both schemes, we expect a simplified and streamlined admission process for Bond Connect.

Trading is also more efficient. Under the CIBM Scheme, an investor has to trade Chinese bonds through the use of an onshore bond settlement agent bank. The settlement agent bank will then negotiate with onshore dealers on the client's behalf. Under Bond Connect, an offshore investor can trade directly with eligible onshore dealers through electronic request-for-quote and native interfaces of established electronic bond trading platforms. This is expected to enhance the price discovery and liquidity of trading Chinese bonds.

Finally, investors have to open onshore settlement accounts with ChinaBond and the Shanghai Clearing House under the CIBM Scheme, while Bond Connect will allow investors to continue using offshore global custodians via the Central Moneymarkets Unit through the nominee structure. There are also mechanisms under Bond Connect to allow flexibility to use CNH or foreign currencies for payment.

In sum, the scheme brings many benefits. Foreign investors account for just 2 per cent of China's bond market, but make up about 30 per cent of the US market. There is substantial room for growth in China's bond market, and Bond Connect could be a catalyst that unlocks that potential. Over time, Bond Connect will also internationalise onshore trading and settlement infrastructure, and help the onshore and offshore RMB

The launching ceremony for Bond Connect.

markets converge.

With Bond Connect, Hong Kong gets an immediate and strong relevance in the fixed income space while HKEX expands the Mutual Market programme that began in 2014 from stocks into a new asset class. Bond Connect won't bring HKEX significant revenue in the short term, but is a key strategic initiative that will see Hong Kong grow further beyond its traditional equities business and put an important foundation in place for HKEX's further developments in FIC, particularly in derivatives.

Even more importantly, Bond Connect is a great example of the value that Hong Kong can bring to China's development. China has built its own capital markets over the past 20 something years. With their deep capital, huge market size and late mover advantage, they have learned from the experience and evolution of major international markets and built their market top-down with a very different design. It is a unique market that operates in a distinct way. Meanwhile international markets have long histories and their own market expectations and practices. Therefore as China continues to open and integrate with the global community, Hong Kong has a vital role to play to bring these two sides together, helping them connect efficiently and interact in a way in which both sides feel comfortable.

Bond Connect is more evidence that China is accelerating the liberalisation of its capital market. It is also a great example of the benefits Hong Kong can enjoy when it creates value and builds connections between China and the international community.

The launching ceremony for Bond Connect.

We have no doubt that China's opening up will continue to bring Hong Kong new opportunities, although the pace will remain uncertain as the Mainland authorities do what they feel is best for the country, at the time they are most confident. We will continue to stay on our toes and work towards creating lasting value for international, Hong Kong and Mainland market users.

8 June 2017

Chapter 5

Commodities Market

42 | Towards LME Week Asia 2013 and beyond

We're four weeks away from London Metal Exchange (LME) Week Asia, and we've been busy gearing up to welcome the global metals trading community to Hong Kong for the first time. LME Week Asia is the Asian edition of LME Week, which is a major event held in London each October. LME Week brings together members of the metals community to network, discuss trends, and learn about growth opportunities, and it has become a cornerstone event for the LME each year over the past several decades.

We're extremely excited to be holding LME Week Asia in Hong Kong this year, and are thrilled with the response. It will be a key opportunity for commodities traders here and around Asia to learn more about the LME's business, and for metals traders around the world to learn more about current trends in the Asian commodities sector. To countdown to the opening of LME Week Asia, I'll be writing a series of blogs over the next few weeks to update you on the developments in our commodities business and how the business fits in with what we're doing here in Hong Kong.

Commodities trading is unfamiliar to many people in Hong Kong and developing a successful commodities business is a huge challenge. Our acquisition of the LME put us firmly on the leadership map of global commodities exchanges. What I'd like to share with you today is how to make it work from here.

Simply put, our commodities strategy has three key objectives:

(1) To make the LME's existing base metals business grow faster by **lowering barriers** to access**,** particularly from Asia;

(2) To extend our commodities platform from London to Asia by launching **new capabilities and new products**, particularly into Hong Kong;

(3) To extend our base metals business into other commodities by developing strategic **partnerships and relationships**, particularly in China.

How will we achieve these objectives? I'd like to explain by way of the house analogy

I used in my last blog. There, I showed how the investments and strategic steps we have been taking for the HKEX business as a whole over the last few years are like laying the foundations of a house, building firewalls, renovating the interior structures and finding ways to attract people to visit and live in our house. The acquisition of the LME allowed us to add an entirely new section to our house — a new commodities extension. I'd like to tell you how we plan to make this extension an integral part of our house.

The new LME section of our house already has a strong "West Wing" — its leading position in base metals worldwide out of London. But although it already has a solid Asian business, the LME could further build its "East Wing" for its Asia-based clients that is well aligned with the rest of the house:

- Trading in LME's existing products by Asian users, particularly from China, is still not free from regulatory and other barriers, both in London and in Asia;
- The LME's infrastructure and product development capability are not well-adapted to support successful extension into Asia; and
- The LME is yet to find strategic partner among key players in China.

So the first stage of our strategy is to take the LME's existing Asian business and make it work better. For this stage it is not about monetary investment or infrastructure build-out. It is about finding ways to make trading on the LME easier, more accessible and more attuned to Asian clients' needs so that they trade more. The first thing we're going to do is to enhance the Asian time zone price discovery mechanism to serve certain Asian users. We are making good progress there and I hope to be able to share more details with you in another blog shortly. A second barrier to entering the "East Wing" for Asian guests is that to become an LME member you have to have an office in London. We are exploring to see whether there can be a better way to connect with Asian players. We are also working on Renminbi (RMB) clearing in London and expansion of the warehouse network in Asia. Last but not the least, we have to promote the LME to Asian users so we can attract new guests to stay in our "East Wing", and that's what LME Week Asia is about.

The second stage of our strategy is to give our commodity extension a good infrastructure. This stage will require significant investment of time and money, but it will be worth it and the payback will not be only in financial terms but also in strategic value. One of the big investments we are making is to set up LME Clear. At present, the LME relies on a third-party clearing house to clear and settle its trades, which means that we don't fully control our own destiny, particularly with regard to speed and flexibility of product launch. Taking clearing in-house through the establishment of LME Clear will speed up time-to-market for new products as well as enable us to capture the related revenue. Financially, LME Clear will be generating meaningful income from its first day of operation, so we would expect it to recover its own set-up costs in the near term. LME also presently outsource their IT operation, and again we are taking that in-house to improve efficiency and develop IT capability as a strategic asset. Finally, we will invest to make our Hong Kong-based trading and clearing systems commodities-enabled.

That brings me to the third strategy we have for our commodities "East Wing" — bringing in more guests. This is about developing mutual product listing/ licensing arrangements and forming strategic partnerships with leading exchanges. Especially Mainland China institutions would be key partners for us given the sheer size of the Mainland market and the Mainland's need to internationalise. These partnerships would ultimately take us beyond metals and into other commodities. We are exploring products related to metals, such as coking coal and iron ore, but we would also like to extend eventually to soft commodities and agriculture as opportunity permits. This will in turn make our market more attractive to existing and potential new users, ensuring that our "East Wing", once expanded, is fully utilised. We're confident we can achieve this because our "East Wing" will be designed to help our potential partners achieve *their* aspirations of internationalisation and global leadership.

That's our plan for taking the LME forward in Asia. What does it mean for Hong Kong? We think it will mean a lot because Hong Kong will be at the centre of our expansion plans — our "East Wing" is located right here. New users and existing users will be channeled from the Mainland to Hong Kong or through Hong Kong *en route* to London, and vice versa. Asian time zone pricing, and ultimately Asian time zone clearing, will involve Hong Kong although the details are yet to be worked out. And Hong Kong and Asian investors and intermediaries may find it interesting to explore the growing range of commodity products we will be bringing here.

There's a lot to talk about as we gear up for LME Week Asia next month. I can't wait to get the conversation started, and hear your ideas as we team up with the LME to build a successful commodities market in Hong Kong.

27 May 2013

43 | Strengthening the LME's Asia benchmark price

In my blog last week, I talked about our plan to build out the "East Wing" of the London Metal Exchange (LME) business here in Asia. And just today, we are delighted to announce one of these "renovations" — an improvement in the way the LME's Asian Benchmark Price is determined. We hope that changes like this will make the LME more accessible to our "guests" from China and other parts of Asia, and I'd like to explain to you what the changes to the benchmark are about and why we're making them.

One of the driving forces behind our acquisition of the LME is the growing relevance that Asia, and particularly China, has in global commodity markets. Even though the LME operates in London in an entirely different time zone, Asian trading on the bourse — through its 18-hour electronic trading system — has been increasing year-on-year. With greater trading volume from Asia comes greater need for Asian dynamics to be reflected in the price discovery in that time zone. It is for this reason that we have come to focus on the Asian Benchmark Price. But before I get into details, let me first provide some background on the "pricing of commodities" on the LME, which is a relatively new topic in Hong Kong.

The LME is the world's leading price formation venue for base metals, with more than 80 per cent market share of global non-ferrous metals futures trading on its platforms. Pricing metals is at the core of what the LME does. The key benchmark prices, including the LME Official Prices and LME Closing Prices, are "discovered" through buying and selling on the Ring, the open-outcry trading floor in London, at set points throughout the day. Open-outcry sessions in the Ring have a long and storied history, dating back to 1877, and still attracting vigorous trading to this day.

The prices "discovered" on the LME's platforms are used as the global benchmark and basis for physical trading as well as in the valuation of portfolios, in commodity indices, and in metal ETFs. The price of the aluminium used to construct your iPad or the copper in the wire inside your home was very likely based on the price discovered on the LME.

The LME's global network of licensed warehouses helps ensure that its prices reflect the reality of supply and demand on the ground.

While this benchmark pricing system has worked for generations and continues to work well, the rise of China as the world's second largest economy has created a vast new market for metals. Today, China is the world's biggest producer and consumer of non-ferrous metals, accounting for about 40 per cent of the global total. This, together with the market growth in other parts of Asia, has direct impact on LME trading patterns. Over the past few years, trading on LMEselect, the LME's electronic trading system, has been increasing in the Asian time zones, which is early morning in London. In 2010, trading on LMEselect before 7 am London time, the afternoon in Asia, doubled from the year before, prompting the LME to launch its Asian Benchmark Price for LME Aluminium, LME Copper and LME Zinc to enhance its market for LME users in Asia. Alongside the global LME price, we believe the Asian Benchmark Price could more accurately reflect the supply and demand of metals in the Asian time zones. In 2012, three-month futures trading on LMEselect in Asian time zones increased by another 13 per cent year-on-year, leading to today's announcement.

The announced improvement is two-fold. Firstly, the timing of the publication of the LME Asian Benchmark Price will be adjusted to the closing of other key commodity futures markets in Asia. Secondly, the LME will improve the Asian Benchmark discovery process by reducing the pricing window from 15 minutes to a new five-minute pricing window, thus focusing liquidity into a shorter trading period.

The effectiveness of the LME Asian Benchmark will become more meaningful over time as the LME gets more traction with wholesale customers in China, who are increasingly exposed to international price risk through the import of base metals, in conjunction with an expected increase in LME membership from Asia. Further, as we look to develop new products tailored to Asian users, the products could be benchmarked or settled against the LME Asian Benchmark. This is all part of polishing up the "East Wing" of the LME's business to make it more attractive to Asian users.

The enhancement of the LME Asian Benchmark is just one step the LME is taking in Asia. Over the coming months and years, we will be working with the LME and its users to grow their business further in China and the rest of Asia, finding new ways to capture the huge surge of physical trading and hedging of metals in the region.

3 June 2013

44 | Why LME volumes are surging

The London Metal Exchange (LME) announced recently that trading volumes on the bourse have reached record highs. The LME said record volumes were transacted in May, beating the previous all-time high from April. Aluminium volumes were up 7 per cent, copper was up 10 per cent, zinc by 13 per cent, nickel 19 per cent, and cobalt trading volumes were up a whopping 30 per cent compared to May 2012.

The record trading volume on the LME is a very encouraging sign, particularly as we look to implement plans to further grow the business in Asia and globally. But it also begs the question: what drives the LME's trading volumes? And why are volumes rising when commodity prices are falling?

First, the factors driving trading volume on the LME have evolved over time. Metal companies have always used the LME to manage volatility. Forward sales on the basis of a fixed price — the traditional contract type — involve the risk of prices moving before delivery, and hence need to be hedged. In fact, that was the principal reason that drove the founding of the LME 135 years ago. Many companies negotiating today agree that the delivery price will be the LME cash average for the month of delivery, for example. As neither party knows today what that price will be, they hedge to avoid disadvantageous price movements; and they hedge via LME contracts. Increasing regulatory pressure alongside pressure from banks and shareholders to reduce credit risk has resulted in further pushes to hedge using LME contracts. We believe there is plenty of room for further growth in this area.

For other players, metals have always been seen as a possible hedge against inflation. And with interest rates low, metals have offered the possibility of an additional return. They have also been seen as a China and BRICS play, as developing countries consume metal for infrastructure and manufacturing.

In addition, electronic trading is playing a bigger role in boosting volumes. Growth in electronic trading on the LME Select system now accounts for a larger share of total volume on the LME.

All these factors have contributed to the long-term growth in volumes on the LME in recent years. Volumes have been rising even more lately because of the volatility of commodity prices.

Yes, that's right — despite the recent fall in commodity prices, volumes on the LME are actually rising. This may not be straightforward for those used to the equity market, but the impact of falling metal prices on the LME is not the same as that of falling stock prices in the equity market. Let me explain.

In the equity market, daily turnover is largely driven by market sentiment; for instance, if the market turns bearish it may result in fewer people wanting to invest and thus lower turnover. Metal trading on the LME works differently, however. Here, if traders and investors believe the price is too high they sell futures, anticipating lower prices. If they think the price is too low, they buy. This means that as the price of LME metals rise and fall, new business is always being attracted to the exchange. In addition, volatile conditions mean that companies involved in producing, shipping and manufacturing will feel a greater need to hedge using LME futures contracts because their price risk goes up. This is one of the key factors that has led to an increase in volumes.

The most exciting thing about the LME's record high volumes is that we're just getting started with our plans for the exchange. Our long-term goal is to further commercialise the business, introduce new members from Asia, launch new products, and build our own clearing platform. As we execute our plans step by step, the full potential of the LME is yet to be unleashed.

13 June 2013

45 | New LME-licensed warehouse locations in Asia approved

As momentum builds towards London Metal Exchange (LME) Week Asia next week, we are very pleased to announce today that Taiwan's Kaohsiung port has been approved by the LME Board as a good delivery point. This means that warehouses in Kaohsiung can apply to become LME-licensed warehouses, making Kaohsiung the ninth LME delivery point in Asia, after locations in Singapore, Malaysia, Korea and Japan.

Warehouses in Kaohsiung and the other locations in Asia are important to the LME despite the fact most futures contracts, including those traded on the LME, are not held to delivery but are closed out prior to expiry for settlement in cash.

The reason is that the LME's warehouse network lies at the core of the LME's value proposition to the metals industry. The requirement for delivery to LME-licensed warehouses helps keep the prices of LME futures contracts "honest". Before I elaborate on this, it's important to understand why the LME's warehouse network is such a key competitive asset.

The LME has licensed over 700 warehouses in 36 locations in 14 countries around the world, predominantly in North America, Northern Europe, and now, increasingly, East Asia. It is the most extensive global warehouse network among futures exchanges, yet each of them has to go through a strict application and approval process. To get approved by the LME as a good delivery point, a port has to meet several requirements relating to capacity, transport access and more. After the port is approved, the warehouse operators in the port need to apply separately for a warehouse license — again, specific standards need to be met. The continuing qualifications of the port are reviewed by LME staff on an annual basis. This is a system that is carefully designed, maintained, and reviewed.

To establish and maintain this huge network requires considerable effort from the LME. Why has the LME made this commitment? Because the warehouse network is pivotal for making the futures price converge with the physical price as the futures contracts get closer to expiry dates. Although only a small percentage of futures contracts are eventually settled in physical delivery, the possibility of physical delivery prevents futures prices from diverging too far from the price of the physical metal.

I explained in a previous blog that it is important for futures prices to closely track the physical, as it is the basis for effective hedging of metal prices fluctuation for both producers and consumers. Therefore, the warehouse network helps tie the LME to the metals industry and the metals industry to the LME.

Now, you may wonder why the LME has warehouses in the locations it does, and not for example in Africa or Latin America, which are major suppliers of metals. The answer is that the warehouse as delivery point has to be as near as possible to the locations of consumption, that is, manufacturing. North America and Northern Europe were the main global centres of manufacturing in the past. But now, increasingly, manufacturing is being done in Asia. In fact, a large proportion of global manufacturing is being done in China, which is now the world's second-largest economy and still growing rapidly. That's why having more LME-approved locations in Asia, such as Kaohsiung, provides some help, especially to China-based users since warehouses in these locations are more accessible for them than warehouses elsewhere. Ideally, the LME would be able to eventually license warehouses in Mainland China itself.

You will probably hear a lot more about warehouses from speakers and commentators at LME Week Asia. Our plan to extend the LME warehouse network in Asia is building out the LME "East Wing", as I described in my earlier blog post. While we have big plans for the LME over the long term, there's one thing we won't change: the fundamental role warehouses play in LME pricing.

17 June 2013

46 | LME Week Asia and the future of commodities in London, Hong Kong, and Mainland China

We've just wrapped up an incredible couple of days filled with events to mark the very first London Metal Exchange (LME) Week Asia in Hong Kong. Producers, consumers, metals traders, Mainland exchanges, and LME members from around the world converged in Hong Kong, with more than 900 tickets selling out for the LME Week Asia gala dinner on Tuesday night. It was a big moment for Hong Kong, especially considering our city hasn't traditionally been a major player in the commodities sector.

We had a marvellous time this week, but we also know that much work lies ahead. When we acquired the LME, it was because we had a broad vision. Our vision, which I outlined at the LME Week Asia seminar on Tuesday morning, involves a three-pronged strategy that focuses on London, Hong Kong, and Mainland China. It boils down to what I like to call the three Ps: Products, Platform, and Participants.

First of all, the LME is equipped with the expertise, intellectual property, brand, and membership necessary to help us develop into a true global leader in commodities. It provides us with the key foundation for developing commodities products. Today, the LME is already a global leader in metals trading and the pricing centre for the global commodities market. In fact, it's extremely rare to find a business in a free market with a market share of over 80 per cent, which the LME has achieved. This success proves the LME already has an incredible business model. As the new owners of the LME, we want to further build on this leadership. We will do this by building LME Clear, insourcing LME's IT and building a trade repository. We want to make access to LME's electronic trading platform as easy as possible for global investors, traders, producers and consumers. All these together will create greater flexibility to develop a broader range of products.

Secondly, we have Hong Kong. We aim to position ourselves as a trading and clearing platform for the new commodities products to be launched in Asia. Our goal in the short term is to ensure the market in Hong Kong is ready for commodities. Developing the necessary infrastructure for physical settlement is a long-term and complex task, so we plan to initially focus on cash settled monthly commodities contracts. That way, we can leverage our existing trading and clearing platforms for equity derivatives, and with some upgrades and enhancements, use them for commodities futures. We are looking at new base metals products that may supplement those in other markets, and also seek to launch commodities products other than base metals when appropriate. Over the long term, we aim to offer physical settlement for certain commodities that are appropriate for our market.

Thirdly, mutual access between HKEx and the Mainland market mainly refers to liquidity and participants. We want to develop Hong Kong into a commodities platform where international products can meet Mainland liquidity, and Mainland liquidity can connect with international liquidity. In other words, we want it to be a market where there is mutual access between international and Mainland liquidity.

This is no doubt an ambitious plan and an ideal situation for Hong Kong. As an optimist, I firmly believe we can make it happen. But I also understand not everybody is convinced. Specifically, London, Hong Kong and Mainland China have raised the following concerns:

(1) LME members in London fear we may fundamentally **change** the business model of the LME;

(2) Our friends in Hong Kong may lack **confidence** in our grand vision; and

(3) Mainland exchanges may fear our plan will lead to **competition**, or wonder if there is room to **cooperate**.

Regarding our stakeholders in London, I want to stress that we are not intending to change the LME's core business model, which has proven to be extremely successful. We are all aware the LME is not an exchange in the traditional sense of the word. Before the acquisition, it was a trading platform mutually owned by market participants in the fields of base metal production, trading, financing and warehousing. The entire LME ecosystem is highly competitive and sustainable. In fact, it is one of the world's most successful business models. Each of you may hold a different view about the LME's ecosystem, but it is undeniable that it has been widely accepted by the global base metal industry, and LME prices have become benchmark prices for global base metals.

These strengths are what made the LME so attractive to HKEx, and what compelled us to make an offer for the bourse. However, the price we offered was based on our intention to turn the LME from a mutually-owned non-profit organisation into a fully commercial exchange. We fully recognise the LME works well and is widely accepted, but it's important that HKEx is also able to benefit from it. We have no intention of fundamentally changing the internal operating structure of the LME ecosystem, but we will seek a reasonable and fair return on our investment. That's why we will introduce clearing services along commercial lines in September 2014, and review fees after January 2015. While we seek a return on our investment, we also aim to invest in LME services to provide better clearing and IT services. We are not here for value destruction or value redistribution — we are here for value creation. I believe we can create a larger pie that benefits the LME membership community as well as the LME.

Over time, we also plan to use the LME to offer base metals trading in Asia and move into other commodities. The LME model works great in London, and we intend to preserve its DNA, but we may make changes to Asian-based trading to meet local needs. These new opportunities will create new value, and we hope they capture the imagination of the current LME membership and they join us as we grow together in this new frontier.

I also wish to address some of our Hong Kong friends' common queries here. Why am I confident Hong Kong will be the first market for mutual access between the Mainland and the world? My rationale is simple. To the Mainland, we are one country. To international investors, Hong Kong has a system different from the Mainland. "One Country, Two Systems" is what makes us unique from anywhere else in the world.

When the Mainland opens up its markets, it is not without risk. Therefore it needs a relatively controlled and safe environment before it's ready to proceed. As a part of China, Hong Kong is an ideal venue and trusted partner that can provide assistance. Hong Kong has served Mainland China well, particularly in securities, which has shown that we can be relied upon.

As there are "Two Systems", Hong Kong has a fair and transparent regulatory environment, the rule of law, sound economics and all the talents and capabilities required for financial development. It is these factors and the protection and convenience they offer that attract international investors to our city.

"One Country, Two Systems" is our core competitive advantage. Without the "One Country", Hong Kong would not have the opportunity to support the Mainland. Without the "Two Systems", Hong Kong would be no different from other Mainland cities. We must be confident in our own unique advantages. We must not waste these opportunities and responsibilities that these times have bestowed on us.

Finally, I know our Mainland exchange peers are concerned about potential competition. I want to be clear here: if we only talk about competition but not collaboration, it demonstrates that we lack vision. But if we talk only about collaboration and not competition, we are not being honest. From HKEx's perspective, our strategy is clear. We do not intend to take away business from our Mainland peers. Rather, we are interested in cooperation and creating new markets and new opportunities together. This is the only way we can have a win-win situation. We will not interfere in what Mainland exchanges are doing. Our hope is to join hands with them for the benefit of us all.

Here, I would like to touch a little on another issue I get asked about, which is what I see as a price-setting centre. In my view, a genuine price-setting centre must be one in which both the buyers and sellers can take part. Over the years, China has been producing and consuming at high volumes, but has been failing to transform this quantity effectively into influence in pricing. In other words, it has always been a price taker. If we can build an ecosystem here in Hong Kong, on Mainland China's doorstep, where both the buyers and sellers can participate and exert their influence, it would definitely help the Mainland's internationalisation and be a step towards achieving mutual access between the Mainland and international markets. Once we have a practicable Hong Kong model, we will be able to replicate it and extend it to other international markets, too. We hope that Hong Kong can become the starting point in China's internationalisation, not the finish line. China could then head to New York, London and Chicago. As the market develops, the influence on pricing will return to the hands of consumers. In the process, Hong Kong will be able to play its historical role while continuing to have room for further development.

All in all, it is history that is giving Hong Kong its unique opportunity, and I am deeply confident about our future.

Before I finish, I just want to thank all those who attended the first LME Week Asia in Hong Kong. I enjoyed meeting many of you and hope you had a wonderful time and enjoyed Hong Kong hospitality. I'm sure this will be the first of many LME Week Asia events in our city, and I believe they will get bigger and better with each passing year.

28 June 2013

47 | LME warehouse queues: Perception and reality

As I expected, the global warehouse network of the London Metal Exchange (LME) was a key talking point at last week's LME Asia events. I was very pleased by the strong market response to our approval of Kaohsiung as the LME's ninth Asian good delivery location. Together with the revised LME Asian Benchmark Price, I'm sure that this will make the LME even more relevant for Asian users, while preserving the global focus which has underpinned the market since its foundation.

You may have heard some criticism of the LME's warehouse network, in particular the warehouse queues. This has been the subject of an ongoing market debate for quite a while. Today, after much deliberation, I'm pleased to report that the LME Board has announced a market consultation to address this important issue for our market.

Our stakeholders in the LME community are much more familiar with this issue and have a much deeper understanding of it. While people may strongly disagree with one another or with the premise and key components of the consultation, the issues and proposed solutions are well understood and the feedback will ultimately help the LME find the best outcome for the industry.

For our new stakeholders in Asia, the warehouse debate is new and the issues are not often easy to understand. Although not all of our Asian stakeholders are deeply affected, they do care about how it affects HKEX's reputation as the new owner of the LME. As readers have different levels of understanding of this issue, I've decided to present this as a series of questions and answers. Let's begin.

What is a warehouse queue?

Because of the LME's prominent role in the metals industry, many owners of metal like to hold their metal in an LME-licensed warehouse. They receive a warrant to certify their holding of a particular lot of metal in a given warehouse. If the owner wants to withdraw the metal, he or she needs to ask the warehouse operator to load it out and cancel the warrant. The LME requires its licensed warehouses to load-out a certain amount of metal — at least 3,000 tonnes per day at the largest locations. But, in some warehouses, a lot of metal has historically been loaded-in — as explained below — so when a lot of users want to cancel warrants, a queue will form. These queues have grown longer at some locations — to hundreds of days in a few cases — as the stock of cancelled warrants waiting for delivery out of the warehouse has grown.

We should note that much of the metal sitting in queues isn't being withdrawn by industrial users, but by financing players who are moving it from one warehouse to another to take advantage of lower rents. To be clear, the vast majority of LME Warehouses do not have queues — only five LME good delivery locations have any warehouses with queues. But the queues at these locations have drawn some attention from metals users and the media.

What macro-economic factors lead to warehouse queues?

In an economic recession, such as the world experienced in 2008, market demand for metal declines. To accommodate reduced demand, producers can cut production. But another solution, particularly for an aluminium smelter which is difficult to shut down, is to carry on producing for inventory. Normally, demand recovers after a while, and the market absorbs the excess inventory. But the slowdown of recent years has been unusually prolonged, while interest rates have been so low so that producers could afford to carry inventory for longer. The contango in aluminium, where the futures price exceeds the current cash price, makes it profitable for financial players to invest in metal to capture an arbitrage profit. All these factors have led to unusually large inventories of metal, particularly aluminium, building up. And the best place to hold inventory of metal is on warrant in an LME-licensed warehouse, where it is available for participation in the world's key market for metal price formation.

Are the queues bad, and if so, for whom?

The market has raised three main concerns with regard to the queues. They are:

(1) that queues are resulting in metal not being accessible to industrial users in a timely manner and hence negatively affecting the real economy;

(2) that queues are pushing the overall cost of metal, particularly aluminium, too high; and

(3) that the premium charged for delivery at a particular warehouse has become too large relative to the underlying cost of the metal.

Let me deal with these questions one by one.

Regarding the first question, it's very important to note that warehouse queues do not stop real-world metals users from getting access to metal. In general, businesses needing metals do not take that metal from LME warehouses — instead, they buy their metal directly from producers, and use the LME to hedge the price. In other words, anybody who wants metal can get metal. Metal stocks in LME warehouses are simply "supplies of last resort", allowing consumers to address small shortfalls in their metal requirements, and keeping the LME price "honest". Nobody is seriously telling us that they can't get access to metal — it's readily available from the producers — but, when consumers buy it from those producers, they do have to pay the producer's premium. I'll elaborate on this in a minute.

Previously, I was quoted saying I would take a "bazooka" to the warehouse queue issue if getting access to metal were a big problem. But our assessment is that buyers have access, so we are not facing a fundamental problem with our market and a bazooka is unnecessary.

Regarding the second concern, that queues are resulting in the underlying price of metal being pushed too high, this is not a problem either. The price of aluminium is currently about 47 per cent below its historical high, reflecting oversupply by aluminium producers, as I mentioned above. The prices of other metals are also below historical peaks. It is likely true that if it weren't for producer stockpiling, supported by metals financiers, prices would be even lower. But prices would shoot up again as soon as market conditions changed and the big fluctuation of metals prices would not necessarily be in the long term interests of the market. In any case, the metal price has nothing to do with the queues, since it is the availability of metal on the open market that determines price and there is plenty of metal available there. I had talked last year about a surgical operation if the price of metal was being distorted, but we've determined that this is not a problem either.

Now, the third issue. Earlier in this blog I mentioned the producer's premium. This premium, which is usually a small percentage on the LME price, is charged to those who want immediate access to metal in a particular location. The size of the premium depends on supply and demand in that particular location. Premiums have always been a feature of the market and can fluctuate wildly, but the lengthening of queues in recent years may have increased the size of that premium. To give you an example, to get immediate delivery of aluminium in the mid-west of the United States, you'll currently have to pay a premium of around 12 per cent above the LME aluminium price, which is a much higher level than has historically been the case.

The metal premiums and warehouse queues are closely related because, historically, the LME warehouse network has acted as a limit on metal premiums; if the producer premium became too high, then arbitrageurs would buy metal via LME contracts (at the LME price) and withdraw metal from the LME warehouses on settlement, rather than paying the premium to the producer. This had the effect of raising the LME price and hence "tightening" the premium. However, because of the queues at certain key warehouses, this arbitrage is more difficult, so the premium has become higher. The premium is widely tracked and reported, so the "all-in" cost of metal (the LME price plus premium) is easily calculated, but we fully understand the frustrations of users who want the LME price to be as close to the "all-in" price as possible, especially since the premium cannot easily be hedged.

I view a persistently high premium as an unhealthy physical condition, such as swelling, that is not life threatening but is highly irritating and has the potential to cause long term harm. This can't be solved with either a "bazooka" or a surgical operation. Instead, a modest approach, such as "Chinese medicine", might be the cure. That's what this consultation is all about.

What has the LME done, to date, to address market concerns?

The consensus at the LME has been that warehouse queues are the result of macro-economic forces and that it is not appropriate for the LME to intervene. Although some market participants have suggested that if the LME had applied stronger measures a few years ago, the size of problem could be reduced. While we acknowledge that possibility with the benefit of hindsight, we strongly believe the LME has applied its rules consistent with its historical practices over many years. The most important thing is to look forward and find ways to resolve the issues we face today as best we can.

While this consultation is about the future, it's also important to note steps the LME has taken in recent years to respond to concerns over warehouse queues.

In 2010, the LME commissioned Europe Economics to prepare an independent assessment of the LME's warehouse rules. The report made several proposals of potential remedial action to ease the pressure of queues at warehouses, and also recommended the LME carry out routine reviews of delivery out rates every six months.

The LME has adopted this recommendation and has made changes to LME warehouse rules since it was published. We doubled the minimum load-out rates at the largest locations in April 2012, added minimum delivery out rates for nickel and tin in April this year, and we introduced additional requirements on warehouse companies to address the effect that queues were having on other metals. That also took effect in April this year.

What is the LME proposing to do in its consultation?

We don't necessarily disagree with the consensus that queues are the result of macro-economic forces. However, as queues continue and even get worse in some cases, we want to take a fresh look at the issue. Therefore the LME Board has put forward a proposal that we believe balances the differing needs of LME members, warehouse operators and the metals industry.

Our goal is to present an alternative to the current arrangement that aims to (1) prevent queues from growing longer through introducing new load-out rates for incremental new load-in; (2) reduce the length of the existing queues by applying new load-out rates for cancelled warrants "recycled back" to the queue; and importantly; (3) avoid penalising players who have invested and adopted business practices in reliance on previous LME rules through the grandfather provision. Finally, we want to ensure the system is flexible enough to respond to potential unintended consequences. If you are interested in the details of the proposal, you can read them here.

In crafting the proposal, the LME Board had to bear in mind the logistical constraints. You cannot require warehouses to increase their load out rates by much as metal is heavy and difficult to move; there just aren't the logistical resources in terms of spare fork-lift trucks, rail capacity and so on to load it out much faster. Nor can the physical layout and distribution of warehouses, which is conditioned by existing LME rules, readily be changed either. Acknowledging these realities, the essence of the proposal is that warehouses with queues of more than 100 days must link their load-in rates to their load-out rates so that the queue problem will not get worse, and over time, actually get better.

I want to reiterate that the Board has not formed a view on whether any changes should be made, only that this proposal is an option we can proceed with should the market indicate a desire for us to do so.

What do we want from the market?

We have undertaken this consultation because it's important we listen to market participants when concerns arise. As the new owner of the LME, we want to be proactive in responding to concerns and addressing perceived problems.

Consultations are only effective if we take them seriously, and we do. We are looking for thoughtful and substantive feedback from market players. If you oppose our proposed solutions, it's particularly important for us to hear why we should not proceed and what unintended consequences it may cause. If you are in favour of this solution, we also would like a detailed explanation of why.

All feedback will be kept confidential unless we are otherwise instructed. We have a long history of public consultations in Hong Kong and we use the feedback received to make further improvements to our market. We look forward to strengthening this tradition in London as well.

I don't yet know whether changes will result from the consultation. But I do know that, whatever decision is made, we can move forward with the confidence that the LME's rules deliver the best physical network for the metals community, and support a truly global LME price.

1 July 2013

48 | LME Week Asia 2013 kicks off with new commodities contracts in Hong Kong

It's amazing to think nearly a full year has passed already since our inaugural London Metal Exchange (LME) Week Asia in 2013, but it's here again and bigger and better than ever.

LME Week Asia brings together producers, consumers, metals traders, Mainland exchanges and LME members from around the world to discuss market trends and seek new opportunities, particularly in China. As the centre of gravity in the commodities market shifts from west to east, Hong Kong is well positioned to connect buyers and sellers, connect Asia with the rest of the world, and connect other markets with Mainland China.

Global connectivity is actually the theme of LME Week Asia this year, and it will be explored in more depth at several events scheduled for this week. I wrote last year that our connectivity plan involves three Ps: Products, Platform and Participants. The LME brings expertise, membership and Products; Hong Kong will be the Platform for products serving the Asian market; and Participants will come from Mainland China. Hong Kong is the nexus that facilitates this connectivity.

Today marked a key step towards fulfilling that vision. Our Asia Commodities team announced four new commodities contracts this afternoon. The London Aluminium Futures, London Copper Futures, and London Zinc Futures unveiled today are based the LME's leading products in terms of volume and value, which is why they will be among the first products traded, cleared and settled here in Hong Kong.

The fourth new contract is the API 8 Thermal Coal Futures contract. The contract is particularly strategic for HKEx because China is one of the largest coal users and Asia, led by China, accounts for more than half of the world's thermal coal use*. We hope to help more consumers use derivatives as a hedging tool.

All four contracts are expected to be launched in the second half of this year, subject to regulatory approval and market readiness. The metals futures will trade in Renminbi (RMB), while the coal futures will trade in US dollars. All four will be cash settled, with clearing and settlement happening in Hong Kong.

We're already seeing signs that the commodities industry is taking root in Hong Kong; for instance, last year Standard Chartered moved its global head of metals trading from London to our city. Other banks are expanding their metals trading desks in Hong Kong, too, in light of our acquisition of the LME, China's growing influence on commodities pricing, and the slow but accelerating opening of China's financial market, as evidenced by a very recent development.

Earlier this month the China Securities Regulatory Commission (CSRC) on the Mainland and Securities and Futures Commission (SFC) in Hong Kong gave preliminary approval to the Shanghai-Hong Kong Stock Connect, a programme that will see investors in the Mainland able to trade eligible shares in Hong Kong and vice versa. The cross-border trading link is a major breakthrough for Mainland China's financial market development and for Hong Kong as a global centre of finance.

If launched successfully, the Shanghai-Hong Kong Stock Connect programme will provide significant insight for the internationalisation of Mainland China's commodities market and the development of Hong Kong's commodities market. With an economy that continues to grow at a robust pace, the Mainland is now a leading consumer and importer of many bulk commodities from all over the world. The Mainland commodities industry shares a common dream of internationalising the Mainland commodities markets and securing for China a level of influence and price-setting authority that matches the country's economic strength and increasing role in the international commodity markets.

There are various ways to realise the internationalisation of Mainland China's commodities market. It can be achieved by exporting the commodities products and

---

* The five largest coal users — China, US, India, Russia and Japan — account for 76 per cent of global coal use, according to the BP Energy Survey 2013 and analysis by a major research and brokerage firm. China is also a major thermal coal user, accounting for more than 45 per cent of global use, according to the National Bureau of Statistics of China. Led by China and the large economies of Japan and South Korea, Asia accounts for more than 50 per cent of global thermal coal use, according to the US Energy Information Agency.

prices in Mainland markets to overseas markets, by importing overseas commodities products and prices to Mainland markets, by inviting international users and investors to Mainland markets (e.g. the oil futures platform in the Shanghai Free Trade Zone), or by giving Mainland users and investors access to overseas market through something like the Shanghai-Hong Kong Stock Connect model. HKEx will try its best to provide more options for the internationalisation of Mainland China's commodities markets through collaboration with our Mainland counterparts.

This year's LME Week Asia is bigger than ever before. More than 600 people from around the world are set to attend the metals seminar on Thursday — one of the week's premier events — and more than 1,400 are expected to attend to the LME Week Asia dinner, nearly 50 per cent more than last year.

At the metals seminar this year, the Financial Secretary of the Hong Kong SAR Government, Mr. John Tsang, will give his thoughts on Hong Kong as a commodities trading hub. Our panel discussions also promise to be lively. The first will focus on what metals trading will look like in 10 years' time, while the second will examine the impact of China's reforms on the metals market. We are also very honoured to have Shanghai Futures Exchange Chairman Yang Maijun, Dalian Commodity Exchange Chairman Liu Xingqiang, Zhengzhou Commodity Exchange Vice President Wu Keli and Peng Gang, Vice Chairman of China Futures Association, for a panel discussion in the afternoon. They will share their insights and visions on the internationalisation of the Mainland commodities markets. As HKEx is a bridge between the Mainland and overseas markets, we hope to work with our Mainland counterparts and come up with a win-win formula for cooperation and mutual support that will advance commodity connectivity between the Mainland and Hong Kong. Our goal is to join hands with the Mainland and turn visions into reality as soon as possible.

On Friday, we will host our first ever Women in Commodities Luncheon. The event will honour women in the commodities industry and provide them with an opportunity to network with other inspiring female leaders.

I've only touched on a few of the exciting things on our agenda this week. There will be many more. LME Week Asia got off to a strong start last year, and we're excited about continuing to build it into an important annual event on Hong Kong's calendar.

22 April 2014

49 | Unlocking the value of the LME

There is a lot of news and discussion today on the changes to the London Metal Exchange (LME) fee structure, so I wanted to share my views about that and how it fits into our overall strategic plan. Ever since we acquired the LME in 2012 we have been talking about our aspirations for the exchange. Not long after the first LME Week Asia in 2013, I wrote on my blog:

Today, the LME is already a global leader in metals trading and the pricing centre for the global commodities market. This success proves the LME already has an incredible business model. As the new owners of the LME, we want to further build on this leadership. We will do this by building LME Clear, insourcing LME's IT and building a trade repository. We want to make access to LME's electronic trading platform as easy as possible for global investors, traders, producers and consumers. All these together will create greater flexibility to develop a broader range of products.

Our vision from day one was to invest in the bourse, commercialise it, and utilise its reputation and Hong Kong's strengths to connect the Mainland and international commodities markets. Rather than change the bourse, we wanted to build on its solid foundation and grow it further. And so far, we've gotten quite a bit done!

We have made a substantial financial investment in the LME since the acquisition. One of the highlights was the in-sourcing of information technology, which enhances the reliability and stability of the trading system and gives us more flexibility for future enhancements. We also reinforced our commitment to the LME Ring, announcing in June that it would stay open as the LME's lively — and historic — trading floor. And last week, LME Clear was launched. With its new technology, LMEmercury, we have delivered a world class, real time risk and clearing platform.

The launch of LME Clear is a tremendous step forward. Now we have not only the trading of metals contracts but also the post-trade processes — the clearing and settlement and risk management of positions. LME is no longer dependent on another clearing house for these vital processes; it is master of its own destiny. The launch of LME Clear marks a new phase for both HKEx Group and the LME, not only as the beginning of the realisation of the LME's commercial potential, but also in giving us essential freedom and leverage to pursue our strategic options, particularly relating to new products, new currencies, and new time zones.

Today's changes to the fee structure fit into the overall objective of commercialising the bourse. It is a key step to unlocking the bourse's value while ensuring that we respect our users and maintain our competitiveness.

It will also enable us to capitalise on our strengths for the benefits of users as well as HKEx shareholders. We believe that the new fees are moderate and fair, and respect all market users while being competitive with other exchanges' tariffs. The fee changes will enable us to continue to invest in new services, products and people, and to continue building out the business globally, particularly here in Asia.

All of this fits into our overall strategy, which is to build HKEx into a leading global multi-asset class exchange in the Asian time zone that connects China and global financial markets. A key step is launching Shanghai-Hong Kong Stock Connect, which is right around the corner. It is a landmark scheme that will open mutual market access in our core cash equity business for the very first time.

Acquiring the LME is the most integral part of our FICC franchise. The strategic vision for our commodity business is threefold: first, we enhance LME and consolidate it into a truly global leader in base metals; second, we pursue an eastward extension of the LME's reach and build a Hong Kong-based commodities platform; and finally we will leverage the Hong Kong platform to pursue mutual market access with our Mainland counterparts.

With our additional investments, the launch of LME Clear and new commercialisation programme, we have largely achieved the first part of our strategy and we are actively moving into the second and third components, which begin with the planned launch of our Asia commodity products in December. Shanghai-Hong Kong Stock Connect, once successfully launched, will become an important model of mutual market access for other asset classes.

While we are making good progress, there is still a long way to go. Our team is continuing to work to realise the full potential of the LME in connecting the international commodities markets with Mainland China. I believe the best is yet to come.

29 September 2014

50 | LME a key contributor to HKEx Group family

London Metal Exchange (LME) Week Asia is right around the corner, so I want to turn my attention to our commodities business.

I was very happy yesterday to report some positive first quarter results for HKEx, boosted by increased trading turnover and, more importantly, fruits of some initiatives we began years ago that are now starting to pay dividends. One of those initiatives was our purchase of the LME in 2012. Back then, people wondered if it was a good move, whether Hong Kong was a natural fit for a commodities exchange, or whether we paid too much.

In hindsight, I can see why there were some skeptics. But all along our team had a clear plan for the LME that is now starting to bear fruit. The LME is a key part of our international strategy, diversifies our business and is already making an important contribution to our bottom line.

First, the numbers. We have seen a significant growth in overall revenue of the LME business, with first quarter revenue more than doubling to HK$647 million from HK$315 million a year ago, equivalent to 23 per cent of HKEx Group's total revenue. This includes HK$447 million from the LME and HK$200 million from LME Clear. This growth comes from our new fee structure introduced in January, steady trading volume, a decrease in litigation costs, and good performance of LME Clear since it launched last September.

So things are looking up in London, and Hong Kong is coming along, too. It's amazing to think that in 2012 Hong Kong wasn't anywhere on the world commodities map, and today we're just days away from the largest LME Week Asia yet, with representatives from more than 300 companies and about 100 journalists scheduled to join the Metals Seminar alone. More than 130 tables have already been reserved for about 1,600 guests at the LME Gala Dinner this year, the most we've ever had.

Outside our seminar and dinner, we are also seeing a budding commodities community in Hong Kong with more networking events and important discussions about the future of metals trading and hedging in Asia.

While the commodities eco-system is still nascent here, the potential is great. Last December we launched three London Metal Mini contracts on aluminium, copper and zinc, the first metal contracts traded and cleared in Hong Kong, and may well look to add more markets soon. While trading has been light on the first three contracts so far, we view them as an important breakthrough. They are like a single flame, and once more parts of the commodities eco-system are in place in Hong Kong, that flame will become a fire.

We're also starting to see more Asian firms become members of the LME: GF Financial Markets became the first Chinese-owned company able to trade in the LME Ring when it became a category 1 member last year. China Merchants Securities (UK) was approved as a category 2 member in November, joining Bank of China International and ICBC, which recently purchased Standard Bank, while Sorin Corporation from Korea and Lee Kee Group from Hong Kong became category 5 members recently. In fact, the CEO of Lee Kee, Clara Chan, sits on the LME's zinc and lead committees plus LME Clear's Risk Committee.

China is also taking more — and faster — steps to open up. Just last week it was announced that 100 more state-owned enterprises will be allowed to trade commodities derivatives overseas without regulatory approval, three-times more than before. It's another sign of China's opening up, which will create new opportunities for everyone.

Even more importantly, the LME gave us the credibility we needed with our Mainland counterparts to begin the Mutual Market programme. Shanghai-Hong Kong Stock Connect was the first step to connect the stock markets across the border, and a connection with Shenzhen is in the pipeline. We said at the outset that the "Mutual Market" can be extended from stocks to other asset classes from equity derivatives to fixed income, currencies and commodities. Even more importantly, it allows both sides to invest in a familiar regulatory and legal environment, so there are many possibilities for future growth.

Specifically, in terms of our commodities strategy, there are essentially three things that we wish to achieve over time: (1) connect Chinese products to international liquidity and vice versa; (2) connect international liquidity with Chinese liquidity; and (3) better connect China's physical markets to international futures market.   Looking forward, by leveraging on the mutual market access breakthroughs and LME, we are in an excellent position to accelerate the internationalisation of China's commodities markets.   An internationalised commodities market not only would allow greater access by investors and users, it could also better serve the Chinese real economy.   In this process, we will continue to play a humble but essential role: to build and connect.

So I'd like to welcome everyone arriving in Hong Kong for LME Week Asia. There are a number of exciting events lined up this week that should provide excellent opportunities for in-depth discussions and insights into the global metals market.   I look forward to getting started next week, and to building a stronger commodities eco-system in Hong Kong.

14 May 2015

51 | Brexit and HKEX's commodities strategy for China

There has been a lot of coverage of "Brexit" since citizens of the UK narrowly voted to leave the European Union. The fallout from the decision still remains to be seen, but many are asking me if it will have an impact on the London Metal Exchange (LME) and implementation of our commodities strategy. I'm happy to say our plans are on track, but I'd like to take a closer look at our growth plans, particularly pertaining to the Mainland.

Brexit

Will Brexit have any impact on the LME?

Many people are still in shock about voters' decision to leave the European Union, with analysts and pundits speculating on what impact the decision might have on global markets and the future of London as a financial centre. We are confident the impact on the LME will be limited, because trading on the LME is global and not confined to the UK or even Europe. Members and customers are from around the world, with future growth concentrated in Asia. Furthermore, LME trading is basically done entirely in US dollars, with operating income primarily in US dollars as well. Only operating expenses are denominated in pounds.

Finally, LME has been operating for almost 140 years, long before a European Union existed, and has thrived through all kinds of change and adversity. It is the world's premier metals market and has proven to be strong, robust, and resilient.

Will Brexit affect plans for a "London-Hong Kong Connect" in commodities or other initiatives?

London-Hong Kong Connect was designed to clear and settle LME trades in Hong Kong, and was subject to Hong Kong, UK and European regulatory authority. With Britain withdrawing from the EU, there is some uncertainty about the policy developments in the UK. Therefore, we will wait and monitor the development of the UK and Europe's regulatory policy before making further plans to connect the commodities markets in London and Hong Kong.

However, we don't expect Brexit to have an impact on any other initiatives, such as establishing LME-certified warehouses in the Mainland. We are also still on track to launch our trading platform in Qianhai, Shenzhen to "physicalise" the Mainland's commodities market.

Physical delivery

Why does the LME want to set up warehouses in China, and how can they serve the country's real economy?

Warehouses are important because when buyers can receive physical delivery, the futures and spot prices of any given metals contract will converge at the contract expiration date. The LME doesn't own or operate any warehouses, but has a network of more than 600 certified warehouses in 35 locations. That means physical delivery is convenient and efficient, and futures prices truly reflect the real economy and global base metals supply and demand.

China is already the world's largest consumer of raw materials, and many of the country's metals users need to manage risk via LME futures contracts. Chinese traders account for more than 30 per cent of total LME trading, but because there are no certified warehouses in the Mainland, customers must make or receive delivery at a warehouse located elsewhere. That means the metal will then be shipped to or from overseas, with the trades stuck with high shipping costs and lengthy delivery times. This is a barrier to China participating in the global metals market.

Having LME-certified warehouses in the Mainland would not only reduce the cost of delivery for Mainland companies and give them greater access to LME trading and pricing, it would also help China gain influence in international commodities pricing over the long term that is commensurate with its size.

Finally, we are making good progress with LMEshield, our secure way to manage warehouse receipts. We recently announced we will list warehouses for the LMEshield repository in countries along China's "Belt and Road" routes.

Pricing

China's influence on pricing is weak, despite the fact it's the world's largest consumer of raw materials. How can this be resolved? As things stand today, Chinese enterprises aren't completely free to participate in international futures markets to hedge transactions because of foreign exchange restrictions, which means they can't influence prices in overseas markets. At the same time, foreign exchanges and bonded warehouses are restricted in the Mainland, artificially increasing delivery costs for Chinese companies and resulting in them buying or selling at global prices without the ability to protect themselves via physical delivery or hedging.

On the other hand, trading is active on Chinese domestic exchanges but international users don't participate in the Mainland price discovery process, so they naturally won't accept prices discovered there because they have no influence. As a result, both international and Mainland users end up using overseas market pricing.

The key is creating a place where Chinese and overseas buyers and sellers can transact. It doesn't matter where that platform is, as long as Chinese enterprises are able to contribute and it is trusted by both Chinese and international market participants.

What can be done to improve China's pricing influence in the commodities sector?

One option is for China to speed up the opening of its commodity futures market and welcome foreign investors to participate. If this isn't feasible because of capital controls and other restrictions, it might make sense in the interim to allow Mainland enterprises and investors to go out and participate in international commodities markets.

Qianhai trading platform

There are thousands of Mainland commodity spot trading platforms; why is HKEX launching its own in Shenzhen?

The Mainland has more than 1,000 commodity trading platforms, but most lack credibility, supervision, or a strong regulatory structure, with others exploiting regulatory rules by promising high-yield, low-risk products. We hope to build a standardised, transparent, credible spot commodities trading platform in Shenzhen for physical delivery that will serve China's real economy.

We are confident because we have three major advantages: first, HKEX has a long track record as a trusted partner to the Mainland. After the acquisition of LME, we were able to combine this with expertise and brand recognition in the commodities space. Second, we have the power, determination and resources to provide reliable warehousing, logistics and other facilities to serve the real economy effectively. Finally, we have the credibility of HKEX. We believe we can create a new price benchmark and use it to develop indices, futures, and other derivatives products in Hong Kong.

What challenges do you expect to face?

Our biggest challenge is not from Mainland regulatory policy, which is designed to protect investors and build a robust market in China, but from finding a way to adapt and customise global practices to the Chinese market. We believe in the merits of our objectives, which are highly consistent with China's regulatory prerogatives. Although LME has a long track record in global commodities markets, the Chinese market is unique and we can't expect to copy the international model and paste it in China. We will need to consider the local conditions and develop a commodities platform that serves the particular needs and characteristics of the Mainland market.

20 July 2016

52 | Taking the first step towards building a long-term commodities platform in China

We are in the midst of a lot of festivities and insightful discussions this week for London Metal Exchange (LME) Asia Week, the fifth time we've brought together market players for the biggest annual commodities gathering in Asia. LME Asia Week has grown every year since we started it in 2013, and this year is even more special, because today we are hosting hundreds of guests at our new trading platform in Mainland China: the Qianhai Mercantile Exchange, or QME.

We decided long ago that it was important to build out our commodities business, first by purchasing the LME in 2012, launching new metals contracts on our Hong Kong platform, and now by building a new spot commodities market in Qianhai, Shenzhen. Our team has been working diligently for many months to make sure everything is ready, and we are stepping up our preparations to get the market ready for trading.

The QME has been a hot topic this week, with many people asking me about its core mission. To better understand the QME, it's important to first get a snapshot of the Mainland's commodities markets. China's three futures markets are highly liquid and freely-traded, but are geared towards retail investors and speculators. They have very few contracts that get physically settled. Spot market users are spread out among hundreds of disparate, poorly-regulated and highly-fragmented trading platforms around the country. This is inefficient, and ultimately results in higher costs for industrial users. Furthermore, there is little convergence between the spot and futures prices, an imbalance between the OTC and exchange-traded markets and a disconnect between the physical market and financial services. On top of this, while China is the world's largest producer and consumer of many different metals, it lacks commensurate pricing power in global markets.

We see an opportunity to address some of these issues, which can be a win-win for Mainland China and us at HKEX. We think we can do it because, at HKEX, and especially with our Shanghai and Shenzhen Stock Connects, we have deep experience working with Mainland authorities, regulators and market players. In addition, we have the LME, which has a rich heritage as the global leader in non-ferrous metals trading. We believe we can leverage these two key advantages to operate a successful spot commodities market that serves China's real economy.

But we can't just blindly copy the LME, because China has its own unique characteristics and challenges I mentioned earlier. So we want to leverage the LME to begin tackling some of China's unique structural issues by building a new model that both fills a market vacuum and meets market needs. It will be a fair, open, transparent and well-regulated market that, unlike many other commodities platforms in China, will be geared towards institutions rather than retail investors or speculators, giving them more hedging options. We want to serve the physical economy, and eventually establish benchmark prices for China.

The long-term objective is to operate an efficient marketplace with streamlined operations to enable physical users to manage their risk and get commodities at a lower cost. It will also help narrow the gap between onshore and offshore pricing and begin giving China a larger influence in global commodities prices.

We have chosen to build our onshore commodities platform in Qianhai, a modern service industry zone in southern China's Guangdong Province, which is only 50 kilometres away from our headquarters in Hong Kong. On top of its close proximity, the long-term cooperative relationship between Hong Kong and Guangdong businesses and absence of a central commodities market in southern China contributed to our decision to open in Qianhai.

Any new effort faces some challenges. I have been asked many times if this is a good time to enter the Mainland commodities market, because China is in the midst of a crackdown on commodities trading platforms. We have just gone through a period where a lot of leverage has been put into China's markets, risk has grown, and some unscrupulous players have been taking advantage. Mainland authorities are now taking a growing interest in this area, and have been looking closely at the models of various trading venues.

As the Mainland evolves and opens, it will always be two steps forward and one step back. The liberalisation process takes time, and we need to be patient. Nevertheless, we welcome and fully support the crackdown, as the long-term effect will be a much healthier and robust market. We intend to continue working alongside the regulators in Mainland China, and hope to make QME the standard-bearer for a fair, transparent and efficient marketplace that can support the next phase of China's development.

As for the exact launch time frame, we aren't too fussed. Whether it's a limited launch earlier or a larger and more comprehensive launch later, our main goal is to have this platform fulfill our long-term vision. Our initial success will be judged in many years, not in a few months.

With our credibility, resources, knowhow and risk management capability, we believe we can begin helping facilitate the convergence of the spot market in China, form a solid and reliable commodities platform to supplement the futures market, and build a stronger connection between the commodities market and the financial services market in the Mainland.

Over the long term, we will look at ways to connect the LME and QME, bringing the liquidity of the China and global markets together to create even more opportunities. A Commodities Connect with Mainland China is a matter of "when", not "if". The current situation in the Mainland commodities markets makes a Commodities Connect more sensitive for Chinese regulators, but we will continue our conversations and remain hopeful we'll secure a breakthrough.

QME is also an important step in our commodities strategy. Internationally, on the spot market front, we purchased the LME to give us instant credibility and know-how. In China, the QME is a project we are building from scratch to address that particular market's needs. In time, we'll also explore the cross-listing of commodities derivatives as part of the Connect programme.

True global financial centres have multi-asset class capabilities, and the QME is our next big step in diversifying our business and is a strategic step in strengthening our city's future as China's international financial centre. As the 20th anniversary of the establishment of the Hong Kong Special Administrative Region approaches this summer, we've been doubling down on our efforts to grow Hong Kong beyond our core equities business.

Today was the first time our market partners got to see the QME Ring. In time, we hope the Ring becomes as important and influential to the commodities market in Mainland China as the LME Ring is to the rest of the world.

11 May 2017

53 | Going for gold

It has been a golden summer for HKEX, because just days after launching the ground-breaking Bond Connect we've turned our attention to one of the world's most precious metals: gold.

Today we launched a dual-currency gold futures contract in Hong Kong that is physically-settled, and a gold contract as part of LMEprecious on the London Metal Exchange (LME). They mark a number of firsts: the first time we've launched a contract in two currencies, the first time we've offered physical delivery in Hong Kong, and LME Gold is the first and only on-exchange, spot to five-year loco London contract in the world.

Why gold?

It's no secret that gold is popular in Asia, and particularly in China, which is the largest consumer of gold in the world. The vast majority of the gold Mainland China imports comes through Hong Kong.

Gold is different from other commodities, because while it's used to make things like watches and bracelets, it's also widely recognised as a form of currency. The price of gold is heavily influenced by macro-economic trends, in addition to things like supply and demand.

Currently the price of gold is set through the OTC market in London and futures trading in New York, both of which are many time zones away from the largest consumer of gold in the world: China. Asia is severely under-represented when it comes to influencing gold pricing.

Why two cities?

China is already the world's largest consumer of gold, and consumption could grow exponentially in the future as the economy continues to expand. This creates new demand for gold trading in an Asian time zone, a need for physical settlement, and demand for hedging and risk management tools.

Hong Kong already has many important ingredients to become a gold trading hub: investors have traded gold here for more than 100 years and we have an active physical gold trading market that is one of the major bullion markets in the world. We are perfectly positioned to be a robust trading hub in the Asian time zone for both Mainland and international gold traders, leveraging the city's existing gold trading market and the massive demand from China to develop new gold benchmarks and provide more hedging opportunities, which will further develop the existing physical gold market in Hong Kong. It also brings other benefits, such as attracting investors who want to add gold to their portfolio and providing more investment options for offshore RMB deposits.

Our long-term plan is to build Hong Kong into a commodities, currencies and risk management centre, and gold ticks all three of those boxes.

London is also the perfect place to introduce on-exchange gold futures trading. We are leveraging the LME's long history and expertise in non-ferrous metals trading to move into precious metals with LMEprecious, which is a team effort involving the World Gold Council as well as Goldman Sachs, ICBC Standard Bank, Morgan Stanley, Natixis, OSTC and Societe Generale.

LMEprecious will bring greater transparency to precious metals trading by bringing more transactions onto the exchange-traded market, support and aid ongoing regulatory change, provide additional robustness to the market, broaden market access, make trading more capital efficient and trade lifecycle management easier. It will accommodate the interests of the full range of market stakeholders and reinforce the strengths of the London market.

While the contracts are not fungible, we are launching in Hong Kong and London at the same time to promote synergy and create on-exchange liquidity to complement the OTC markets and provide further risk management and arbitrage opportunities across different time zones.

What makes these products unique?

The Hong Kong gold contract is unique because HKEX is the first exchange to simultaneously offer dual-currency gold derivatives contracts in both US dollars and offshore RMB (CNH). It's significant because investors can now arbitrage between CNH futures, and the synthetic currency prices derived from dual currency gold contracts.

Physical settlement is important, too. If these contracts were purely cash-settled, there could be a price discrepancy between the US dollar and CNH tranches. But with physical settlement, smart investors will be able to arbitrage the price differences in the two liquidity pools, which will then lead to price convergence. This physical delivery option keeps the implied exchange rate between the two liquidity pools honest.

This is unique, and gives professional investors the option to use the new dual-currency gold futures as well as CNH options to create a powerful derivatives strategy.

In London, which is already a global gold centre, LMEprecious will meet the needs of investors who want access to exchange-traded, transparent and centrally-cleared precious metals. The contracts will include spot, daily, monthly and quarterly futures out to five years for both gold and silver.

With these two products launching today, investors will be able to trade gold products on HKEX's platforms 24 hours a day.

This is also another key step in our goal to build out a multi-asset class exchange with a focus on China. Gold is an important contract that diversifies our portfolio of CNH-denominated products, moves Hong Kong forward as an RMB pricing centre, and gives investors even more options in the Hong Kong and London markets.

10 July 2017

Chapter 6
Others

54 | Letter to Santa

Christmas is just around the corner, and it is usually the time to reflect on the year past and ponder on the year to come. But I've never felt entirely comfortable writing year-end reviews. This year, though, I decided to do something I've never done before: write a letter to Santa, just like my three children used to do at this time of year when they were little. They used to talk about how good they've been during the year, and ask Santa for gifts. I think it's a great idea, so here is my humble first try.

Dear Santa,

What a year it has been, eh? It has been so busy in our home at HKEx, I forgot to send you my Christmas wish list! Now that things have begun to quieten down, I have a few moments to tell you what good boys and girls we have all been this year and what we are hoping to receive under the tree on Christmas Day.

First off, I'm really pleased to say that I have shed some weight this year! I am down almost 2 full pounds, mainly thanks to playing soccer. That might not seem like a lot, but you know those noodles and fried rice dishes are just too hard for me to resist, Santa. Did you know we played the Shenzhen Stock Exchange earlier this month, and held them to a 1-1 draw? I played the entire second half, but hardly touched the ball. Maybe you could put a few goals under the tree for me this year.

In fact, Santa, everyone at our HKEx home deserves a great Christmas. The whole team has been working like diligent elves to make our house bigger and more beautiful, and improve the neighbourhood.

Our IT elves, for instance, have kept us safe from any electrical outages and ensured there were no leaks in our roof. I'm really proud of them. And don't worry Santa, the chimney has been cleaned and is ready for your arrival! Next year, we'll start making our home even stronger and more resilient through upgrades like a closing auction and Volatility Control Mechanism.

Santa, the two new commodities wings of our home are also coming along quite nicely. We worked hard this year to continue renovating and improving the new LME restaurant we bought two years ago, which is now connected to West Wing of the home. We announced we would raise the food prices a bit from the start of next year, but we're confident people will still fill the seats because our food quality is getting even better. The increase in revenues will also help us introduce delicious new items to the menu. Aside from that, we installed a whole new plumbing system which we named LME Clear. It has already become the envy of the neighborhood, so we're really proud of it.

The new East Wing to our commodity restaurant has also come a long way from last year. It's a big job building an entirely new wing from scratch, but we made some good progress in 2014. After you have the milk and cookies we're putting out for you, please do take a look. You'll notice the three new mini contracts we installed just this month. It will take some time, but I know it will soon grow into a beautiful new addition to the home.

The biggest new development in the neighbourhood this year, Santa, was a sparkling new bridge we built to connect us with the big market on the other side of the river. It took more than a year to build, and nearly everyone in the house pulled together to get it done. If that's not teamwork, Santa, then I don't know what is! We thought there might be a few more cars making use of the bridge, but we'll keep working on it next year, I promise. We want to install more and better road signs to the bridge, and show people the benefits and convenience of using it. I have no doubt traffic will increase.

But Santa, I'm afraid not everything was perfect this year. A special guest was planning to come to stay with us, but he decided to have a party at someone else's house instead. He was asking for some pretty major renovations to make room for his arrival, and we just couldn't change our house that fundamentally without consulting everyone in the neighbourhood. The good news is the community is having a good discussion about whether or not we should make those renovations in the future. In the meantime, our home is still beautiful and very attractive to other visitors — including you, I'm sure, Santa!

Before you visit, though, I hope your sleigh stops by the house of our friendly neighbourhood regulator. The SFC has done a fantastic job this year of keeping the neighbourhood secure, and will work even more closely in the years to come with its equally friendly regulator, the CSRC, across the river. They deserve excellent Christmas gifts under the tree and we are going to build more bridges and roads for them to police!

Also, please treat our HKEx family well. Without each and every person working together in our home, we wouldn't have had such a breakthrough year. They have all been good boys and girls, and deserve a wonderful holiday season. Please bring them whatever they are asking for (within reason, Santa, as I'm sure not even you have an unlimited budget). Maybe bring a little something extra for the 21 couples who got married this year, and the 28 babies born to HKEx staff who will be celebrating their first Christmas.

Finally Santa, I'd be most grateful if you could bring each and every one of us health, prosperity, and happiness in the year ahead. I can't think of a better present than that.

Sincerely,

Charles

22 December 2014

55 | Looking to Santa

Readers of my blog might remember that last December I wrote a letter to Santa — and he gave us a pretty good year. So I was thinking whether or not to write to Santa again this year, and considered asking my daughter for her thoughts. After all, at HKEx we are all superheroes now, superheroes having been the theme of our annual dinner last Thursday. But I knew my daughter would just say I should get better at soccer first before thinking I'm a superhero. And she would probably be right. Although my colleagues at HKEx are very hardworking and capable they don't have super powers, and anyway, superheroes don't give gifts.

So, I'd better get my letter off right away before the sleigh leaves the North Pole!

Dear Santa,

It's so good to write to you again after all the wonderful gifts you gave to us last year. In fact, it was a pretty great year for HKEx overall.

The entire staff at HKEx were good boys and girls, and worked hard during the first full year of Shanghai-Hong Kong Stock Connect (can you believe it's a year old already?). We are now busy improving this important bridge we built by working on access roads, signs, lighting and more while getting ready for the opening of another bridge. This is also the first full year the London Metal Exchange (LME) has operated as a truly commercial exchange, and we are celebrating the first anniversary of the launch of LME Clear, a brand new clearing house we built in Europe. LME and LME Clear are now important contributors to the financial performance of the group this year.

Of course, like in any year, there were some trials and tribulations. 2015 was a year of two halves for the domestic A share market, which has become increasingly important to our own market after Stock Connect. The first half saw the most bullish run in A shares since the global financial crisis and the second half witnessed one of the most dramatic market routs in recent years. The good news is the Hong Kong market withstood the volatility, proving again that it is a mature, reliable, and sophisticated market. It's really great to be working in such a strong and respected financial centre, Santa, so I'm very grateful about that.

We also rolled out several new products this year, which was only the beginning of our plan to build a critical mass of new products primarily in derivatives across different asset classes, with an increasing focus on attracting new investors from the Mainland. We do not expect these products to generate immediate trading success, though. I think of it a little like a carnival: people won't show up to a carnival with just one or two shows, it needs to have a wide variety of performances, games and fun that draw people from near and far. Our new products represent the first couple of shows at the carnival, but pretty soon I hope we can grow into one of the biggest and best carnivals around! We just launched our second batch of LME mini contracts, as well as 34 new stock futures contracts. As we roll out more games, we expect more crowds to come and enjoy them, particularly from the north.

But back to my wish list. This year, I'm hoping to find under the Christmas tree vigorous market support as we roll out our Volatility Control Mechanism and Closing Auction Session next year. These enhancements will strengthen our market structure and align ourselves with international market practices. On top of this, I wish that we work hard to strengthen our market quality. You know the theme of our annual staff dinner was superheroes. In reality we can't count on heroic characters to protect our market though, so we all need to pitch in to ensure our market regulation is robust. And Santa, don't forget we will soon be announcing our next three year strategic plan, which we're really excited about. We'll accept any gifts that help us implement our plans! And just one more thing that would mean a lot... HKEx lost to the Shenzhen Stock Exchange 1-0 in our soccer match about a week ago. The captain tapped me on the shoulder and told me to get into the game in the second half but I only touched the ball once. Maybe you could give us a goal or two next year? :)

Finally, Santa, I want you to be good to all my colleagues at HKEx. Each year they work tirelessly to ensure HKEx remains one of the top exchanges in the world, and they are among the reasons Hong Kong is a respected global centre of finance. They deserve a lot of generosity this year, so I hope you are good to them. If you are still thinking about gifts for them, perhaps I could offer a suggestion - toy trains are good, but the trains coming over the Connect bridges are fabulous!

I'd like to thank them too, Santa, and everybody else who has supported HKEx this year. We have a busy year ahead, but now's the time to kick-back, relax, and enjoy good times with family and friends over the holiday season. I wish everyone good health and prosperity in the year ahead.

Sincerely,

Charles

20 December 2015

56 | Talking weather on a typhoon day

This article is only available in Chinese.　(It is on p.265.)

57 | Year-end reflections (2016)

Things have finally started slowing down as we approach Christmas, with colleagues getting ready to leave to spend the holidays with their families. As the office hallways quiet down and the piles of paper on my desk begin to disappear, I found myself leaning back in my chair and reflecting on 2016.

According to the Chinese calendar, 2016 is the year of the monkey. I can't help but be amazed at the great wisdom and foresight of our ancestors who predicted that 2016 would be one of the most unpredictable years ever, just like the monkey. At the beginning of the year, nobody thought Britain would vote to exit the European Union, but it did; nobody thought Donald Trump would be nominated for President of the United States, let alone elected, but he was; here in Hong Kong, nobody thought 2016 would have so much drama, but it did.

Closer to home at HKEX, 2016 has not been easy for us either. Our volumes and revenues are both down. But some very important initiatives, such as Shenzhen Connect, were launched as anticipated. So we thank the monkey for allowing us to achieve at least some of our big dreams. In these difficult times, it is important to continue to harbour hopes and dreams, and strive to work together if we want to progress and turn our collective dreams into reality.

Speaking of dreams, I couldn't help but look back at some dreams from my early years. One of my first memories was when I was a young oil worker wondering outside the gates of Zhongshan University in Guangzhou one day, where our crew was passing through after a training camp. Watching students coming and going through the gate proudly with their student passes, I so desperately wished that someday I could carry my backpack through those gates too.

When I finally graduated from college, I began to dream to go to America to study, but my student visa application was rejected three times! When I was finally approved, I remember screaming with joy and celebration outside the US Embassy in Beijing, scaring people walking by. I also remember, while a student in Alabama, how much I dreamed of being able to walk into a McDonald's without ever having to worry about the cost, or have a car that didn't constantly break down, or have enough money to fix it if it did. When I look back at these dreams, I chuckle because they seem quaint now. But at the time, they were big dreams, dreams so vast that it seemed like they might never come true. I am so thankful that with hard work, lots of help from family and friends and with lots of luck, I have come a long way.

I approach my work at HKEX the same way. Just a few years ago, market commentators in Hong Kong said the "through train", as it was known then, would be difficult to realise because of technical and regulatory hurdles. Having the Mainland and Hong Kong markets open for trading in both directions seemed like a fantasy. But after the launch of Shanghai Connect in November 2014 and the start of Shenzhen Connect earlier this month, it's almost taken for granted that the Hong Kong, Shanghai and Shenzhen markets are connected and accessible on both sides of the boundary.

Lots have been said about Hong Kong being an unlikely commodity market, but we purchased London Metal Exchange (LME) and we are on our way to expand from that. We are also preparing to establish a commodities trading platform in Qianhai, Shenzhen. So no matter how big your dreams, stay hopeful, keep your chin up, and most importantly, keep working hard!

I want to share some of my plans for 2017, but recognising that this has been a long and exhausting year, I'll put them on hold until January. Let's take a rest. Now's the time to reflect, and I hope everyone is able to do that over the holidays with their loved ones.

I want to thank everyone again for their support this year: the regulators, our exchange participants and brokers who had to put in long hours to prepare and test for initiatives like Stock Connect, the greater financial market, and of course the staff at HKEX. Without your support our dreams for 2016 wouldn't have become a reality.

I wish everyone a Merry Christmas and a happy and prosperous 2017.

19 December 2016

58 | Years of hard work begin to show signs of harvest

Happy New Year!

I know I'm a little bit late wishing everyone a happy 2018, but we've been busy at HKEX preparing to unveil our plans for the year. This year marks the final year of our 2016-2018 Strategic Plan and I have a lot of exciting things to share, but before diving into our initiatives let's take a moment to reflect on 2017 — a truly breakthrough year for HKEX in a number of areas.

A look back at 2017

Perhaps the biggest news came towards the end of the year: we announced the conclusions to our consultation paper on listing reform which will usher in the most significant change to Hong Kong's listing regime since 1993. Hong Kong has collectively decided to take a big step forward as a financial centre and welcome innovative, new economy companies that use non-standard share structures and pre-revenue companies from the biotech sector. The market had vigorously debated the topic but we are glad that we have come forward together on a path that will ultimately underscore Hong Kong's attractiveness as a capital raising centre for a new generation of companies. We expect companies from the new economy to begin applying for listing under the new rules by the end of June this year.

We strengthened the Connect programme last year as well, by extending it into a new asset class: bonds. The Northbound channel under Bond Connect will have a far-reaching impact on the orderly opening of Mainland China's capital market and the internationalisation of the Renminbi (RMB).

On the fixed-rate and currency side, our turnover of RMB Currency Futures continued to increase, which helped pave the way for the launch of RMB Currency Options. This puts more tools in the hands of investors for hedging RMB exchange rate risk.

In commodities, we launched the first dual-currency and physically-settled gold futures contracts in Hong Kong alongside the launch of our first ferrous metals product, iron ore futures. The commercialisation of the London Metal Exchange (LME) is basically complete and we remain focused on launching our spot trading platform in the Mainland, the Qianhai Mercantile Exchange.

Actually, there are too many things to list in this blog! Cumulatively, the efforts of the entire market over the past few years was reflected in the market performance of 2017: the average daily turnover in our securities market reached HK$88 billion, an increase of over 30 per cent from 2016. The average daily volume of our derivatives market hit a new high of 873,000 contracts, with the momentum sustained into 2018: so far in January derivatives trading volume has passed one million contracts in 11 of the first 16 trading days. The Hong Kong market's total market cap sat at HK$34 trillion, a record high, at the end of 2017 and it has crossed over HK$36 trillion for the first time last week.

Stock Connect has continued to grow as well. The total volume of Northbound transactions to Shanghai and Shenzhen reached RMB 2.266 trillion, up 194 per cent over the same period last year and demonstrating international investor appetite for A shares. Southbound transactions from those two cities to Hong Kong hit HK$2.259 trillion, up 170 per cent over 2016. The benefits of the scheme are being shared by all three exchanges, with net inflows of RMB 200 billion Northbound and HK$340 billion Southbound. International and Hong Kong investors now cumulatively hold nearly RMB 600 billion in A shares, while Mainland investors hold over HK$1 trillion in stocks listed in Hong Kong via Stock Connect.

Motivation for success

The green shoots of success we witnessed last year were not because of work we did in 2017 — the work began much earlier, and that work involved strong support from the market. Our predecessors laid the groundwork for us to take the mantle, and they instilled in us the courage and foresight to work on programmes like Connect, listing reform and the launch of new asset classes.

Our media lunch in 2016 was the first time we put our connectivity plans into words:

to connect China with the world and reshape the global market landscape. It has been our primary goal to be the first choice for Mainland investors seeking international diversification and international investors seeking exposure to China. We have been consumed with this goal over the past two years, and are working diligently to continue improving the Connect scheme by expanding it and including other asset classes. At the same time, we want to make sure Hong Kong has products that are attractive to investors everywhere, which is why we have undertaken listing reform and aim to enrich our product offering across the major asset classes of equities, commodities, fixed income and currency. We must be proactive and anticipate the changing needs of the market.

Two years have passed since we unveiled our Strategic Plan, and many of our objectives have been achieved — but not all. It's easy to focus on the successes, but we know some challenges remain ahead of us, particularly in the commodities sector.

We have embarked on a difficult path to strengthen Hong Kong's traditionally robust equities business while growing into new asset classes that make HKEX, and the Hong Kong financial market in general, more diversified and resilient, and more prosperous in the long term. This task can't be completed overnight, or even in a year or two — it takes time.

Looking to the future

The emergence of the new economy, particularly in Mainland China, is a fantastic catalyst for us to take a closer look at our business and see if we're as competitive and attractive as we can be. We need to look at Connect to make sure we are continuing to work with regulators and our Mainland counterparts to nurture and grow the scheme. We have already begun laying the groundwork for new products in new asset classes that appeal to investors in Mainland China and around the world. Diversifying our market will ensure our business is less susceptible to factors beyond our control, like market sentiment and trading volumes.

If we continue to persevere and work hard in cooperation with the market, I believe that over time Hong Kong will reap the benefits of being a comprehensive financial centre across asset classes and with ground-breaking, innovative connections and access to Mainland China. If we don't try, we'll never reach our goal!

In 2018, we will continue doing our best to plant the seeds of growth for tomorrow and make our business as diverse and resilient as we can.

It is an exciting time. We are starting to see some of our efforts bear fruit, which is an indication that we mustn't stop pushing forward and strengthening our market. Our team at HKEX will continue working hard in 2018 and beyond to ensure we can have a reliable, bountiful harvest in the years to come.

24 January 2018

免責聲明

本書所載資料及分析只屬資訊性質，概不構成要約、招攬、邀請或推薦買賣任何證券、期貨合約或其他產品，亦不構成提供任何形式的建議或服務。書中表達的意見不一定代表香港交易及結算所有限公司（「香港交易所」）的立場。書中內容概不構成亦不得被視為投資或專業建議。儘管本書所載資料均取自認為是可靠的來源或按當中內容編備而成，香港交易所及其附屬公司、董事及僱員概不就有關資料（就任何特定目的而言）的準確性、適時性或完整性作任何保證。香港交易所及其附屬公司、董事及僱員對使用或依賴本書所載的任何資料而引致任何損失或損害概不負責。

Disclaimer

All information and views contained in this book are for informational purposes only and does not constitute an offer, solicitation, invitation or recommendation to buy or sell any securities, futures contracts or other products or to provide any advice or service of any kind. The views expressed in this book do not necessarily represent the position of Hong Kong Exchanges and Clearing Limited ("HKEX"). Nothing in this book constitutes or should be regarded as investment or professional advice. While information contained in this book is obtained or compiled from sources believed to be reliable, neither HKEX nor any of its subsidiaries, directors or employees guarantee its accuracy, timeliness or completeness for any particular purpose. Neither HKEX nor any of its subsidiaries, directors or employees shall be responsible for any loss or damage arising from the use of, or reliance upon, any information contained in this book.